Discrete Multivariate Distributions

WILEY SERIES IN PROBABILITY AND STATISTICS

Established by WALTER A. SHEWHART and SAMUEL S. WILKS

A complete list of the titles in this series appears at the end of this volume.

Discrete Multivariate Distributions

NORMAN L. JOHNSON
University of North Carolina
Chapel Hill, North Carolina

SAMUEL KOTZ
University of Maryland
College Park, Maryland

N. BALAKRISHNAN
McMaster University
Hamilton, Ontario, Canada

A Wiley-Interscience Publication
JOHN WILEY & SONS, INC.
New York • Chichester • Weinheim • Brisbane • Singapore • Toronto

This text is printed on acid-free paper.

Library of Congress Cataloging in Publication Data:

Johnson, Norman Lloyd.
Discrete multivariate distributions / Norman L. Johnson, Samuel Kotz, N. Balakrishnan.
p. cm. — (Wiley series in probability and statistics. Applied probability and statistics)
"A Wiley-Interscience publication."
Includes bibliographical references and index.
ISBN 0-471-12844-9 (cloth : alk. paper)
1. Distribution (Probability theory) I. Kotz, Samuel. II. Balakrishnan, N., 1956– . III. Title. IV. Series.
QA273.6.J617 1996
519.2′4—dc20 96-23727

Printed in the United States of America

10 9 8 7 6 5 4 3 2 1

To
Regina Elandt-Johnson
Rosalie Kotz
Colleen Cutler, Sarah and Julia Balakrishnan

Contents

Preface

This volume constitutes an expanded and updated version of Chapter 11 of *Distributions in Statistics: Discrete Distributions*, co-authored by Norman L. Johnson and Samuel Kotz, and published in 1969. This is also the fourth volume in a series, jointly forming second editions of the original first three volumes of *Distributions in Statistics*. The earlier three publications in this series are

Univariate Discrete Distributions (with A. W. Kemp)	1992
Continuous Univariate Distributions—1 (with N. Balakrishnan)	1994
Continuous Univariate Distributions—2 (with N. Balakrishnan)	1995

In a review of the 1969 volume, Sir David R. Cox [*Biometrika* (1970), **57**, p. 468] remarked that "relatively little has been done on multivariate discrete distributions." The situation has changed considerably since then, though "little" might still only be replaced by "less."

As compared with Chapter 11 of the original 1969 volume, there is more than a sixfold increase in the number of pages, reflecting a welcome and unprecedented growth in publications on discrete multivariate distributions during the past quarter of a century. Although a substantial part of this increase has been of a theoretical nature, it has been accompanied by a substantial increase of applications (in number as well as variety). This can be ascribed (at least partly) to the wide availability of computer programs facilitating the execution of the detailed calculations often associated with the use of discrete multivariate distributions. We have endeavored in this volume to give due attention to representative examples of such applications derived from extensive searches of available literature.

The theoretical advances recorded in this volume have their own intrinsic value. We particularly draw readers' attention to results concerning distributions other than the commonly used multinomial and multivariate negative binomial, hypergeometric, and Poisson distributions.

As in the other volumes of this series, we have generally omitted papers restricted to significance tests and have also omitted some theoretical results which are of use in some current developments in the theory of statistical inference. While the former has been a rule with all the volumes in this series, the latter is due to the fact that many of these results require advanced mathematical tools not commonly available to the majority of statisticians engaged in problems of practical nature. This is in accordance with the originally stated policy of providing "a collection of the more important facts about the commonly used distributions in statistical theory and practice." There will inevitably be omissions of some papers containing important results that could have been included in this volume. These should be considered as consequences of our ignorance and not of personal nonscientific antipathy.

Our sincere thanks go to Professors W. J. Ewens and S. Tavaré for providing a write-up of Chapter 41 on their initiative. We are also very thankful to an anonymous reviewer who provided valuable suggestions which led to a considerable improvement in the organization and presentation of the material. We express our sincere thanks to the authors from all over the world, too numerous to be cited individually, who were kind enough to provide us with copies of their published papers and other reports. We are also indebted to the librarians at Bowling Green State University, Ohio; Catholic University of America and George Washington University, Washington, D.C.; St. Petersburg University of Finance and Economics, St. Petersburg; Hong Kong Baptist University, Hong Kong; Institute of Statistics at Academia Sinica, Taipei; University of North Carolina, Chapel Hill; and McMaster University, Hamilton, who assisted us in our extensive literature search. Special thanks are due to Ms. Cyndi Patterson (Department of Mathematics and Statistics, Bowling Green State University) for the initial typing of some early chapters of this volume, as well as to Mrs. Debbie Iscoe (Hamilton, Ontario, Canada) for typing the entire volume. Overcoming logistic and organizational problems was facilitated by the hospitality of the Department of Mathematics and Statistics, Bowling Green State University; the Department of Statistics, University of Hong Kong; and the Division of Quality Technology and Statistics, University of Luleå, Sweden.

We are happy to acknowledge the support and encouragement of Ms. Margaret Irwin, Ms. Kate Roach, and Mr. Steve Quigley of John Wiley & Sons throughout the course of this project. The managerial and editorial help provided by Ms. Angela Volan and Ms. Kimi Sugeno are also gratefully acknowledged.

As always we welcome readers to comment on the contents of this volume and are grateful in advance for informing us of any errors, misrepresentations or omissions that they may notice.

Norman L. Johnson
Samuel Kotz
N. Balakrishnan

Chapel Hill, North Carolina
College Park, Maryland
Hamilton, Ontario
November 1996

Postscript

Although the preparation of the second edition of *Continuous Multivariate Distributions* (originally published in 1972) is in advanced planning stages, it may somewhat be delayed for a while; the present volume marks the completion of the first and major phase in compilation of the second editions of the four volume series on *Distributions in Statistics* (originally published during 1969-1972). The second author—the initiator of this project—takes this opportunity to express his most sincere thanks to the first author—Professor Norman L. Johnson, a doyen of Statistics in this century—for a most fruitful and harmonious collaboration spanning over close to 35 years. He is also very thankful to a distinguished pair of scholars and dear friends of many years, Dr. A. W. Kemp and Professor C. D. Kemp, for paving the way and providing erudite leadership for the second edition of *Univariate Discrete Distributions*. (I particularly enjoyed our regular Sunday 4 a.m. telephone conversations, Freda!) Special thanks are due to Professor N. Balakrishnan, the co-author of the last three volumes, for sticking with us through thick and thin during the last 5 years and carrying with great devotion more than his share related to production aspects of this undertaking (while being an equal partner in all the scientific matters). Last but not least, I acknowledge with gratitude the letters of encouragement and appreciation I received during the last 27 years from a multitude of statisticians throughout the world which provided a convincing testimony of the usefulness and desirability of this comprehensive and difficult project. It is these letters coupled with a formal recognition from the University of Athens (Greece) and more recently from Bowling Green State University, two prominent centers of research in the area of statistical distribution theory, that sustained me during all these years and prevented the abandonment of this labor of love.

SAMUEL KOTZ

College Park, Maryland
December 1996

List of Tables

Glossary

$a^{(b)}$	bth descending factorial of a $= a(a-1)\cdots(a-b+1)$
$a^{[b]}$	bth ascending factorial of a $= a(a+1)\cdots(a+b-1)$
$a^{[b,c]}$	$= a(a+c)\cdots(a+\overline{b-1}\,c)$
$s(n,i)$	Stirling numbers of the first kind $=$ coefficient of x^i in $\{x!/(x-n)!\}$, $i=0,1,\ldots,n$
$S(n,i)$	Stirling numbers of the second kind $=$ coefficient of $\{x!/(x-i)!\}$ in x^n, $i=0,1,\ldots,n$
$\bar{s}(n,i)$	Stirling numbers of the third kind $=$ coefficient of x^i in $x^{[n]}$, $i=0,1,\ldots,n$
$\bar{S}(n,i)$	Stirling numbers of the fourth kind $=$ coefficient of $x^{[i]}$ in x^n, $i=0,1,\ldots,n$
$\Gamma(\alpha)$	complete gamma function $= \int_0^\infty e^{-z}z^{\alpha-1}dz$, $\alpha>0$
$B(\alpha,\beta)$	complete beta function $= \Gamma(\alpha)\Gamma(\beta)/\Gamma(\alpha+\beta)$, $\alpha,\beta>0$
$B(\alpha_1,\ldots,\alpha_k)$	multiple beta function $= \Gamma(\alpha_1)\cdots\Gamma(\alpha_k)/\Gamma(\alpha_1+\cdots\alpha_k)$, $\alpha_1,\ldots,\alpha_k>0$
${}_2F_1(a,b;c;x)$	Gaussian hypergeometric function $= \sum_{j=0}^{\infty}\frac{a^{[j]}b^{[j]}}{j!c^{[j]}}x^j$, $c\neq 0,-1,-2,\ldots$
$\binom{n}{i}$	binomial coefficient (n combination i) $= \frac{n!}{i!(n-i)!}$
$\binom{n}{i_1,i_2,\ldots,i_k}$	multinomial coefficient $= \frac{n!}{i_1!\cdots i_k!(n-\sum_{j=1}^k i_j)!}$ $= \frac{\Gamma(n+1)}{\Gamma(i_1+1)\cdots\Gamma(i_k+1)\Gamma(n-\sum_{j=1}^k i_j+1)}$
$\lvert A\rvert$	determinant of the matrix A
$\mathrm{Diag}(a_1,\ldots,a_k)$	diagonal matrix with $a_1,\ldots,a_k$ as the diagonal

I	identity matrix
$\nabla_x f(x)$	backward difference operator $= f(x) - f(x-1)$
$\Delta_x f(x)$	forward difference operator $= f(x+1) - f(x)$
$\nabla_x^n f(x)$	$= \nabla_x \left(\nabla_x^{n-1} f(x)\right),\ n = 2, 3, \ldots$
$\Delta_x^n f(x)$	$= \Delta_x \left(\Delta_x^{n-1} f(x)\right),\ n = 2, 3, \ldots$
$I(x \in T)$	indicator function taking on the value 1 if $x \in T$ and 0 if $x \notin T$
$F_1(\boldsymbol{x};\boldsymbol{\theta}) \bigwedge_{\boldsymbol{\theta}} F_2(\boldsymbol{\theta};\boldsymbol{\alpha})$	distribution $F_1(\boldsymbol{x};\boldsymbol{\theta})$ is compounded (by $\boldsymbol{\theta}$) with distribution $F_2(\boldsymbol{\theta};\boldsymbol{\alpha})$
$\otimes$	Kronecker product operator defined by $\boldsymbol{A}_{r_1\times s_1} \otimes \boldsymbol{B}_{r_2\times s_2} = \begin{pmatrix} a_{11}\boldsymbol{B} & a_{12}\boldsymbol{B} & \cdots & a_{1s_1}\boldsymbol{B} \\ \vdots & \vdots & & \vdots \\ a_{r_1 1}\boldsymbol{B} & a_{r_1 2}\boldsymbol{B} & \cdots & a_{r_1 s_1}\boldsymbol{B} \end{pmatrix}_{r_1 r_2 \times s_1 s_2}$

Notations

$\boldsymbol{X}$	random k-dimensional vector $= (X_1, \ldots, X_k)'$
$P_{\boldsymbol{X}}(\boldsymbol{x})$	probability mass function of $\boldsymbol{X}$ $= \Pr\left[\bigcap_{i=1}^{k}(X_i = x_i)\right]$
$F_{\boldsymbol{X}}(\boldsymbol{x})$	cumulative distribution function of $\boldsymbol{X}$ $= \Pr\left[\bigcap_{i=1}^{k}(X_i \le x_i)\right]$
$G_{\boldsymbol{X}}(\boldsymbol{t})$	probability generating function of $\boldsymbol{X}$ $= E\left[\prod_{i=1}^{k} t_i^{X_i}\right]$
$\varphi_{\boldsymbol{X}}(\boldsymbol{t})$	moment generating function of $\boldsymbol{X}$ $= E[e^{\boldsymbol{t}'\boldsymbol{X}}]$
$K_{\boldsymbol{X}}(\boldsymbol{t})$	cumulant generating function of $\boldsymbol{X}$ $= \log \varphi_{\boldsymbol{X}}(\boldsymbol{t})$
$E[e^{i\boldsymbol{t}'\boldsymbol{X}}]$	characteristic function of $\boldsymbol{X}$
$\mu'_{\boldsymbol{r}}(\boldsymbol{X})$	$\boldsymbol{r}$th mixed moment (about zero) of $\boldsymbol{X}$ $= E\left[\prod_{i=1}^{k} X_i^{r_i}\right]$
$\mu_{\boldsymbol{r}}(\boldsymbol{X})$	$\boldsymbol{r}$th central mixed moment of $\boldsymbol{X}$ $= E\left[\prod_{i=1}^{k}(X_i - E[X_i])^{r_i}\right]$
$\kappa_{\boldsymbol{r}}(\boldsymbol{X})$	$\boldsymbol{r}$th mixed cumulant of $\boldsymbol{X}$ $=$ coefficient of $\prod_{i=1}^{k}(t_i^{r_i}/r_i!)$ in $K_{\boldsymbol{X}}(\boldsymbol{t})$
$\mu'_{(\boldsymbol{r})}(\boldsymbol{X})$	descending $\boldsymbol{r}$th factorial moment of $\boldsymbol{X}$ $= E\left[\prod_{i=1}^{k} X_i^{(r_i)}\right]$
$\mu'_{[\boldsymbol{r}]}(\boldsymbol{X})$	ascending $\boldsymbol{r}$th factorial moment of $\boldsymbol{X}$ $= E\left[\prod_{i=1}^{k} X_i^{[r_i]}\right]$
$\log G_{\boldsymbol{X}}(\boldsymbol{t} + \boldsymbol{1})$	factorial cumulant generating function of $\boldsymbol{X}$ $= \log E\left[\prod_{i=1}^{k}(t_i + 1)^{X_i}\right]$

$\kappa_{(\boldsymbol{r})}(\boldsymbol{X})$	descending $\boldsymbol{r}$th factorial cumulant of $\boldsymbol{X}$ $= \text{coefficient of } \prod_{i=1}^{k}(t_i^{r_i}/r_i!) \text{ in } \log G_{\boldsymbol{X}}(\boldsymbol{t}+\boldsymbol{1})$
$E[X_i \mid \boldsymbol{X_a} = \boldsymbol{x_a}]$	conditional expectation of X_i on $\boldsymbol{X_a}$ $= E\left[X_i \mid \bigcap_{j=1}^{p}(X_{a_j} = x_{a_j})\right]$
$P_{\boldsymbol{X}\mid z}(\boldsymbol{x} \mid \boldsymbol{Z} = z)$	conditional probability mass function of $\boldsymbol{X}$, given $\boldsymbol{Z} = z$ $= \Pr\left[\bigcap_{i=1}^{k}(X_i = x_i) \mid \bigcap_{j=1}^{h}(Z_j = z_j)\right]$
$F_{\boldsymbol{X}\mid z}(\boldsymbol{x} \mid \boldsymbol{Z} = z)$	conditional cumulative distribution function of $\boldsymbol{X}$, given $\boldsymbol{Z} = z$ $= \Pr\left[\bigcap_{i=1}^{k}(X_i \le x_i) \mid \bigcap_{j=1}^{h}(Z_j = z_j)\right]$
$E[\boldsymbol{X}]$	mean vector of $\boldsymbol{X}$
$\text{var}(\boldsymbol{X})$	variance–covariance matrix of $\boldsymbol{X}$
$P_{\boldsymbol{X}}^{T}(\boldsymbol{x})$	probability mass function of the truncated distribution of $\boldsymbol{X}$
$\mu'_{(\boldsymbol{r})T}(\boldsymbol{x})$	descending $\boldsymbol{r}$th factorial moment of the truncated distribution of $\boldsymbol{X}$
$\text{cov}(X_i, X_j)$	covariance of X_i and X_j $= E[X_iX_j] - E[X_i]E[X_j]$
$\text{corr}(X_i, X_j)$	correlation coefficient between X_i and X_j $= \text{cov}(X_i, X_j)/\sqrt{\text{var}(X_i)\text{var}(X_j)}$

CHAPTER 34

General Remarks

1 INTRODUCTION

As mentioned in the Preface, this book is a considerably expanded version of Chapter 11 of the volume by Johnson and Kotz (1969). A very substantial revision of Chapters 1–10 of that volume has already appeared in Johnson, Kotz, and Kemp (1992).

As in the earlier volume, most of the distributions discussed here are closely related to the discrete univariate distributions described in Johnson, Kotz, and Kemp (1992). In particular, the *marginal distributions* of individual random variables are, for the most part, simple binomial, Poisson, geometric, negative binomial, hypergeometric or logarithmic series distributions, or distributions obtained by compounding these distributions.

Relevant chapters from Johnson, Kotz, and Kemp (1992) and Johnson, Kotz, and Balakrishnan (1994, 1995) should be used as references while reading the present volume. From now on, these chapters will be referred to simply by number without giving the authors and the publication date. In fact, it is only for this purpose that the chapters in the above-mentioned three volumes (and the present volume) have been arranged in a consecutive numerical order. Since we are using a great deal of prior material available in these volumes, it has been possible to condense the treatment and also include all fairly important multivariate discrete distributions in a book of moderate size. The amount of details provided will depend on the importance of the distribution as well as on the existence of easily available alternative sources of information. For example, in regard to the first point, the multinomial and the negative multinomial distributions receive rather extensive consideration due to the significant role they play in both statistical theory and practice. In regard to the second point, we have attempted to avoid unnecessary overlap with the information on the bivariate distributions already available in the volume by Kocherlakota and Kocherlakota (1992), and that on the mixtures of multivariate distributions available in Titterington, Smith, and Makov (1985) and Prakasa Rao (1992). A concise review of multivariate discrete distributions prepared recently by Papageorgiou (1997) and that of bivariate discrete distributions by Kocherlakota and Kocherlakota (1997) will provide some supporting information to the content of this chapter as well as subsequent chapters.

Many of the mathematical symbols used in this book are commonly used ones. However, there are some which are less commonly encountered, and hence for the convenience of the readers we have prepared a *Glossary* in which we have defined them. We have also presented a list of *Notations* used consistently throughout this book.

2 FEATURES OF MULTIVARIATE DISTRIBUTIONS

On passing from univariate to multivariate distributions, some essentially new features require attention. These are connected with relationships among *sets* of random variables and include regression, correlation, and, more generally, conditional distributions. These aspects of the discrete multivariate distributions will be given special attention in this volume. There will be somewhat less attention paid to methods of estimation. This is not because estimation methods are unimportant, but is due to the fact that there has been a notably slower historical development in estimation methods for discrete multivariate distributions. In spite of the rapid penetration of computer technology into statistical theory and practice, this slower growth in estimation methods may perhaps be ascribed to (a) the greater expense of obtaining multivariate data and (b) the heavier calculations involved in applying estimation techniques.

In the volume by Johnson and Kotz (1969), it was somewhat optimistically predicted that there would be "a rapid development of multivariate estimation techniques in the near future." Unfortunately, the trend over the past 27 years has been more toward the construction of elaborate (hopefully more realistic) models rather than toward the development of convenient and effective methods of estimation. An important exception, as described for example by Kocherlakota and Kocherlakota (1992), is in the development of Bayesian estimation procedures.

For the convenience of the readers we now present some definitions of special relevance to the description and analysis of discrete multivariate distributions. [See also Chapter 1 for additional information.]

2.1 Probabilities and Probability Generating Function

For a set of random variables $\boldsymbol{X} = (X_1, X_2, \ldots, X_k)'$, the joint *probability mass function* (pmf) of $\boldsymbol{X}$ is

$$P_{\boldsymbol{X}}(\boldsymbol{x}) = P_{X_1,X_2,\ldots,X_k}(x_1, x_2, \ldots, x_k) = \Pr\left[\bigcap_{i=1}^{k}(X_i = x_i)\right]. \tag{34.1}$$

(It is often referred to as the joint *probability function.*) The joint distribution of a subset of $(X_1, \ldots, X_k)'$, say $(X_1, \ldots, X_m)'$ for $m < k$, is a *marginal* distribution.

The joint *probability generating function* (pgf) of $\boldsymbol{X}$ is

$$G_{\boldsymbol{X}}(\boldsymbol{t}) = E\left[\prod_{i=1}^{k} t_i^{X_i}\right] = \sum_{x_1} \cdots \sum_{x_k} P_{X_1,X_2,\ldots,X_k}(x_1, x_2, \ldots, x_k) \prod_{i=1}^{k} t_i^{x_i}, \tag{34.2}$$

where $\boldsymbol{t} = (t_1, t_2, \ldots, t_k)'$. The pgf of the marginal distribution of $(X_1, \ldots, X_m)'$ is $G_{\boldsymbol{X}}(t_1, \ldots, t_m, 1, \ldots, 1)$. If $\boldsymbol{X}$ and $\boldsymbol{Y}$ are mutually independent k-dimensional random vectors, then

$$G_{\boldsymbol{X}+\boldsymbol{Y}}(\boldsymbol{t}) = G_{\boldsymbol{X}}(\boldsymbol{t})G_{\boldsymbol{Y}}(\boldsymbol{t})\,. \tag{34.3}$$

Also

$$G_{-\boldsymbol{X}}(\boldsymbol{t}) = G_{\boldsymbol{X}}(\boldsymbol{t}^{-1})\,, \tag{34.4}$$

where $\boldsymbol{t}^{-1} = (t_1^{-1}, t_2^{-1}, \ldots, t_k^{-1})'$, and so

$$G_{\boldsymbol{X}-\boldsymbol{Y}}(\boldsymbol{t}) = G_{\boldsymbol{X}}(\boldsymbol{t})G_{\boldsymbol{Y}}(\boldsymbol{t}^{-1})\,. \tag{34.5}$$

Further, with $G_{\boldsymbol{X}}(\boldsymbol{t})$ as the joint probability generating function of $\boldsymbol{X}$, upon setting $t_1 = t_2 = \cdots t_k = t$ one readily obtains

$$G_{\boldsymbol{X}}(t, \ldots, t) = E\left[t^{X_1+X_2+\cdots+X_k}\right] \equiv G_{\sum_{i=1}^k X_i}(t)$$

which is the probability generating function of the variable $\sum_{i=1}^k X_i$.

2.2 Moments and Moment-Type Generating Functions

The mixed $\boldsymbol{r} = (r_1, r_2, \ldots, r_k)'$th moment about zero is

$$\mu_{\boldsymbol{r}}'(\boldsymbol{X}) = \mu_{r_1,r_2,\ldots,r_k}'(X_1, X_2, \ldots, X_k) = E\left[\prod_{i=1}^k X_i^{r_i}\right]. \tag{34.6}$$

The corresponding central mixed moment is

$$\mu_{\boldsymbol{r}}(\boldsymbol{X}) = \mu_{r_1,r_2,\ldots,r_k}(X_1, X_2, \ldots, X_k) = E\left[\prod_{i=1}^k (X_i - E[X_i])^{r_i}\right]. \tag{34.7}$$

When there is no risk of confusion, the parenthetical $\boldsymbol{X}$ or $(X_1, X_2, \ldots, X_k)$ will be omitted.

The *characteristic function* of $\boldsymbol{X}$ is

$$E\left[e^{i\boldsymbol{t}'\boldsymbol{X}}\right] = E\left[e^{i(t_1X_1+t_2X_2+\cdots+t_kX_k)}\right], \tag{34.8}$$

where $i = \sqrt{-1}$ is the imaginary number and $\boldsymbol{t} = (t_1, t_2, \ldots, t_k)'$. The *moment generating function* (mgf) of $\boldsymbol{X}$ is

$$\varphi_{\boldsymbol{X}}(\boldsymbol{t}) = E\left[e^{\boldsymbol{t}'\boldsymbol{X}}\right] = E\left[e^{t_1X_1+t_2X_2+\cdots+t_kX_k}\right], \tag{34.9}$$

provided that this is finite. The central moment generating function is

$$\varphi_{\boldsymbol{X}}(\boldsymbol{t})e^{-\boldsymbol{t}'E[\boldsymbol{X}]}\,. \tag{34.10}$$

The mixed $\boldsymbol{r}$th moment about zero [central moment] may be determined from the moment [central moment] generating function in (34.9) [(34.10)] as the coefficient of $\prod_{i=1}^{k}(t_i^{r_i}/r_i!)$.

The *cumulant generating function* of $\boldsymbol{X}$ is

$$K_{\boldsymbol{X}}(\boldsymbol{t}) = \log \varphi_{\boldsymbol{X}}(\boldsymbol{t}) \tag{34.11}$$

and the mixed $\boldsymbol{r}$th cumulant, $\kappa_{\boldsymbol{r}}(\boldsymbol{X}) = \kappa_{r_1,r_2,\ldots,r_k}(X_1, X_2, \ldots, X_k)$, is the coefficient of $\prod_{i=1}^{k}(t_i^{r_i}/r_i!)$ in the cumulant generating function in (34.11).

We may also define the *factorial cumulant generating function* as

$$\log G_{\boldsymbol{X}}(\boldsymbol{t}+\boldsymbol{1}) = \log E\left[\prod_{i=1}^{k}(t_i+1)^{X_i}\right],$$

where $\boldsymbol{1} = (1, \ldots, 1)'_{1\times k}$, from which the *descending factorial cumulant*, $\kappa_{(\boldsymbol{r})} = \kappa_{(r_1,r_2,\ldots,r_k)}$, can be defined as the coefficient of $\prod_{i=1}^{k}(t_i^{r_i}/r_i!)$.

Remark. It should be noted that some authors define the bivariate factorial cumulant generating function as $\log G(t_1, t_2)$ instead of $\log G(t_1+1, t_2+1)$.

We will use the notations

$$a^{(b)} = a(a-1)\cdots(a-b+1) \tag{34.12}$$

for the bth *descending factorial of* a and use

$$a^{[b]} = a(a+1)\cdots(a+b-1) \tag{34.13}$$

for the bth *ascending factorial of* a, as in Johnson and Kotz (1969). [However, different notations are used by many authors including $a^{[b]}$ instead of $a^{(b)}$ for the bth descending factorial of a.] Pochhammer's symbol $(a)_b$ [Chapter 1, Eq. (1.6)] will sometimes be used in place of $a^{(b)}$ for the bth descending factorial of a.

The descending $\boldsymbol{r} = (r_1, r_2, \ldots, r_k)'$th factorial moment of $\boldsymbol{X}$ is

$$\mu'_{(\boldsymbol{r})}(\boldsymbol{X}) = \mu'_{(r_1,r_2,\ldots,r_k)}(\boldsymbol{X}) = E\left[\prod_{i=1}^{k} X_i^{(r_i)}\right]. \tag{34.14}$$

Note that

$$\mu'_{(\boldsymbol{r})}(\boldsymbol{X}) = d^{\sum_{i=1}^{k} r_i} G_{\boldsymbol{X}}(\boldsymbol{t}) \Big/ \left\{\prod_{i=1}^{k} dt_i^{r_i}\right\}\Bigg|_{\boldsymbol{t}=\boldsymbol{1}}. \tag{34.15}$$

Similarly, the ascending $\boldsymbol{r} = (r_1, r_2, \ldots, r_k)'$th factorial moment of $\boldsymbol{X}$ is

$$\mu'_{[\boldsymbol{r}]}(\boldsymbol{X}) = \mu'_{[r_1,r_2,\ldots,r_k]}(\boldsymbol{X}) = E\left[\prod_{i=1}^{k} X_i^{[r_i]}\right]. \tag{34.16}$$

We use $\mu'_{[\boldsymbol{r}]}$ to denote the ascending $\boldsymbol{r}$th factorial moment of $\boldsymbol{X}$ throughout this volume. Unfortunately, this notation has been used in the literature sometimes to denote the $\boldsymbol{r}$th descending factorial moment of $\boldsymbol{X}$.

2.3 Stirling Numbers and Moments

Recalling that $s(n, i)$ and $S(n, i)$ denote the Stirling numbers of the first kind and the second kind, respectively [see Chapter 1, p. 10], we have the following relationships:

$$\mu'_{(\boldsymbol{r})}(\boldsymbol{X}) = E\left[\prod_{i=1}^{k} X_i^{(r_i)}\right] = E\left[\prod_{i=1}^{k} \frac{X_i!}{(X_i - r_i)!}\right]$$

$$= \sum_{i_1=0}^{r_1} \cdots \sum_{i_k=0}^{r_k} s(r_1, i_1)s(r_2, i_2)\cdots s(r_k, i_k)\mu'_{\boldsymbol{i}}(\boldsymbol{X}) \tag{34.17}$$

where $\boldsymbol{i} = (i_1, \ldots, i_k)'$, and

$$\mu'_{\boldsymbol{r}}(\boldsymbol{X}) = E\left[\prod_{i=1}^{k} X_i^{r_i}\right]$$

$$= \sum_{i_1=0}^{r_1} \cdots \sum_{i_k=0}^{r_k} S(r_1, i_1)S(r_2, i_2)\cdots S(r_k, i_k)\mu'_{(\boldsymbol{i})}(\boldsymbol{X}). \tag{34.18}$$

It is easy to observe that the Stirling numbers of the first kind and the second kind satisfy the following relations:

$$s(n, 1) = (-1)^{n-1}(n-1)!,\ s(n, i) = s(n-1, i-1) - (n-1)s(n-1, i)$$
$$\text{for } i = 1, 2, \ldots, n-1, \text{ and } s(n, n) = 1; \tag{34.19}$$

$$S(n, 1) = 1,\ S(n, i) = S(n-1, i-1) + i\,S(n-1, i)$$
$$\text{for } i = 1, 2, \ldots, n-1, \text{ and } S(n, n) = 1\,. \tag{34.20}$$

For convenience, some values of the Stirling numbers of the first kind and the second kind are presented in Tables 34.1 and 34.2, respectively.

Table 34.1. Selected Values of Stirling Numbers of the First Kind, $s(n, i)$

	i							
n	1	2	3	4	5	6	7	8
1	1							
2	−1	1						
3	2	−3	1					
4	−6	11	−6	1				
5	24	−50	35	−10	1			
6	−120	274	−225	85	−15	1		
7	720	−1764	1624	−735	175	−21	1	
8	−5040	13068	−13132	6769	−1960	322	−28	1

Table 34.2. Selected Values of Stirling Numbers of the Second Kind, $S(n, i)$

	i							
n	1	2	3	4	5	6	7	8
1	1							
2	1	1						
3	1	3	1					
4	1	7	6	1				
5	1	15	25	10	1			
6	1	31	90	65	15	1		
7	1	63	301	350	140	21	1	
8	1	127	966	1701	1050	266	28	1

Next, with $x^{[n]}$ denoting the nth ascending factorial of x, we define the Stirling numbers of the third kind by

$$x^{[n]} = \sum_{i=0}^{n} \bar{s}(n, i)\, x^i \tag{34.21}$$

and define the Stirling numbers of the fourth kind by

$$x^n = \sum_{i=0}^{n} \bar{S}(n, i)\, x^{[i]} . \tag{34.22}$$

Then, with $\mu'_{[\boldsymbol{r}]}$ denoting the ascending $\boldsymbol{r}$th factorial moment of $\boldsymbol{X}$ as defined in (34.16), we have the following relationships:

$$\begin{aligned} \mu'_{[\boldsymbol{r}]}(\boldsymbol{X}) &= E\left[\prod_{i=1}^{k} X_i^{[r_i]}\right] = E\left[\prod_{i=1}^{k}\{X_i(X_i+1)\cdots(X_i+r_i-1)\}\right] \\ &= \sum_{i_1=0}^{r_1}\cdots\sum_{i_k=0}^{r_k} \bar{s}(r_1, i_1)\bar{s}(r_2, i_2)\cdots\bar{s}(r_k, i_k)\mu'_{\boldsymbol{i}}(\boldsymbol{X}) , \end{aligned} \tag{34.23}$$

where $\boldsymbol{i} = (i_1, i_2, \ldots, i_k)'$, and

$$\begin{aligned} \mu'_{\boldsymbol{r}}(\boldsymbol{X}) &= E\left[\prod_{i=1}^{k} X_i^{r_i}\right] \\ &= \sum_{i_1=0}^{r_1}\cdots\sum_{i_k=0}^{r_k} \bar{S}(r_1, i_1)\bar{S}(r_2, i_2)\cdots\bar{S}(r_k, i_k)\mu'_{[\boldsymbol{i}]}(\boldsymbol{X}) . \end{aligned} \tag{34.24}$$

It is easy to observe that the Stirling numbers of the third kind and the fourth kind satisfy the following relations:

Table 34.3. Selected Values of Stirling Numbers of the Third Kind, $\bar{s}(n, i)$

	i							
n	1	2	3	4	5	6	7	8
1	1							
2	1	1						
3	2	3	1					
4	6	11	6	1				
5	24	50	35	10	1			
6	120	274	225	85	15	1		
7	720	1764	1624	735	175	21	1	
8	5040	13068	13132	6769	1960	322	28	1

Table 34.4. Selected Values of Stirling Numbers of the Fourth Kind, $\bar{S}(n, i)$

	i							
n	1	2	3	4	5	6	7	8
1	1							
2	−1	1						
3	1	−3	1					
4	−1	7	−6	1				
5	1	−15	25	−10	1			
6	−1	31	−90	65	−15	1		
7	1	−63	301	−350	140	−21	1	
8	−1	127	−966	1701	−1050	266	−28	1

$$\bar{s}(n, 1) = (n-1)!, \quad \bar{s}(n, i) = \bar{s}(n-1, i-1) + (n-1)\bar{s}(n-1, i)$$
$$\text{for } i = 1, 2, \ldots, n-1, \text{ and } \bar{s}(n, n) = 1; \tag{34.25}$$

$$\bar{S}(n, 1) = (-1)^{n-1}, \quad \bar{S}(n, i) = \bar{S}(n-1, i-1) - i\,\bar{S}(n-1, i)$$
$$\text{for } i = 1, 2, \ldots, n-1, \text{ and } \bar{S}(n, n) = 1\,. \tag{34.26}$$

For convenience, some values of the Stirling numbers of the third kind and the fourth kind are presented in Tables 34.3 and 34.4, respectively.

2.4 Recurrence Relations

Note that the central moment generating function of $\boldsymbol{X}$ in (34.10) can also be written as

$$E[e^{\boldsymbol{t}'\boldsymbol{X}}]e^{-\boldsymbol{t}'\boldsymbol{\mu}} = \varphi_{\boldsymbol{X}}(\boldsymbol{t})\, e^{-\boldsymbol{t}'\boldsymbol{\mu}} = \exp\{K_{\boldsymbol{X}}(\boldsymbol{t}) - \boldsymbol{t}'\boldsymbol{\mu}\}, \tag{34.27}$$

where $\boldsymbol{\mu} = E[\boldsymbol{X}]$. Utilizing binomial expansions we readily obtain the relationships

(similar to those in the univariate case presented in Chapter 1, p. 42)

$$\mu_{\boldsymbol{r}}(\boldsymbol{X}) = \sum_{i_1=0}^{r_1} \cdots \sum_{i_k=0}^{r_k} (-1)^{i_1+\cdots+i_k} \binom{r_1}{i_1} \cdots \binom{r_k}{i_k} (E[X_1])^{i_1} \cdots (E[X_k])^{i_k} \cdot \mu'_{\boldsymbol{r}-\boldsymbol{i}}(\boldsymbol{X}) \tag{34.28}$$

and

$$\mu'_{\boldsymbol{r}}(\boldsymbol{X}) = \sum_{i_1=0}^{r_1} \cdots \sum_{i_k=0}^{r_k} \binom{r_1}{i_1} \cdots \binom{r_k}{i_k} (E[X_1])^{i_1} \cdots (E[X_k])^{i_k} \mu_{\boldsymbol{r}-\boldsymbol{i}}(\boldsymbol{X}). \tag{34.29}$$

To simplify the notation, denote $\mu'_{r_1,r_2,\ldots,r_\ell,0,\ldots,0}$ by $\mu'_{r_1,r_2,\ldots,r_\ell}$ and denote $\kappa_{r_1,r_2,\ldots,r_\ell,0,\ldots,0}$ by $\kappa_{r_1,r_2,\ldots,r_\ell}$. Smith (1995) has established the following two recurrence relations, which are convenient for computation:

$$\mu'_{r_1,r_2,\ldots,r_{\ell+1}} = \sum_{i_1=0}^{r_1} \cdots \sum_{i_\ell=0}^{r_\ell} \sum_{i_{\ell+1}=0}^{r_{\ell+1}-1} \binom{r_1}{i_1} \cdots \binom{r_\ell}{i_\ell} \binom{r_{\ell+1}-1}{i_{\ell+1}} \cdot \kappa_{r_1-i_1,r_2-i_2,\ldots,r_{\ell+1}-i_{\ell+1}}\, \mu'_{i_1,i_2,\ldots,i_{\ell+1}} \tag{34.30}$$

and

$$\kappa_{r_1,r_2,\ldots,r_{\ell+1}} = \sum_{i_1=0}^{r_1} \cdots \sum_{i_\ell=0}^{r_\ell} \sum_{i_{\ell+1}=0}^{r_{\ell+1}-1} \binom{r_1}{i_1} \cdots \binom{r_\ell}{i_\ell} \binom{r_{\ell+1}-1}{i_{\ell+1}} \cdot \mu'_{r_1-i_1,r_2-i_2,\ldots,r_{\ell+1}-i_{\ell+1}}\, \mu'^{*}_{i_1,i_2,\ldots,i_{\ell+1}}. \tag{34.31}$$

In the above relationship, $\mu'^{*}_{\boldsymbol{i}}$ denotes the formal "mixed $\boldsymbol{i}$th moment about zero" of a "distribution" with the "cumulant generating function" $-K_{\boldsymbol{X}}(\boldsymbol{t})$ or, equivalently, with the "moment generating function"

$$\varphi^*(\boldsymbol{t}) = \exp\{-K_{\boldsymbol{X}}(\boldsymbol{t})\}, \tag{34.32}$$

that is,

$$\varphi^*(\boldsymbol{t}) = \sum_{r_1=0}^{\infty} \cdots \sum_{r_k=0}^{\infty} \mu'^{*}_{\boldsymbol{r}} \prod_{i=1}^{k} (t_i^{r_i}/r_i!). \tag{34.33}$$

Note that the negative sign in (34.32) is nonstandard and Smith (1995) has introduced it more for computational convenience.

Employing (34.32), Balakrishnan, Johnson, and Kotz (1995) have established the recurrence relations

$$\mu_{r_1,r_2,\ldots,r_{\ell+1}} = \sum_{i_1=0}^{r_1} \cdots \sum_{i_\ell=0}^{r_\ell} \sum_{i_{\ell+1}=0}^{r_{\ell+1}-1} \binom{r_1}{i_1} \cdots \binom{r_\ell}{i_\ell} \binom{r_{\ell+1}-1}{i_{\ell+1}} \cdot \{\kappa_{r_1-i_1,r_2-i_2,\ldots,r_{\ell+1}-i_{\ell+1}}\, \mu_{i_1,i_2,\ldots,i_{\ell+1}} - E[X_{\ell+1}]\, \mu_{i_1,i_2,\ldots,i_\ell,i_{\ell+1}-1}\} \tag{34.34}$$

and

$$\kappa_{r_1,r_2,\dots,r_{\ell+1}} = \sum_{i_1,j_1=0}^{r_1} \cdots \sum_{i_\ell,j_\ell=0}^{r_\ell} \sum_{i_{\ell+1},j_{\ell+1}=0}^{r_{\ell+1}-1} \binom{r_1}{i_1, j_1, r_1 - i_1 - j_1}$$

$$\cdots \binom{r_\ell}{i_\ell, j_\ell, r_\ell - i_\ell - j_\ell} \binom{r_{\ell+1} - 1}{i_{\ell+1}, j_{\ell+1}, r_{\ell+1} - i_{\ell+1} - j_{\ell+1} - 1}$$

$$\cdot \mu_{r_1-i_1-j_1,\dots,r_{\ell+1}-i_{\ell+1}-j_{\ell+1}} \mu^*_{i_1,\dots,i_{\ell+1}} \prod_{i=1}^{\ell+1} (\mu_i + \mu_i^*)^{j_i}$$

$$+ \mu_{\ell+1} I\{r_1 = \cdots = r_\ell = 0, r_{\ell+1} = 1\}, \qquad (34.35)$$

where μ_i^* corresponds to the "distribution" defined by (34.32), and $I\{\cdot\}$ denotes the indicator function. In (34.34) and (34.35), $\mu_{r_1,r_2,\dots,r_{\ell+1}}$ and $\kappa_{r_1,r_2,\dots,r_{\ell+1}}$ have been used as simplified notations for $\mu_{r_1,\dots,r_{\ell+1},0,\dots,0}$ and $\kappa_{r_1,\dots,r_{\ell+1},0,\dots,0}$, respectively. Furthermore, $\sum_{i,j=0}^{r}$ denotes the summation over all non-negative integers i and j such that $0 \le i + j \le r$, and as usual

$$\binom{r}{i, j, r - i - j} = \frac{r!}{i!j!(r-i-j)!} .$$

Balakrishnan and Johnson (1997) recently established a recurrence relation for the mixed moments of $\boldsymbol{X}$ when the probability generating function $G_{\boldsymbol{X}}(\boldsymbol{t})$ assumes a special form. Specifically, let

$$G_{\boldsymbol{X}}(\boldsymbol{t}) = \frac{A(\boldsymbol{t})}{B(\boldsymbol{t})} , \qquad (34.36)$$

where $A(\boldsymbol{t})$ and $B(\boldsymbol{t})$ are k-variate polynomials in $\boldsymbol{t}$ given by

$$A(\boldsymbol{t}) = \sum_{\ell_1=0}^{p_1} \cdots \sum_{\ell_k=0}^{p_k} a(\ell_1, \dots, \ell_k) t_1^{\ell_1} \cdots t_k^{\ell_k} \qquad (34.37)$$

and

$$B(\boldsymbol{t}) = \sum_{\ell_1=0}^{q_1} \cdots \sum_{\ell_k=0}^{q_k} b(\ell_1, \dots, \ell_k) t_1^{\ell_1} \cdots t_k^{\ell_k} . \qquad (34.38)$$

Then, the following recurrence relation holds for the joint probability mass function:

$$\sum_{\ell_1=0}^{q_1} \cdots \sum_{\ell_k=0}^{q_k} b(\ell_1, \dots, \ell_k) P_{\boldsymbol{X}}(\boldsymbol{x} - \boldsymbol{\ell}) = a(x_1, \dots, x_k) I(\boldsymbol{x} \in T) , \qquad (34.39)$$

where $I(\boldsymbol{x} \in T)$ is the indicator function taking on the value 1 if $\boldsymbol{x} \in T$ and 0 if $\boldsymbol{x} \notin T$, $\boldsymbol{\ell} = (\ell_1, \dots, \ell_k)'$, and

$$T = \{(\ell_1, \dots, \ell_k) : 0 \le \ell_i \le p_i \text{ for } i = 1, 2, \dots, k\} . \qquad (34.40)$$

Further, the following recurrence relation holds for the joint moments:

$$\sum_{\ell_1=0}^{q_1} \cdots \sum_{\ell_k=0}^{q_k} b(\ell_1, \ldots, \ell_k) E\left[(X_1 + \ell_1)^{r_1} \cdots (X_k + \ell_k)^{r_k}\right]$$

$$= \sum_{x_1=0}^{p_1} \cdots \sum_{x_k=0}^{p_k} a(x_1, \ldots, x_k) x_1^{r_1} \cdots x_k^{r_k} . \qquad (34.41)$$

From (34.41), upon using binomial expansions, we readily obtain the recurrence relation

$$\sum_{\ell_1=0}^{q_1} \cdots \sum_{\ell_k=0}^{q_k} b(\ell_1, \ldots, \ell_k) \mu_{\boldsymbol{r}}'(\boldsymbol{X})$$

$$= \sum_{x_1=0}^{p_1} \cdots \sum_{x_k=0}^{p_k} a(x_1, \ldots, x_k) x_1^{r_1} \cdots x_k^{r_k}$$

$$- \sum_{\ell_1=0}^{q_1} \cdots \sum_{\ell_k=0}^{q_k} \underset{\boldsymbol{i} \neq \boldsymbol{r}}{\sum_{i_1=0}^{r_1} \cdots \sum_{i_k=0}^{r_k}} \binom{r_1}{i_1} \cdots \binom{r_k}{i_k} \ell_1^{r_1 - i_1} \cdots \ell_k^{r_k - i_k} \mu_{\boldsymbol{i}}'(\boldsymbol{X}) . \qquad (34.42)$$

In addition, for the special case when $a(\ell_1, \ldots, \ell_k) = 0$ for all $(\ell_1, \ldots, \ell_k) \neq (p_1, \ldots, p_k)$ and $a(p_1, \ldots, p_k) = a$, (34.41) gives

$$\sum_{\ell_1=0}^{q_1} \cdots \sum_{\ell_k=0}^{q_k} b(\ell_1, \ldots, \ell_k) E\left[(X_1 + \ell_1)^{r_1} \cdots (X_k + \ell_k)^{r_k}\right]$$

$$= a\, p_1^{r_1} p_2^{r_2} \cdots p_k^{r_k} \qquad (34.43)$$

and, if $p_i = 0$ for some i, then

$$\sum_{\ell_1=0}^{q_1} \cdots \sum_{\ell_k=0}^{q_k} b(\ell_1, \ldots, \ell_k) E\left[(X_1 + \ell_1)^{r_1} \cdots (X_k + \ell_k)^{r_k}\right]$$

$$= a \left\{ \prod_{\substack{j=1 \\ j \neq i}}^{k} p_j^{r_j} \right\} I(r_i = 0) , \qquad (34.44)$$

where $I(r_i = 0)$ is the indicator function taking on the value 1 if $r_i = 0$ and 0 if $r_i > 0$.

2.5 Incomplete Moments

Analogous to the case of univariate distributions, it is sometimes useful to determine *incomplete moments*. While the complete moments are of the general form

$$\sum_{\boldsymbol{x} \in T} \cdots \sum g(\boldsymbol{x}) P_{\boldsymbol{X}}(\boldsymbol{x}), \qquad (34.45)$$

the incomplete moments are of the same form with T replaced by a subset T^*. For example, while the mixed $\boldsymbol{r}$th moment about zero is

$$\mu'_{\boldsymbol{r}}(\boldsymbol{X}) = \sum_{\boldsymbol{x}\in T} \cdots \sum \left(\prod_{i=1}^{k} x_i^{r_i} \right) P_{\boldsymbol{X}}(\boldsymbol{x}), \tag{34.46}$$

the mixed $\boldsymbol{r}$th incomplete moment about zero is [Gerstenkorn (1980)]

$$\sum_{\boldsymbol{x}\in T^*} \cdots \sum \left(\prod_{i=1}^{k} x_i^{r_i} \right) P_{\boldsymbol{X}}(\boldsymbol{x}), \tag{34.47}$$

which is just a partial sum from (34.46).

2.6 Conditional Distributions and Moments

The conditional probability generating function $G_{X_2,\ldots,X_k|X_1=x_1}(t_2,\ldots,t_k)$ is given by

$$G_{X_2,\ldots,X_k|X_1=x_1}(t_2,\ldots,t_k) = \frac{G^{(x_1,0,\ldots,0)}(0,t_2,\ldots,t_k)}{G^{(x_1,0,\ldots,0)}(0,1,\ldots,1)}, \tag{34.48}$$

where $G^{(x_1,\ldots,x_k)}(t_1,\ldots,t_k) = \frac{\partial^{x_1+\cdots+x_k}}{\partial t_1^{x_1}\cdots\partial t_k^{x_k}} G(t_1,\ldots,t_k)$. A simple extension of this formula to the case of the joint conditional probability generating function of m components conditional on l components has been derived by Xekalaki (1987). The conditional expected value of X_i, given $X_j = x_j$, denoted by

$$E[X_i \mid X_j = x_j], \tag{34.49}$$

is the *regression (function) of* X_i *on* X_j. It is a function of x_j (**not** a random variable). The symbol

$$E[X_i \mid X_j] \tag{34.50}$$

will also be used, which, in general, is a random variable. If X_j is replaced by a set of random variables $\boldsymbol{X_a} = (X_{a_1}, X_{a_2}, \ldots, X_{a_p})'$, then the conditional expected value

$$E[X_i \mid \boldsymbol{X_a} = \boldsymbol{x_a}] = E\left[X_i \mid \bigcap_{j=1}^{p} (X_{a_j} = x_{a_j}) \right] \tag{34.51}$$

is the *multiple regression (function) of* X_i *on* $\boldsymbol{X_a}$. As above, we emphasize that this is a function of $(x_{a_1}, x_{a_2}, \ldots, x_{a_p})$ and not a random variable, while $E[X_i \mid \boldsymbol{X_a}]$ is, in general, a random variable.

The joint *cumulative distribution function* (cdf) of $\boldsymbol{X}$ is

$$\begin{aligned} F_{\boldsymbol{X}}(\boldsymbol{x}) &= F_{X_1,X_2,\ldots,X_k}(x_1,x_2,\ldots,x_k) = \Pr\left[\bigcap_{i=1}^{k} (X_i \le x_i) \right] \\ &= \sum_{\{y_i \le x_i\}} \cdots \sum P_{X_1,X_2,\ldots,X_k}(y_1,y_2,\ldots,y_k). \end{aligned} \tag{34.52}$$

The joint *conditional distribution* of $\boldsymbol{X}$, given that some other random variables $\boldsymbol{Z} = (Z_1, Z_2, \ldots, Z_h)'$ have values $\boldsymbol{Z} = \boldsymbol{z}$, has probability mass function

$$P_{\boldsymbol{X}|\boldsymbol{z}}(\boldsymbol{x} \mid \boldsymbol{Z} = \boldsymbol{z}) = P_{X_1,X_2,\ldots,X_k}(x_1, x_2, \ldots, x_k \mid z_1, z_2, \ldots, z_h)$$
$$= \Pr\left[\bigcap_{i=1}^{k}(X_i = x_i) \mid \bigcap_{j=1}^{h}(Z_j = z_j)\right] \qquad (34.53)$$

and cumulative distribution function

$$F_{\boldsymbol{X}|\boldsymbol{z}}(\boldsymbol{x} \mid \boldsymbol{Z} = \boldsymbol{z}) = \Pr\left[\bigcap_{i=1}^{k}(X_i \le x_i) \mid \bigcap_{j=1}^{h}(Z_j = z_j)\right]$$
$$= \sum_{\{y_i \le x_i\}} \cdots \sum P_{\boldsymbol{X}|\boldsymbol{z}}(y_1, y_2, \ldots, y_k \mid z_1, z_2, \ldots, z_h)\,. \qquad (34.54)$$

Sometimes, (34.54) is called the *array distribution* of $\boldsymbol{X}$ given $\boldsymbol{Z} = \boldsymbol{z}$. This term is more commonly used when $k = 1$—that is, when there is only a single X_1 so that the array distribution is univariate. The variance–covariance matrix of the array distribution is called the *array variance–covariance matrix*. [Naturally, it is called the *array variance* when $k = 1$.] If the array distribution of $\boldsymbol{X}$ given $\boldsymbol{Z} = \boldsymbol{z}$ does not depend on $\boldsymbol{z}$, then the regression of $\boldsymbol{X}$ on $\boldsymbol{Z}$ is said to be *homoscedastic*. The term "homoscedastic" is sometimes restricted to those cases wherein the array distributions for different values of $\boldsymbol{z}$ differ only in respect to location, and cases in which only the variance–covariance matrix remains constant are referred to as "weakly homoscedastic." We will not follow this usage in this volume.

3 DISCRETE MULTIVARIATE FAMILIES

In Chapter 2 (and also in Chapter 12, for *continuous* variables), some examples are presented of "families" of distributions embracing wide varieties of special distributions. We shall now proceed with a similar course of action for discrete multivariate distributions. However, we will present here relatively less material reflecting, to some degree, the less fully developed work that exists on this topic for discrete joint distributions; furthermore, one will also find quite broad families of distributions among generalizations of the specific distributions that are discussed in the later chapters of this volume.

3.1 Multivariate Discrete Exponential Family

First, we describe a family originally proposed by Khatri (1983a,b) which he termed a *multivariate discrete exponential family (MDEF)*. This family includes all joint distributions of $\boldsymbol{X} = (X_1, X_2, \ldots, X_k)'$ with probability mass functions of the form

$$P_{\boldsymbol{X}}(\boldsymbol{x}) = g(\boldsymbol{x}) \exp\left\{\sum_{i=1}^{k} R_i(\boldsymbol{\theta}) x_i - Q(\boldsymbol{\theta})\right\}, \qquad (34.55)$$

where $\boldsymbol{\theta} = (\theta_1, \theta_2, \ldots, \theta_k)'$ and $\boldsymbol{x}$ takes on values in a region T. Since

$$\sum_{\boldsymbol{x} \in T} \cdots \sum P_{\mathbf{X}}(\boldsymbol{x}) = 1\,, \tag{34.56}$$

it readily follows that

$$Q(\boldsymbol{\theta}) = \log\left[\sum_{\boldsymbol{x} \in T} \cdots \sum g(\boldsymbol{x}) \exp\left\{\sum_{i=1}^{k} R_i(\boldsymbol{\theta}) x_i\right\}\right]. \tag{34.57}$$

Often, T includes only integer values of the $\boldsymbol{x}$'s, but this need not be the case. Among the many classes of distributions included in this family, we note the following:

1. "Nonsingular" multinomial

$$\text{with } p = k;\ g(\boldsymbol{x}) = \left\{\left(n - \sum_{i=1}^{k} x_i\right)! \prod_{i=1}^{k} x_i!\right\}^{-1},$$

$$R_i(\boldsymbol{\theta}) = \log\left\{\theta_i \Big/ \left(1 - \sum_{j=1}^{k} \theta_j\right)\right\},$$

$$\text{and } Q(\boldsymbol{\theta}) = -n \log\left(1 - \sum_{i=1}^{k} \theta_i\right)$$

(see Chapter 35)

2. Multivariate Poisson (see Chapter 37)

3. Multivariate hypergeometric (see Chapter 39)

4. Multivariate power series (see Chapter 38)

5. Multivariate Lagrangian Poisson (see Chapter 37)

6. Multivariate negative binomial (see Chapter 36)

7. Lagrangian multinomial

8. Multivariate Lagrangian logarithmic

9. Multivariate Lagrangian negative binomial (see Chapter 36)

We believe that, in general, separate discussions focused on each specific class of distributions provide a more thorough understanding of their particular features than would an attempt to provide a broad picture based on the membership of a large family such as MDEF. However, there are some common general features that can usefully be developed. For example, Khatri (1983a) has presented the following formulas, valid for all MDEF distributions. They all involve the $(k \times k)$ matrix

$$\boldsymbol{M} = \big((m_{ij})\big) = \big((\partial R_i(\boldsymbol{\theta})/\partial \theta_j)\big)\,, \quad i, j = 1, 2, \ldots, k, \tag{34.58}$$

and the element $((m^{ij}))$ of its inverse $\boldsymbol{M}^{-1}$. With $\boldsymbol{e}_i$ denoting the ith column of the $(k \times k)$ identity matrix $\boldsymbol{I}_k$, we have

$$E[\boldsymbol{X}] = \boldsymbol{M}^{-1} \cdot \frac{\partial Q(\boldsymbol{\theta})}{\partial \boldsymbol{\theta}}, \tag{34.59}$$

$$\frac{\partial}{\partial \boldsymbol{\theta}} \Pr[\boldsymbol{X} = \boldsymbol{x}] = \boldsymbol{M}(\boldsymbol{x} - E[\boldsymbol{X}]) \Pr[\boldsymbol{X} = \boldsymbol{x}], \tag{34.60}$$

$$\frac{\partial}{\partial \boldsymbol{\theta}} \mu'_{\boldsymbol{r}} = \boldsymbol{M} \left\{ \sum_{i=1}^{k} \boldsymbol{e}_i \, \mu'_{\boldsymbol{r}+\boldsymbol{e}_i} - \mu'_{\boldsymbol{r}} E[\boldsymbol{X}] \right\}$$

$$\text{(for moments about zero)}, \tag{34.61}$$

$$\mu_{\boldsymbol{r}+\boldsymbol{e}_i} = \sum_{j=1}^{k} \left\{ m^{ij} + r_j \, \mu_{\boldsymbol{r}-\boldsymbol{e}_j} \text{cov}(X_i, X_j) \right\}$$

$$\text{(for central moments)}, \tag{34.62}$$

$$\kappa_{\boldsymbol{r}+\boldsymbol{e}_i} = \sum_{j=1}^{k} m^{ij} \, \partial \kappa_{\boldsymbol{r}} / \partial \theta_j \quad \text{(for cumulants)}, \tag{34.63}$$

and, in particular,

$$\text{cov}(X_i, X_j) = \sum_{\ell=1}^{k} m^{j\ell} \frac{\partial}{\partial \theta_\ell} E[X_i] \quad \text{for } 1 \leq i \neq j \leq k.$$

In the case when $Q(\boldsymbol{\theta})$ and $R(\boldsymbol{\theta})$ are infinitely differentiable with respect to $\boldsymbol{\theta}$ and $\boldsymbol{\theta} = \boldsymbol{0} \in \boldsymbol{D}$, a simpler representation of (34.55) can be obtained as

$$P_{\boldsymbol{X}}(\boldsymbol{x}) = g(\boldsymbol{x}, \boldsymbol{p}) \{ \boldsymbol{\theta}^{\boldsymbol{x}} (f(\boldsymbol{\theta}))^{\boldsymbol{x}} \} \exp(-\boldsymbol{p}'\boldsymbol{\theta}), \quad \boldsymbol{x} \in T, \tag{34.64}$$

where $g(\boldsymbol{x}, \boldsymbol{p})$ can be determined from

$$(\boldsymbol{x}!) g(\boldsymbol{x}) = [\boldsymbol{D}^{\boldsymbol{x}} \{ \exp(Q(\boldsymbol{\theta}))(f(\boldsymbol{\theta}))^{\boldsymbol{x}} | \boldsymbol{I} - \boldsymbol{A}(\boldsymbol{\theta}) |]_{\boldsymbol{\theta}=\boldsymbol{0}}. \tag{34.65}$$

In the above equations,

$$f_i(\boldsymbol{\theta}) = \theta_i \exp(-R_i(\boldsymbol{\theta})) \quad \text{for } i = 1, 2, \ldots, k,$$

$$f(\boldsymbol{\theta}) = (f_1(\boldsymbol{\theta}), \ldots, f_k(\boldsymbol{\theta}))',$$

$$D_i = \partial / \partial \theta_i,$$

$$\boldsymbol{D}^{\boldsymbol{x}} = \prod_{i=1}^{k} D_i^{x_i},$$

$$\boldsymbol{x}! = \prod_{i=1}^{k} (x_i!),$$

$$(f(\boldsymbol{\theta}))^{\boldsymbol{x}} = \prod_{i=1}^{k} (f_i(\boldsymbol{\theta}))^{x_i},$$

$$\boldsymbol{A}(\boldsymbol{\theta}) = \left((a_{ij}(\boldsymbol{\theta}))\right), \qquad \text{where } a_{ij}(\boldsymbol{\theta}) = \theta_i D_j \log f_i(\boldsymbol{\theta}), \quad 1 \le i, j \le k,$$

and

$$Q(\boldsymbol{\theta}) = \sum_{i=1}^{k} p_i \theta_i = \boldsymbol{p}'\boldsymbol{\theta},$$

where $\boldsymbol{p} \in D_1$ – the parametric space of $\boldsymbol{p}$ such that $g(\boldsymbol{x}) \equiv g(\boldsymbol{x}, \boldsymbol{p}) \ge 0$ for all $\boldsymbol{x} \in T$ and $\boldsymbol{p} \in D_1$.

The distribution in (34.64) possesses additive property; that is, if $\boldsymbol{X}$ is distributed according to (34.64) with $\boldsymbol{p} \in D_1$ and $\boldsymbol{Y}$ is independently distributed according to (34.64) with $\boldsymbol{q} \in D_1$, then $\boldsymbol{Z} = \boldsymbol{X} + \boldsymbol{Y}$ is distributed as multivariate discrete exponential in (34.64) with $\boldsymbol{p} + \boldsymbol{q} \in D_1$. Thus,

$$\Pr[\boldsymbol{X} = \boldsymbol{x} | \boldsymbol{Z} = \boldsymbol{z}] = \frac{g(\boldsymbol{x}, \boldsymbol{p}) g(\boldsymbol{z} - \boldsymbol{x}, \boldsymbol{q})}{g(\boldsymbol{z}, \boldsymbol{p} + \boldsymbol{q})} \tag{34.66}$$

for all $\boldsymbol{x} \in T_1(\boldsymbol{z}) = \{\boldsymbol{x} : x_i = 0, 1, \ldots, z_i \text{ for } 1 \le i \le k;\ \boldsymbol{p}, \boldsymbol{q}, \boldsymbol{p} + \boldsymbol{q} \in D_1\}$.

Khatri (1983b) also introduced *multivariate Markov–Pólya distribution* by choosing

$$f_i(\boldsymbol{\theta}) = c_i \theta_i \{1 - \exp(-c_i \theta_i)\}^{-1} \exp\left(\sum_{j=1}^{k} d_{ij} \theta_j\right), \qquad i = 1, 2, \ldots, k \tag{34.67}$$

in (34.64). This is a generalization of the corresponding univariate distribution with probability mass function

$$\Pr[X = x] = g(x, p)\{(1 - \exp(-c\theta))/(c\theta)^x\}(1 + d\theta)^{-bx} \exp(-p\theta), \tag{34.68}$$

which also satisfies the property in (34.66). In this case, an explicit (but rather complicated) expression for $g(x, p)$ is obtained from (34.65) as

$$\begin{aligned} x! g(x, p) &= \left[D^x \{\exp(p\theta)(f(\theta))^x (1 - \theta D \log f(\theta))\}\right]_{\theta=0} \\ &= p \left[D^{x-1} \exp(p\theta)(f(\theta))^x\right]_{\theta=0} \\ &= p \left[D^{x-1} \left\{ \left(\frac{c\theta}{1 - \exp(-c\theta)} \right)^x \exp(p\theta)(1 + d\theta)^{bx} \right\} \right]_{\theta=0} \\ &= p \sum_{j=0}^{x-1} \binom{x-1}{j} \left[D^{x-1-j} \left\{ \left(\frac{c\theta}{1 - \exp(-c\theta)} \right)^x \exp(p\theta) \right\} \right]_{\theta=0} \end{aligned}$$

$$\times \left[D^j(1+d\theta)^{bx}\right]_{\theta=0}$$

$$= p\sum_{j=0}^{x-1}\binom{x-1}{j}\binom{bx}{j} j!d^j(1+pc)^{[x-1-j,c]}, \tag{34.69}$$

where $a^{[0,c]} = 1$ and $a^{[y,c]} = a(a+c)\cdots(a+\overline{y-1}c)$.

If we set $\phi_i(\boldsymbol{\theta}) = \exp(R_i(\boldsymbol{\theta}))$ for $i = 1, 2, \ldots, k$, we can write from Eq. (34.57)

$$\begin{aligned}\exp(Q(\boldsymbol{\theta})) &= \sum_{\boldsymbol{x}\in T}\cdots\sum g(\boldsymbol{x})\exp\left\{\sum_{i=1}^{k} R_i(\boldsymbol{\theta})x_i\right\}\\ &= \sum_{\boldsymbol{x}\in T}\cdots\sum g(\boldsymbol{x})\prod_{i=1}^{k}(\phi_i(\boldsymbol{\theta}))^{x_i}\\ &= h(\boldsymbol{\theta}), \qquad \text{say.}\end{aligned} \tag{34.70}$$

Therefore, if the transformations $\phi_i(\boldsymbol{\theta}) = \exp(R_i(\boldsymbol{\theta}))$ for $i = 1, 2, \ldots, k$ are invertible, then the multivariate discrete exponential family in (34.55) becomes a power series distribution (see Chapter 38) given by

$$P_{\boldsymbol{X}}(\boldsymbol{x}) = g(\boldsymbol{x})\prod_{i=1}^{k}(\phi_i(\boldsymbol{\theta}))^{x_i}/h(\boldsymbol{\theta}) \qquad \text{for } \boldsymbol{x}\in T\,. \tag{34.71}$$

3.2 Multivariate Lagrangian Distributions

The multivariate discrete exponential family in (34.55) includes *multivariate Lagrangian distributions* when

$$R_i(\boldsymbol{\theta}) = \log\theta_i - \log g_i(\boldsymbol{\theta}) \quad \text{with } g_i(\boldsymbol{0}) \neq 0, \qquad i = 1, 2, \ldots, k\,.$$

In addition, it includes the following distributions:

1. *Multivariate Lagrangian Poisson distribution* with joint probability mass function [Khatri (1962)]

$$\begin{aligned}P_{\boldsymbol{X}}(\boldsymbol{x}) = |\boldsymbol{I}-\boldsymbol{A}(\boldsymbol{x})|\prod_{j=1}^{k}\Bigg\{&\left(\sum_{i=0}^{k} c_{ji}x_i\right)^{x_j}\\ &\times\exp\left(-\sum_{i=0}^{k} c_{ji}x_i\theta_i\right)\theta_j^{x_j}/j!\Bigg\},\end{aligned} \tag{34.72}$$

where $x_0 = 1$, $x_j = 0, 1, \ldots$ for all $j = 1, 2, \ldots, k$, and c_{ji} are non-negative constants for $j = 1, 2, \ldots, k$ and $i = 0, 1, \ldots, k$. Further, let $\boldsymbol{A}(\boldsymbol{x}) = \big((a_{ij}(\boldsymbol{x}))\big)$

with

$$a_{ij}(\boldsymbol{x}) = \frac{x_i}{\sum_{q=1}^{k} c_{iq} x_q} c_{ij} \quad \text{for } i, j = 1, 2, \ldots, k \,. \tag{34.73}$$

In this case

$$\begin{aligned} Q(\boldsymbol{\theta}) &= \textstyle\sum_{j=1}^{k} c_{j0}\theta_j \,, \\ R_i(\boldsymbol{\theta}) &= \log\theta_i - \textstyle\sum_{\ell=1}^{k} c_{\ell i}\theta_\ell, \quad 1 \le i \le k \,, \end{aligned} \tag{34.74}$$

and $\boldsymbol{M} = \big((m_{ij})\big)$, where

$$\begin{aligned} m_{ij} &= \tfrac{1}{\theta_i} - c_{ii} && \text{if } i = j \\ &= -c_{ji} && \text{if } i \neq j \,. \end{aligned}$$

Hence, $E[\boldsymbol{X}] = \boldsymbol{M}^{-1}\boldsymbol{c}_0$, where $\boldsymbol{c}_0 = (c_{10}, c_{20}, \ldots, c_{k0})'$, and

$$\operatorname{cov}(X_i, X_j) = \sum_{\ell=1}^{k} \left(m^{j\ell}\, m^{i\ell}/\theta_\ell^2\right) E[X_\ell] \quad \text{for all } i, j.$$

2. *Zero-truncated multivariate Lagrangian Poisson distribution* with joint probability mass function

$$\Pr[\boldsymbol{X}_0 = \boldsymbol{x}] = P(\boldsymbol{x}|\boldsymbol{\theta})/\{1 - \exp(-Q(\boldsymbol{\theta}))\} \tag{34.75}$$

for all $x_i = 0, 1, \ldots$, excluding $x_i = 0$ for all $i = 1, 2, \ldots, k$, where $P(\boldsymbol{x}|\boldsymbol{\theta})$ is given by (34.72) and $Q(\boldsymbol{\theta})$ is as defined in (34.74). Evidently,

$$E[\boldsymbol{X}_0] = E[\boldsymbol{X}]/\{1 - \exp(-Q(\boldsymbol{\theta}))\} \tag{34.76}$$

and

$$\operatorname{var}(\boldsymbol{X}_0) = \boldsymbol{M}^{-1}\boldsymbol{D}_{\eta_0}(\boldsymbol{M}^{-1})' - \{E[\boldsymbol{X}_0]\}\{E[\boldsymbol{X}_0]\}'\exp(-Q(\boldsymbol{\theta})), \tag{34.77}$$

where $\boldsymbol{D}_{\eta_0} = \operatorname{diag}(\eta_{01}, \eta_{02}, \ldots, \eta_{0k})$ and $\eta_{0i} = E[X_{0i}]/\theta_i^2$ for $i = 1, 2, \ldots, k$.

3. *Lagrangian negative multinomial distribution* with joint probability mass function [Mohanty (1966)]

$$\begin{aligned} \Pr[\boldsymbol{X} = \boldsymbol{x}] = {} & \frac{a_0}{a_0 + \sum_{i=1}^{k} a_i x_i} \binom{a_0 + \sum_{i=1}^{k}(a_i + 1)x_i}{x_1, \ldots, x_k,\ a_0 + \sum_{i=1}^{k} a_i x_i} \\ & \times \left\{\prod_{i=1}^{k} \theta_i^{x_i}\right\} \left\{1 - \sum_{i=1}^{p} \theta_i\right\}^{a_0 + \sum_{i=1}^{k} a_i x_i}, \end{aligned} \tag{34.78}$$

where

$$\binom{n}{y_1, \ldots, y_k} = \frac{n!}{y_1!\, y_2! \cdots y_k!} \qquad \left(\sum_{i=1}^{k} y_i = n\right),$$

$a_i > 0$ for $i = 0, 1, \ldots, k$, and $\sum_{i=1}^{k} \theta_i < 1$ with $0 \le \theta_i < 1$. In this case

$$\exp(-Q(\boldsymbol{\theta})) = \left(1 - \sum_{i=1}^{k} \theta_i\right)^{a_0} \quad \text{and}$$

$$R_i(\boldsymbol{\theta}) = \log \theta_i + a_i \log \left(1 - \sum_{\ell=1}^{k} \theta_\ell\right); \tag{34.79}$$

furthermore,

$$E[\boldsymbol{X}] = a_0 \boldsymbol{\theta}/b \tag{34.80}$$

and

$$\begin{aligned} \text{var}(\boldsymbol{X}) = {} & \frac{a_0}{b} (\boldsymbol{I} + \boldsymbol{D}_\theta \mathbf{1} \boldsymbol{a}'/b) \boldsymbol{D}_\theta (\boldsymbol{I} + \boldsymbol{a} \mathbf{1}' \boldsymbol{D}_\theta / b) \\ & + \left(1 - \sum_{\ell=1}^{k} \theta_\ell\right) a_0 \boldsymbol{D}_\theta \mathbf{1}\, \mathbf{1}' \boldsymbol{D}_\theta / b^3, \end{aligned} \tag{34.81}$$

where $b = 1 - \sum_{\ell=1}^{k} (a_\ell + 1)\theta_\ell$, $\boldsymbol{D}_\theta = \text{diag}(\theta_1, \theta_2, \ldots, \theta_k)$, $\boldsymbol{a} = (a_1, \ldots, a_k)'$, and $\mathbf{1} = (1, 1, \ldots, 1)'$; see Khatri (1962).

4. *Zero-truncated Lagrangian negative multinomial distribution* with joint probability mass function

$$\Pr[\boldsymbol{X}_0 = \boldsymbol{x}] = P(\boldsymbol{x}|\boldsymbol{\theta}) \Bigg/ \left\{1 - \left(1 - \sum_{\ell=1}^{k} \theta_\ell\right)^{a_0}\right\}, \tag{34.82}$$

for all $x_i = 0, 1, \ldots$, excluding $x_i = 0$ for all $i = 1, 2, \ldots, k$, where $P(\boldsymbol{x}|\boldsymbol{\theta})$ is given by (34.78). Evidently,

$$E[\boldsymbol{X}_0] = E[\boldsymbol{X}] \Bigg/ \left\{1 - \left(1 - \sum_{\ell=1}^{k} \theta_\ell\right)^{a_0}\right\} \tag{34.83}$$

and

$$\text{var}(\boldsymbol{X}_0) = \text{var}(\boldsymbol{X}) \left\{1 - \left(1 - \sum_{\ell=1}^{k} \theta_\ell\right)^{a_0}\right\} + \{E[\boldsymbol{X}_0]\}\{E[\boldsymbol{X}_0]\}'. \tag{34.84}$$

5. *Multivariate Lagrangian logarithmic distribution*, obtained from the zero-truncated Lagrangian negative multinomial distribution by taking the limit as $a_0 \to 0$, with joint probability mass function

$$\begin{aligned} P_X(\boldsymbol{x}) = {} & \frac{1}{\sum_{i=1}^{k} a_i x_i} \binom{\sum_{i=1}^{k} (a_i + 1) x_i}{x_1, \ldots, x_k, \sum_{i=1}^{k} a_i x_i} \left\{\prod_{i=1}^{k} \theta_i^{x_i}\right\} \\ & \times \left(1 - \sum_{i=1}^{k} \theta_i\right)^{\sum_{i=1}^{k} a_i x_i} \Bigg/ \left\{-\log\left(1 - \sum_{i=1}^{k} \theta_i\right)\right\} \end{aligned} \tag{34.85}$$

for all $x_i = 0, 1, \ldots,$ except $x_1 = x_2 = \cdots = x_k = 0$ and for $a_i \geq 0$ for all i.

In this case,

$$E[\boldsymbol{X}] = \boldsymbol{\theta} \Big/ \left[\left\{ -\ln\left(1 - \sum_{i=1}^{k} \theta_i\right) \right\} b \right] \tag{34.86}$$

and

$$\text{var}(\boldsymbol{X}) = \frac{(\boldsymbol{I} + \boldsymbol{D}_\theta \mathbf{1}\boldsymbol{a}'/b)\boldsymbol{D}_\theta(\boldsymbol{I} + \boldsymbol{a}\mathbf{1}'\boldsymbol{D}_\theta/b)}{b\{-\log(1 - \sum_{i=1}^{k} \theta_i)\}} + \frac{(1 - \sum_{i=1}^{k} \theta_i)\boldsymbol{D}_\theta \mathbf{1}\mathbf{1}'\boldsymbol{D}_\theta}{b^3\{-\log(1 - \sum_{i=1}^{k} \theta_i)\}} + \{E[\boldsymbol{X}]\}\{E[\boldsymbol{X}]\}', \tag{34.87}$$

where b, $\boldsymbol{a}$, $\mathbf{1}$, and $\boldsymbol{D}_\theta$ are as given above.

3.3 Homogeneous Multivariate Distributions

The idea of homogeneous probability generating functions was introduced by Kemp (1981). Several bivariate probability generating functions belong to this class. The probability generating function $G_{X,Y}(t_1, t_2)$ of a bivariate random variable $(X, Y)'$ is said to be of the *homogeneous type* if

$$G_{X,Y}(t_1, t_2) = H(at_1 + bt_2)$$

with $H(a + b) = 1$. A characterization result can be provided for homogeneous probability generating functions as follows [see Kocherlakota and Kocherlakota (1992, pp. 18–19)]:

> The probability generating function $G_{X,Y}(t_1, t_2)$ is of the homogeneous type if and only if the conditional distribution of X, given $X + Y = z$, is binomial $(z, \frac{a}{a+b})$.

It is clear from the above characterization result that the binomial distribution plays a unique role in discrete bivariate distributions with homogeneous probability generating functions.

This concept can be generalized to discrete multivariate distributions as follows. We say the probability generating function $G_{\boldsymbol{X}}(\boldsymbol{t})$ of a multivariate random variable $\boldsymbol{X}$ is of the *homogeneous type* if

$$G_{\boldsymbol{X}}(\boldsymbol{t}) = H\left(\sum_{i=1}^{k} a_i t_i\right)$$

with $H(\sum_{i=1}^{k} a_i) = 1$. Once again, one may observe that several discrete multivariate distributions discussed in subsequent chapters belong to this class. Further, a characterization result (analogous to the one stated above) in this general case can be stated as follows:

The probability generating function $G_{\boldsymbol{X}}(\boldsymbol{t})$ of a multivariate random variable $\boldsymbol{X}$ is of the homogeneous type if and only if the conditional distribution of $(X_1, X_2, \ldots, X_{k-1})'$, given $X_1 + \cdots + X_k = z$, is multinomial with parameters $\left(z; \frac{a_1}{\sum_{i=1}^k a_i}, \ldots, \frac{a_k}{\sum_{i=1}^k a_i}\right)$.

Unfortunately, nothing general can be said about the conditional distributions involved in nonhomogeneous probability generating functions.

4 MIXED, MODIFIED, AND TRUNCATED DISTRIBUTIONS

As with the discrete univariate distributions, standard discrete multivariate distributions can be modified in different ways. We shall now briefly describe the most common methods.

4.1 Mixture Distributions

Mixtures of discrete multivariate distributions can be generated conveniently by ascribing some distribution(s) to the parameter(s), and then finding the "expected value" of the probability generating function resulting from such imposed distributions. Many examples of this nature will be seen in subsequent chapters of this book; see also Section 2 of Chapter 43 for some additional details.

4.2 "Inflated" Distributions

For a discrete univariate distribution with pmf

$$\Pr[X = x] = p_x, \qquad x = 0, 1, 2, \ldots, \tag{34.88}$$

the distribution obtained by increasing (decreasing) p_{x_0} for some x_0 and then decreasing (increasing) proportionately the values of all other p_x's is called an *inflated* (deflated) distribution (relative to the original distribution, of course). Gerstenkorn and Jarzebska (1979) have described an extension of the same idea to a multivariate setting. They have defined an inflated distribution corresponding to a distribution with pmf $P(x_1, x_2, \ldots, x_k)$ as one having a pmf

$$\begin{aligned} P^*(x_1, x_2, \ldots, x_k) &= \beta + \alpha\, P(x_{10}, x_{20}, \ldots, x_{k0}) \quad \text{for} \bigcap_{i=1}^{k} (x_i = x_{i0}), \\ &= \alpha\, P(x_1, x_2, \ldots, x_k) \qquad \text{otherwise,} \end{aligned} \tag{34.89}$$

where α and β are non-negative numbers such that $\alpha + \beta = 1$. The "inflation" refers to the values of the probability of the event $\bigcap_{i=1}^k (X_i = x_{i0})$ while values of all other probabilities in the distribution are "deflated." It is clear that the (descending) factorial moments corresponding to P and P^* satisfy the relationship (in an obvious notation)

$$\mu^*_{(r)} = \beta \prod_{i=1}^{k} x_{i0}^{r_i} + \alpha\, \mu_{(r)} \,. \tag{34.90}$$

There exists a similar relationship between the moments about zero of the inflated and the original distributions.

It is evident that the above definitions could be extended to modifications involving values of $P_X(\boldsymbol{x})$ for more than one set of x's. The necessary extensions of the formulas are quite straightforward, though quite cumbersome to evaluate.

4.3 Truncated Distributions

As in the case of discrete univariate distributions, truncated forms of discrete multivariate distributions may be introduced. Corresponding to a distribution with the pmf $P(x_1, x_2, \ldots, x_k)$ and support T, one may consider a truncated distribution with $\boldsymbol{x}$ restricted to the region T^*, a subset of T. It is then evident that the pmf of such a truncated distribution is

$$\frac{1}{C}\, P_X(\boldsymbol{x}), \qquad \boldsymbol{x} \in T^*\,, \tag{34.91}$$

where

$$C = \sum \cdots \sum_{\boldsymbol{x} \in T^*} P_X(\boldsymbol{x})\,. \tag{34.92}$$

The moments of this truncated distribution may be defined in a similar manner; in fact, they may be written in terms of incomplete moments of the original distribution [cf. Eq. (34.48)]. Truncated forms of some specific distributions are discussed in subsequent chapters. A general discussion of truncated discrete multivariate distributions is provided in Gerstenkorn (1977).

Single-Variable Truncation

Let X_i, $i = 0, 1, 2, \ldots, k$, be variables taking on integer values from 0 to n so that $\sum_{i=0}^{k} X_i = n$, that is, X_0 is determined as $X_0 = n - \sum_{i=1}^{k} X_i$. Suppose one of the variables, say X_i, is limited by $b < X_i \le c$, where $0 \le b < c \le n$. Then, the variables $X_0, \ldots, X_{i-1}, X_{i+1}, \ldots, X_k$ must satisfy the condition

$$n - c \le \sum_{\substack{j=0 \\ j \ne i}}^{k} X_j < n - b\,.$$

Under this condition, the probability mass function of the doubly truncated distribution (truncation based on X_i) is

$$\begin{aligned} P_{\boldsymbol{X}}^T(\boldsymbol{x}) &= P(x_1, \ldots, x_{i-1}, x_i^T, x_{i+1}, \ldots, x_k) \\ &= \frac{P(x_1, \ldots, x_k)}{F_i(c) - F_i(b)}\,, \end{aligned} \tag{34.93}$$

where $F_i(a) = \sum_{x_i=0}^{a} P_{X_i}(x_i)$. Here, P^T denotes the truncated form of P and x_i^T the truncated form of x_i. In particular, if $b < X_i \le n$, we have the probability mass function of the resulting left truncated distribution as

$$\frac{P(x_1, \dots, x_k)}{1 - F_i(b)} \,. \tag{34.94}$$

Several Variables Truncation

Suppose ℓ of the variables, say $X_1, X_2, \dots, X_\ell$, are limited by $b_i < X_i \le c_i$ ($i = 1, 2, \dots, \ell$), where b_i and c_i are non-negative integers such that $0 \le \sum_{i=1}^{\ell} b_i < \sum_{i=1}^{\ell} c_i \le n$. Then, the variables $X_0, X_{\ell+1}, \dots, X_k$ must satisfy the condition

$$n - \sum_{i=1}^{\ell} c_i \le \sum_{i=0,\ell+1}^{k} X_i < n - \sum_{i=1}^{\ell} b_i \,.$$

Under these conditions, the probability mass function of the doubly truncated distribution (truncation based on $X_1, \dots, X_\ell$) is

$$\begin{aligned} P_{\boldsymbol{X}}^T(\boldsymbol{x}) &= P(x_1^T, \dots, x_\ell^T, x_{\ell+1}, \dots, x_k) \\ &= \frac{P(x_1, x_2, \dots, x_k)}{F_{1,\dots,\ell}(c_1, \dots, c_\ell) - F_{1,\dots,\ell}(b_1, \dots, b_\ell)} \,, \end{aligned} \tag{34.95}$$

where $F_{1,\dots,\ell}(a_1, \dots, a_\ell) = \sum_{x_1=0,\dots,x_\ell=0}^{a_1,\dots,a_\ell} P_{X_1,\dots,X_\ell}(x_1, \dots, x_\ell)$. Here, P^T denotes the truncated form of P and x_i^T the truncated form of x_i ($i = 1, 2, \dots, \ell$). In particular, if $b_i < X_i \le n$ ($i = 1, 2, \dots, \ell$) with $\sum_{i=1}^{\ell} b_i < n$, the probability mass function of the resulting left truncated distribution is

$$\frac{P(x_1, \dots, x_k)}{1 - F_{1,\dots,\ell}(b_1, \dots, b_\ell)} \,. \tag{34.96}$$

5 APPROXIMATIONS

In this section we describe two methods of approximating discrete multivariate distributions and make some comments with regard to their merits and usage.

5.1 Product Approximations

Consider the joint probability mass function

$$P(\boldsymbol{x}), \quad \boldsymbol{x} = (x_1, \dots, x_k)' \in \mathcal{X}, \ \mathcal{X} = \mathcal{X}_1 \times \mathcal{X}_2 \times \cdots \times \mathcal{X}_k \,, \tag{34.97}$$

where $\mathcal{X}_i$ ($i = 1, 2, \dots, k$) is a set of discrete values of the variable X_i. To simplify notations, let us denote the set of indices $\{1, 2, \dots, k\}$ by I, A is a subset of I, and $\bar{A}$ is its complement; that is,

$$I = \{1, 2, \dots, k\}, \quad A = \{i_1, \dots, i_\ell\} \subset I, \quad \bar{A} = \{i_{\ell+1}, \dots, i_k\} = I \setminus A \,. \tag{34.98}$$

Let us use

$$\begin{aligned} \boldsymbol{x}_A &= (x_{i_1}, \ldots, x_{i_\ell}) \in \mathcal{X}_A \equiv \mathcal{X}_{i_1} \times \cdots \times \mathcal{X}_{i_\ell} ; \\ \boldsymbol{x}_{\bar{A}} &= (x_{i_{\ell+1}}, \ldots, x_{i_k}) \in \mathcal{X}_{\bar{A}} \end{aligned} \tag{34.99}$$

for the corresponding subvectors and subspaces. Then, the marginal distribution of $\boldsymbol{X}_A$ is

$$P_{\boldsymbol{X}_A}(\boldsymbol{x}_A) = \sum_{\boldsymbol{x}_{\bar{A}} \in \mathcal{X}_{\bar{A}}} P_{\boldsymbol{X}}(\boldsymbol{x}), \qquad \boldsymbol{x}_A \in \mathcal{X}_A,\ A \subset I\,. \tag{34.100}$$

The subscript $\boldsymbol{X}_A$ in $P(\boldsymbol{x}_A)$ will be omitted whenever no confusion may result. When $A = \emptyset$ (empty set), we obtain from (34.100) that $P(\boldsymbol{x}_\phi) = 1$.

If $\mathcal{A}$ is a partition of the set I, namely,

$$\mathcal{A} = \{A_1, A_2, \ldots, A_L\}, \quad \bigcup_{\ell=1}^{L} A_\ell = I,\ A_\ell \cap A_{\ell'} = \emptyset \qquad \text{for } \ell \neq \ell'\,, \tag{34.101}$$

then, using conditional probability distributions, we have the well-known expansion formula

$$\begin{aligned} P(\boldsymbol{x}) = P(\boldsymbol{x}_{A_L} \mid \boldsymbol{x}_{A_{L-1}}, \ldots, \boldsymbol{x}_{A_1}) P(\boldsymbol{x}_{A_{L-1}} \mid \boldsymbol{x}_{A_{L-2}}, \ldots, \boldsymbol{x}_{A_1}) \\ \cdots P(\boldsymbol{x}_{A_2} \mid \boldsymbol{x}_{A_1})\,. \end{aligned} \tag{34.102}$$

The principle of product approximation follows naturally from (34.102) if one uses only subsets of the conditioning variables.

Let $\mathcal{A}$ be a partition of the set I and let S be a sequence of pairs of subsets with the property

$$S = \{(A_1 \mid B_1), \ldots, (A_L \mid B_L)\}, \quad B_\ell \subset \bigcup_{i=1}^{\ell-1} A_i \subset I \qquad (\ell = 1, \ldots, L)$$

with $B_1 = \emptyset$, and let $P(\boldsymbol{x}_A)$ $(A \subset I)$ be marginal distributions of the distribution P. Then the function

$$P^*(\boldsymbol{x} \mid S) = \prod_{\ell=1}^{L} \frac{P(\boldsymbol{x}_{A_\ell}, \boldsymbol{x}_{B_\ell})}{P(\boldsymbol{x}_{B_\ell})} = P(\boldsymbol{x}_{A_1}) \prod_{\ell=2}^{L} P(\boldsymbol{x}_{A_\ell} \mid \boldsymbol{x}_{B_\ell}), \qquad \boldsymbol{x} \in \mathcal{X}\,, \tag{34.103}$$

which is a valid probability distribution, is called the *product approximation* of the distribution P [cf. Grim (1984)]; S is called the *dependence structure* of this approximation.

The closeness of this approximation may be measured by means of the relative entropy (which is called the *Shannon entropy*)

$$H(P, P^*) = \sum P(\boldsymbol{x}) \ln \left\{ \frac{P(\boldsymbol{x})}{P^*(\boldsymbol{x} \mid S)} \right\} \geq 0 \tag{34.104}$$

which will equal 0 only if P and P^* are identical. Alternative forms of the relative entropy in (34.104) are

$$-H(I) + \sum_{\ell=1}^{L} H(A_\ell \mid B_\ell) = -H(I) + \sum_{\ell=1}^{L} H(A_\ell) - \sum_{\ell=1}^{L} I(A_\ell, B_\ell)\,, \qquad (34.105)$$

where

$$H(A \mid B) = H(A \cup B) - H(B)\,,$$
$$H(A) = -\sum_{\boldsymbol{x}_A \in \mathcal{X}_A} P(\boldsymbol{x}_A) \ln P(\boldsymbol{x}_A) = H(P_A), \qquad A \subset I\,,$$

and

$$I(A, B) = H(A) + H(B) - H(A \cup B), \qquad A \cup B \subset I\,.$$

Since $H(I)$ is a constant for a fixed distribution P, we can optimize the product approximation (34.103) by maximizing

$$Q(S) = -\sum_{\ell=1}^{L} H(A_\ell) + \sum_{\ell=1}^{L} I(A_\ell, B_\ell) = -\sum_{\ell=1}^{L} H(A_\ell \mid B_\ell) \qquad (34.106)$$

independently of P [*cf.* Lewis (1959)].

Statistically oriented formulation of the approximation problem is useful when a direct estimation of the unknown multivariate distribution is difficult because of a limited sample size. In this case, the estimates of marginal distributions up to a given order r may be expected to be more reliable. For this reason, Grim (1984) confined the general dependence structure S by the inequality

$$|A_\ell \cup B_\ell| \le r, \qquad \ell = 1, 2, \ldots, L, \ 1 \le r \le k\,. \qquad (34.107)$$

Nevertheless, the underlying problem of discrete optimization, despite this constraint, is extremely difficult.

One way to simplify the optimization problem is to set $B_\ell = A_{\ell-1}$ ($\ell = 2, 3, \ldots, L$) in S. We then obtain a particular *Markovian* dependence structure

$$S = \{(A_1 \mid \emptyset), (A_2 \mid A_1), \ldots, (A_L \mid A_{L-1})\}\,. \qquad (34.108)$$

The corresponding solution, however, has the complexity of the traveling-salesman problem [see Grim (1979)] and is therefore intractable in higher dimensions.

It can be shown that if $P^*(\cdot \mid S)$ is a product approximation with property (34.108), then there exists a product approximation $P^*(\cdot \mid S^*)$ with the dependence structure

$$S^* = \{(A_1^* \mid B_1^*), \ldots, (A_\ell^* \mid B_\ell^*)\}, \quad |A_\ell^*| = 1, \ |B_\ell^*| = \min\{r-1, \ell-1\}\,,$$
$$\ell = 1, 2, \ldots, k\,, \qquad (34.109)$$

which is better in the sense of the inequality

$$Q(S^*) \geq Q(S) . \tag{34.110}$$

This result simplifies the optimization problem considerably, since, without loss of any generality, the criterion $Q(S)$ may be maximized on the following reduced class of dependence structures:

$$\begin{aligned}
&\overline{S} = \{(i_1 \mid B_1), \ldots, (i_k \mid B_k)\}, \quad \Pi = \Pi(I) = (i_1, i_2, \ldots, i_k) , \\
&B_\ell \subset \{i_1, i_2, \ldots, i_{\ell-1}\}, \quad |B_\ell| = \min(r - 1, \ell - 1), \quad \ell = 1, 2, \ldots, k , \\
&P(\boldsymbol{x} \mid \overline{S}) = \prod_{\ell=1}^{k} P(x_{i_\ell} \mid \boldsymbol{x}_{B_\ell}), \quad \boldsymbol{x} \in X ,
\end{aligned} \tag{34.111}$$

where Π is a permutation of the elements of the set I.

For the case of two-dimensional marginals, Chow and Lin (1968) provided an efficient solution using the concept of *weighted graphs*. Let $\mathcal{G}_k$ be a complete weighted graph over the set of vertices I, namely,

$$\mathcal{G}_k = \{I, E \mid w\}, \quad E = \{(\ell, \ell') : \ell \in I, \ \ell' \in I, \ \ell \neq \ell'\} , \tag{34.112}$$

and the edge-weight function w be defined by equation

$$w : E \longrightarrow \mathbb{R}_1, \quad w(\ell, \ell') = I(\ell, \ell') = H(\ell) + H(\ell') - H(\ell, \ell'). \tag{34.113}$$

Then the maximum weight spanning tree $\tau_k = \{I, F \mid w\}$, $F \subset E$ of the graph $\mathcal{G}_k$ defines a *product approximation*

$$\begin{aligned}
P^*(\boldsymbol{x} \mid \sigma) &= P(x_{i_1}) \prod_{\ell=2}^{k} P(x_{i_\ell} \mid x_{j_\ell}) \\
&= \left\{ \prod_{\ell=1}^{k} P(x_{i_\ell}) \right\} \prod_{\ell=2}^{k} \left\{ \frac{P(x_{i_\ell}, x_{j_\ell})}{P(x_{i_\ell}) P(x_{j_\ell})} \right\}
\end{aligned} \tag{34.114}$$

with the dependence structure σ given by

$$\sigma = \{(i_1 \mid -), (i_2 \mid j_2), \ldots, (i_k \mid j_k)\}, \quad j_\ell \in \{i_1, i_2, \ldots, i_{\ell-1}\} \subset I, \tag{34.115}$$

which maximizes the corresponding criterion $Q(\sigma)$ given by

$$Q(\sigma) = -\sum_{\ell=1}^{k} H(i_\ell) + \sum_{\ell=2}^{k} I(i_\ell, j_\ell) . \tag{34.116}$$

To apply this result, we need to evaluate the mutual information $I(i, j)$ for all pairs of random variables X_i and X_j, which is

$$I(i, j) = \sum_{x_i \in X_i} \sum_{x_j \in X_j} P(x_i, x_j) \ln \left\{ \frac{P(x_i, x_j)}{P(x_i) P(x_j)} \right\}, \quad i, j \in I, \ i \neq j \tag{34.117}$$

and find the maximum-weight spanning tree of the corresponding complete graph. For this purpose, one may use a standard algorithm proposed by Prim (1957).

5.2 Estimation and Structural Approximations

In estimation problems, only a sample of independent observations of a discrete random vector is available:

$$S = \{\boldsymbol{x}^{(1)}, \boldsymbol{x}^{(2)}, \ldots, \boldsymbol{x}^{(N)}\}, \quad \boldsymbol{x}^{(n)} = (x_1^{(n)}, \ldots, x_k^{(n)})' \in \mathcal{X} . \tag{34.118}$$

To construct product approximations in such situations, we could replace the unknown marginal distributions by their respective sample estimates without any difficulty. However, it is more justified to use a *parametric approach* and the maximum-likelihood principle, even when (in the discrete case) both approaches lead to the same optimization procedure [as shown by Chow and Lin (1968)]. In this sense, the product approximation (34.103) may be viewed as a *parametric probability distribution* defined by a dependence structure σ and a set of two-dimensional marginals $\mathcal{P}_2$:

$$P^*(\boldsymbol{x} \mid \sigma, \mathcal{P}_2) = P(x_{i_1}) \prod_{\ell=2}^{k} P(x_{i_\ell} \mid x_{j_\ell}) = P(x_{i_1}) \prod_{\ell=2}^{k} \left\{ \frac{P(x_{i_\ell}, x_{j_\ell})}{P(x_{i_\ell})P(x_{j_\ell})} \right\}, \quad \boldsymbol{x} \in \mathcal{X} , \tag{34.119}$$

$$\sigma = \{(i_1 \mid -), (i_2 \mid j_2), \ldots, (i_k \mid j_k)\}, \quad \mathcal{P}_2 = \{P_{ij} = P(x_i, x_j);\ i, j \in I\} .$$

Grim (1984) provided the following explicit expression for the maximized likelihood function:

$$\begin{aligned} L(\sigma, \mathcal{P}_2) &= \frac{1}{N} \sum_{n=1}^{N} \ln P^*(\boldsymbol{x}^{(n)} \mid \sigma, \mathcal{P}_2) \\ &= \frac{1}{N} \sum_{n=1}^{N} \ln P(x_{i_1}^{(n)}) + \sum_{\ell=2}^{k} \left[\frac{1}{N} \sum_{n=1}^{N} \ln P(x_{i_\ell}^{(n)} \mid x_{j_\ell}^{(n)}) \right], \end{aligned} \tag{34.120}$$

which may be rewritten as

$$\begin{aligned} L(\sigma, \mathcal{P}_2) = &\sum_{\xi \in X_{i_1}} \hat{P}_{i_1}(\xi) \ln P_{i_1}(\xi) \\ &+ \sum_{\ell=2}^{k} \left[\sum_{\eta \in X_{j_\ell}} \hat{P}_{j_\ell}(\eta) \sum_{\xi \in X_{i_\ell}} \frac{\hat{P}_{i_\ell j_\ell}(\xi, \eta)}{\hat{P}_{j_\ell}(\eta)} \ln P_{i_\ell | j_\ell}(\xi \mid \eta) \right], \end{aligned} \tag{34.121}$$

where

$$\hat{P}_i(\xi) = \frac{1}{N} \sum_{n=1}^{N} \delta(\xi, x_i^{(n)}), \quad \hat{P}_{ij}(\xi, \eta) = \frac{1}{N} \sum_{n=1}^{N} \delta(\xi, x_i^{(n)}) \delta(\eta, x_j^{(n)}) ,$$

$$\begin{aligned} i, j \in I,\ \xi \in X_i,\ \eta \in X_j,\ \delta(\xi, \eta) &= 0 \quad \text{if} \quad \xi \neq \eta \\ &= 1 \quad \text{if} \quad \xi = \eta \end{aligned}$$

are the sample estimates of the marginal probabilities. Consequently, for any fixed dependence structure σ, the likelihood function $L(\sigma, \mathcal{P}_2)$ is maximized by

$$P_{i_1}(\xi) = \hat{P}_{i_1}(\xi), \quad P_{i_\ell | j_\ell}(\xi \mid \eta) = \frac{\hat{P}_{i_\ell j_\ell}(\xi, \eta)}{\hat{P}_{j_\ell}(\eta)}, \quad \xi \in X_{i_\ell}, \ \eta \in X_{j_\ell}, \quad \ell = 2, 3, \ldots, k. \tag{34.122}$$

Substituting (34.122) in (34.121) yields

$$L(\sigma, \hat{\mathcal{P}}_2) = \sum_{\ell=1}^{k} \sum_{\xi \in X_\ell} \hat{P}_\ell(\xi) \ln \hat{P}_\ell(\xi) + \sum_{\ell=2}^{k} \sum_{\xi \in X_{i_\ell}} \sum_{\eta \in X_{j_\ell}} \hat{P}_{i_\ell j_\ell}(\xi, \eta) \ln \left\{ \frac{\hat{P}_{i_\ell j_\ell}(\xi, \eta)}{\hat{P}_{i_\ell}(\xi) \hat{P}_{j_\ell}(\eta)} \right\} \tag{34.123}$$

and therefore we may use Chow and Lin's (1968) result to compute the optimal dependence structure $\hat{\sigma}$. Denoting

$$I_0(i, j) = \sum_{\xi \in X_i} \sum_{\eta \in X_j} \hat{P}_{ij}(\xi, \eta) \ln \left\{ \frac{\hat{P}_{ij}(\xi, \eta)}{\hat{P}_i(\xi) \hat{P}_j(\eta)} \right\}, \quad i, j \in I, \tag{34.124}$$

and using the maximum weight spanning tree, we obtain

$$\hat{\sigma} = \arg \max_{\sigma} \left\{ \sum_{\ell=2}^{k} I_0(i_\ell, j_\ell) \right\}. \tag{34.125}$$

Grim (1984) suggested the method of mixtures to improve (34.119) wherein, using product approximations as components of mixtures, a new class of approximations is obtained with dependence structures of qualitatively higher complexity. Grim used the term *structural approximations* in order to differentiate this class of approximate distributions from the standard parametric types. For further details on these structural approximations, one may refer to Grim (1984).

6 LOG-CONCAVITY OF DISCRETE MULTIVARIATE DISTRIBUTIONS

Log-concavity of discrete multivariate distributions has attracted the attention of many researchers over the years. However, the log-convexity has not received similar attention and hence seems to be an area which merits further research.

Let $N = \{0, 1, 2, \ldots\}$. Then, a function $P : N \to (0, \infty)$ is said to be *log-concave* if

$$\{P(x)\}^2 \geq P(x-1)P(x+1), \qquad x = 1, 2, \ldots. \tag{34.126}$$

Many standard discrete univariate distributions on N are log-concave. For example, the binomial distribution with probability mass function

$$P(x) = \binom{n}{x} p^x (1-p)^{n-x}, \qquad x = 0, 1, \ldots, n,$$

and the Poisson distribution with probability mass function

$$P(x) = e^{-\theta}\,\theta^x/x!, \qquad x = 0, 1, 2, \ldots$$

can be easily shown to be log-concave. For more details, one may refer to Chapter 5, p. 209.

Let $N^k = \{\boldsymbol{\ell} = (\ell_1, \ell_2, \ldots, \ell_k)' : \ell_i = 0, 1, 2, \ldots\}$. Then, Bapat (1988) proposed a generalization of the above concept of log-concavity for positive functions defined on a subset of N^k and called it *generalized log-concavity*. This generalization is, incidentally, stronger than a generalization introduced by Karlin and Rinott (1981). To introduce this concept of generalized log-concavity, let $S \subset N^k$, $\boldsymbol{e}_i$ be the ith column of an identity matrix, and P be a function such that $P : S \to (0, \infty)$. Then, P is said to be generalized log-concave on S if, for any $\boldsymbol{x} \in N^k$ such that $\boldsymbol{x} + \boldsymbol{e}_i + \boldsymbol{e}_j \in S$ $(i, j = 1, 2, \ldots, k)$, the symmetric positive $k \times k$ matrix $((P(\boldsymbol{x} + \boldsymbol{e}_i + \boldsymbol{e}_j))$ has exactly one positive eigenvalue.

Bapat (1988) has justified the usage of the term "generalized log-concave" through the result that, for $S \subset N^k, P : S \to (0, \infty)$ and log-concave functions $P_i : N \to (0, \infty)$, $i = 1, 2, \ldots, k$, the function

$$P(\boldsymbol{x}) = \prod_{i=1}^{k} P_i(x_i), \qquad \boldsymbol{x} = (x_1, \ldots, x_k)' \in S \tag{34.127}$$

is generalized log-concave. In other words, if $X_1, X_2, \ldots, X_k$ are independent random variables (each taking values in N) with log-concave probability mass functions, then $\boldsymbol{X} = (X_1, X_2, \ldots, X_k)'$ has a probability mass function which is generalized log-concave. Furthermore, the multinomial distribution with probability mass function (see Chapter 35)

$$P(\boldsymbol{x}) = n! \prod_{i=1}^{k} \frac{p_i^{x_i}}{x_i!}, \quad x_i \geq 0, \quad \sum_{i=1}^{k} x_i = n, \; p_i \geq 0, \quad \sum_{i=1}^{k} p_i = 1\,, \tag{34.128}$$

the negative multinomial distribution with probability mass function (see Chapter 36)

$$P(\boldsymbol{x}) = \frac{(n-1+\sum x_i)!}{(n-1)!}\, p_0^n \prod_{i=1}^{k} \frac{p_i^{x_i}}{x_i!}, \quad x_i \geq 0, \quad \sum_{i=0}^{k} p_i = 1, \tag{34.129}$$

the multivariate hypergeometric distribution with probability mass function (see Chapter 39)

$$P(\boldsymbol{x}) = \frac{1}{\binom{M}{m}} \prod_{i=1}^{k} \binom{M_i}{x_i}, \quad x_i \geq 0, \quad \sum_{i=1}^{k} x_i = m, \quad \sum_{i=1}^{k} M_i = M, \tag{34.130}$$

and the multivariate negative hypergeometric distribution with probability mass function (see Chapter 39)

$$P(\boldsymbol{x}) = \frac{1}{\binom{m+k+M}{m}} \prod_{i=1}^{k} \binom{x_i + M_i + 1}{x_i}, \quad x_i \geq 0, \quad \sum_{i=1}^{k} x_i = m, \quad \sum_{i=1}^{k} M_i = M, \tag{34.131}$$

can all be shown to be generalized log-concave.

In addition, Bapat (1988) has shown that the multiparameter forms of multinomial and negative multinomial probability mass functions are also generalized log-concave.

BIBLIOGRAPHY

Balakrishnan, N., and Johnson, N. L. (1997). A recurrence relation for discrete multivariate distributions and some applications to multivariate waiting time problems, in *Advances in the Theory and Practice of Statistics—A Volume in Honor of Samuel Kotz* (Eds. N. L. Johnson and N. Balakrishnan), New York: John Wiley & Sons (to appear).

Balakrishnan, N., Johnson, N. L., and Kotz, S. (1995). A note on some relationships between moments, central moments and cumulants from multivariate distributions, *Report*, McMaster University, Hamilton, Canada.

Bapat, R. B. (1988). Discrete multivariate distributions and generalized log-concavity, *Sankhyā, Series A*, **50**, 98–110.

Chow, C. K., and Lin, C. N. (1968). Approximating discrete probability distributions with dependence trees, *IEEE Transactions on Information Theory*, **14**, 462–467.

Gerstenkorn, T. (1977). Multivariate doubly truncated discrete distributions, *Acta Universitatis Lodziensis*, 1–115.

Gerstenkorn, T. (1980). Incomplete moments of the multivariate Pólya, Bernoulli (polynomial), hypergeometric and Poisson distributions, *Studia Scientiarum Mathematicarum Hungarica*, **15**, 107–121.

Gerstenkorn, T., and Jarzebska, J. (1979). Multivariate inflated discrete distributions, *Proceedings of the Sixth Conference on Probability Theory*, pp. 341–346, Brasov, Romania.

Grim, J. (1979). On estimation of multivariate probability density functions for situation recognition in large scale systems, in *Proceedings of Third Formator Symposium* (Eds. J. Beneš and L. Bakule), Prague: Academia.

Grim, J. (1984). On structural approximating multivariate discrete probability distributions, *Kibernetika*, **20**, No. 1, 1–17.

Johnson, N. L., and Kotz, S. (1969). *Distributions in Statistics: Discrete Distributions,* New York: John Wiley & Sons.

Johnson, N. L., Kotz, S., and Balakrishnan, N. (1994). *Continuous Univariate Distributions—1*, second edition, New York: John Wiley & Sons.

Johnson, N. L., Kotz, S., and Balakrishnan, N. (1995). *Continuous Univariate Distributions—2*, second edition, New York: John Wiley & Sons.

Johnson, N. L., Kotz, S., and Kemp, A. W. (1992). *Univariate Discrete Distributions*, second edition, New York: John Wiley & Sons.

Karlin, S., and Rinott, Y. (1981). Entropy inequalities for classes of probability distributions. II. The multivariate case, *Advances in Applied Probability*, **13**, 325–351.

Kemp, A. W. (1981). Computer sampling from homogeneous bivariate discrete distributions, *ASA Proceedings of the Statistical Computing Section*, 173–175.

Khatri, C. G. (1962). Multivariate Lagrangian Poisson and multinomial distributions, *Sankhyā, Series B*, **44**, 259–269.

Khatri, C. G. (1983a). Multivariate discrete exponential distributions and their characterization by Rao–Rubin condition for additive damage model, *South African Statistical Journal*, **17**, 13–32.

Khatri, C. G. (1983b). Multivariate discrete exponential family of distributions, *Communications in Statistics—Theory and Methods*, **12**, 877–893.

Kocherlakota, S., and Kocherlakota, K. (1992). *Bivariate Discrete Distributions*, New York: Marcel Dekker.

Kocherlakota, S., and Kocherlakota, K. (1997). Bivariate discrete distributions, in *Encyclopedia of Statistical Sciences—Update Volume 1* (Eds. S. Kotz, C. B. Read, and D. L. Banks), New York: John Wiley & Sons (to appear).

Lewis, P. M. (1959). Approximating probability distributions to reduce storage requirements, *Information and Control*, **2**, 214–225.

Mohanty, S. G. (1966). On a generalized two-coin tossing problem, *Biometrische Zeitschrift*, **8**, 266–272.

Papageorgiou, H. (1997). Multivariate discrete distributions, in *Encyclopedia of Statistical Sciences—Update Volume 1* (Eds. S. Kotz, C. B. Read, and D. L. Banks), New York: John Wiley & Sons (to appear).

Prakasa Rao, B. L. S. (1992). *Identifiability in Stochastic Models: Characterization of Probability Distributions*, London: Academic Press.

Prim, R. C. (1957). Shortest connection networks and some generalizations, *Bell System Technical Journal*, **36**, 1389–1401.

Smith, P. J. (1995). A recursive formulation of the old problem of obtaining moments from cumulants and vice versa, *The American Statistician*, **49**, 217–218.

Titterington, D. M., Smith, A. F. M., and Makov, U. E. (1985). *Statistical Analysis of Finite Mixture Distributions*, New York: John Wiley & Sons.

Xekalaki, E. (1987). A method for obtaining the probability distribution of m components conditional on l components of a random sample, *Rev. Roumaine Math. Pure Appl.*, **32**, 581–583.

CHAPTER 35

Multinomial Distributions

1 INTRODUCTION

The multinomial distribution, like the multivariate normal distribution among the continuous multivariate distributions, consumed a sizable amount of the attention that numerous theoretical as well as applied researchers directed towards the area of discrete multivariate distributions. This lengthy chapter summarizes numerous results on the theory and applications of multinomial distributions. The multitude of results available on this distribution forced us to be somewhat more selective and briefer than we had planned originally. However, the comprehensive list of references at the end of this chapter may compensate partially for some of the likely regrettable omissions in the text.

2 DEFINITION AND GENESIS

Consider a series of n independent trials, in each of which just one of k mutually exclusive events $E_1, E_2, \ldots, E_k$ must be observed, and in which the probability of occurrence of event E_j in any trial is equal to p_j (with, of course, $p_1 + p_2 + \cdots + p_k = 1$). Let $N_1, N_2, \ldots, N_k$ be the random variables denoting the numbers of occurrences of the events $E_1, E_2, \ldots, E_k$, respectively, in these n trials, with $\sum_{i=1}^k N_i = n$. Then the joint distribution of $N_1, N_2, \ldots, N_k$ is given by

$$\Pr\left[\bigcap_{i=1}^{k}(N_i = n_i)\right] = n! \prod_{i=1}^{k} \left(p_i^{n_i}/n_i!\right) = P(n_1, n_2, \ldots, n_k), \text{ (say)}$$

$$n_i \geq 0, \ \sum_{i=1}^{k} n_i = n \,. \qquad (35.1)$$

This is called a *multinomial distribution*, with parameters $(n; p_1, p_2, \ldots, p_k)$. The p_i's are often called *cell probabilities*; n is sometimes called the *index*.

Introducing the multinomial coefficient [cf. (1.13)]

$$\binom{a}{b_1, b_2, \ldots, b_k} = \frac{a!}{\prod_{i=1}^k b_i!} = \frac{\Gamma(a+1)}{\prod_{i=1}^k \Gamma(b_i+1)}, \quad \sum_{i=1}^k b_i = a \ (a, b_i > 0) \tag{35.2}$$

(using the middle expression when a and all b_i's are integers and the last expression when not all are integers), (35.1) can be expressed as

$$P(n_1, n_2, \ldots, n_k) = \binom{n}{n_1, n_2, \ldots, n_k} \prod_{i=1}^k p_i^{n_i}. \tag{35.3}$$

The expression in (35.1) or (35.3) can easily be seen to be the coefficient of $\prod_{i=1}^k t_i^{n_i}$ in the multinomial expansion of

$$(p_1t_1 + p_2t_2 + \cdots + p_kt_k)^n = \left(\sum_{i=1}^k p_it_i\right)^n, \tag{35.4}$$

which is, therefore, the probability generating function (pgf) of the multinomial distribution. [Note that if $k = 2$ the distribution reduces to a binomial distribution (for either N_1 or N_2).]

The marginal distribution of N_i is binomial with parameters (n, p_i). Multinomial distributions can be regarded as multivariate generalizations of binomial distributions [Feller (1957, 1968)]. They appear in the classical work of J. Bernoulli (1713), in addition to his studies of binomial distributions therein. Lancaster (1966) has noted that, as early as in 1838, Bienaymé considered the multinomial distribution in (35.1) and showed that the limiting distribution of the random variable $\sum_{i=1}^k (N_i - np_i)^2/(np_i)$ is chi-square.

Multinomial distribution seems to have been explicitly introduced in the literature in connection with the classical "problem of points for three players of equal skill" by de Montmort in his famous essay in 1708. It was further used by de Moivre (1730) in connection with the same problem; refer to Hald (1990) for more details.

Multinomial distributions can arise in other contexts as well. For example, if $X_1, X_2, \ldots, X_k$ are independent Poisson random variables with expected values $\theta_1, \theta_2, \ldots, \theta_k$, respectively, then the conditional joint distribution of $X_1, X_2, \ldots, X_k$, given that $\sum_{i=1}^k X_i = t$, is multinomial with parameters $(t; \frac{\theta_1}{\theta_\bullet}, \frac{\theta_2}{\theta_\bullet}, \ldots, \frac{\theta_k}{\theta_\bullet})$, where $\theta_\bullet = \sum_{i=1}^k \theta_i$. This result has been utilized in the construction of tests of the hypothesis that the multinomial cell probabilities are all equal–that is, $p_1 = p_2 = \cdots = p_k \ (= 1/k)$. [See Patel (1979) and also Holst (1979), who extends the analysis to cases when zero frequencies are not recorded.]

Sometimes the k-variable multinomial distributions described above are regarded as $(k$-$1)$-*variable* multinomial distributions, while the k-variable distributions are defined as the joint distribution of $N_1, N_2, \ldots, N_k$ when $(N_1, N_2, \ldots, N_k, N_{k+1})$ has the distribution (35.1) with k replaced by $(k+1)$. This is not unreasonable, since in (35.1) the value of any one of the variables is determined by the values of the remaining variables (because $N_{i'} = n - \sum_{i \neq i'} N_i$). We prefer to use our formulation, which is, indeed, employed by many authors—perhaps even a majority. However, it will be a good practice to check which definition is being used when studying any work on this topic.

An appealing interpretation of multinomial distributions is as follows. Let $X_1, X_2, \ldots, X_n$ be independent and identically distributed random variables with probability mass function

$$\Pr[X = x] = p_x, \quad x = 0, 1, \ldots, k, \ p_x > 0, \ \sum_{x=0}^{k} p_x = 1.$$

Thus, these random variables can be viewed as multinomial with index 1. The joint distribution of $\{X_i\}$ is the product measure on the space of $(k+1)^n$ points. The multinomial distribution is then obtained by identifying all points such that

$$\sum_{j=1}^{n} \delta_{ix_j} = y_i, \qquad i = 0, 1, \ldots, k,$$

and

$$\Pr[Y = y] = \frac{n!}{y_0! y_1! \cdots y_k!} p_0^{y_0} p_1^{y_1} \cdots p_k^{y_k}.$$

The number of points in the multinomial distribution is $\binom{n+k}{k}$, with the special case of the binomial distribution having $\binom{n+1}{n} = n+1$ points.

Some authors [e.g., Patil and Bildikar (1967), Patil and Joshi (1968), and Janardan (1974)] refer to (35.1) as a *singular multinomial* distribution, reserving the name "multinomial distribution" for cases when there is an additional category not represented in the N's.

3 MOMENTS AND OTHER PROPERTIES

The mixed factorial $(r_1, r_2, \ldots, r_k)$ moment of (35.1) is

$$\begin{aligned}\mu'_{(r_1, r_2, \ldots, r_k)} &= E\left[N_1^{(r_1)} N_2^{(r_2)} \cdots N_k^{(k)}\right] \\ &= \sum n_1^{(r_1)} n_2^{(r_2)} \cdots n_k^{(r_k)} \binom{n}{n_1, n_2, \ldots, n_k} \prod_{i=1}^{k} p_i^{n_i} \\ &= n^{\left(\sum_{i=1}^{k} r_i\right)} \prod_{i=1}^{k} p_i^{r_i}. \end{aligned} \tag{35.5}$$

From (35.5), we obtain, in particular,

$$E[N_i] = n p_i \tag{35.6}$$

and

$$\text{var}(N_i) = np_i(1 - p_i), \tag{35.7}$$

as indeed follows from the fact that (already noted) N_i has a binomial (n, p_i) distribution. More generally, the subset

$$(N_{a_1}, N_{a_2}, \ldots, N_{a_s}) \text{ of the } N_i\text{'s together with } n - \sum_{i=1}^{s} N_{a_i}$$

has a joint multinomial distribution with parameters

$$\left(n; p_{a_1}, p_{a_2}, \ldots, p_{a_s}, 1 - \sum_{i=1}^{s} p_{a_i}\right).$$

From (35.5), we also readily obtain

$$E[N_i N_j] = n(n-1)p_i p_j$$

whence

$$\text{cov}(N_i, N_j) = -n\, p_i p_j \tag{35.8}$$

and

$$\text{corr}(N_i, N_j) = -\sqrt{\frac{p_i p_j}{(1 - p_i)(1 - p_j)}}. \tag{35.9}$$

Note that since $p_i + p_j \le 1$, $p_i p_j \le (1 - p_i)(1 - p_j)$. The variance–covariance matrix of the multinomial distribution [formed from (35.7) and (35.8)] is singular with rank as $k - 1$.

Ronning (1982) observed that the variance–covariance matrix of the multinomial distribution is of the form

$$\boldsymbol{V} = \text{Diag}(\boldsymbol{p}) - \boldsymbol{p}\boldsymbol{p}', \tag{35.10}$$

where $\boldsymbol{p} = (p_1, \ldots, p_k)'$, $p_i > 0$, and $\sum_{i=1}^{k} p_i = 1$. For matrices of this form, if we order the components

$$p_1 \ge p_2 \ge \cdots \ge p_k > 0$$

and denote by $\lambda_1(\boldsymbol{V}) \ge \lambda_2(\boldsymbol{V}) \ge \cdots \ge \lambda_k(\boldsymbol{V})$ the ordered eigenvalues of $\boldsymbol{V}$, the following inequalities are valid:

$$p_1 \ge \lambda_1(\boldsymbol{V}) \ge p_2 \ge \lambda_2(\boldsymbol{V}) \ge \cdots \ge p_k \ge \lambda_k(\boldsymbol{V}) > 0. \tag{35.11}$$

If $\boldsymbol{p}$ is unknown, this implies the upper bound $\lambda_1(\boldsymbol{V}) \le 1$ derived by Ronning (1982), which has been improved by Huschens (1990) to

$$\lambda_1(\boldsymbol{V}) \le \frac{1}{4} \quad \text{for } k = 2 \qquad \text{and} \qquad \lambda_1(\boldsymbol{V}) \le \frac{1}{2} \quad \text{for } k \ge 3. \tag{35.12}$$

Tanabe and Sagae (1992) and Watson (1996) have given valuable information on the spectral decomposition of the covariance matrix of a multinomial distribution. For example, by using the result that $|\boldsymbol{A} - \boldsymbol{u}\boldsymbol{u}'| = |\boldsymbol{A}|(1 - \boldsymbol{u}'\boldsymbol{A}^{-1}\boldsymbol{u})$ where $\boldsymbol{A}$ is a $(k \times k)$ nonsingular matrix and $\boldsymbol{u}$ is a $(k \times 1)$ vector [see Rao (1973, p. 32)] and setting $\boldsymbol{A} = \text{Diag}(\boldsymbol{p}) - \lambda \boldsymbol{I}_{k \times k}$, when λ is an eigenvalue not equal to any p_i, we obtain [Watson (1996)]

$$\sum_{i=1}^{k} p_i^2/(p_i - \lambda) = 1 .$$

If $p_1 > p_2 > \cdots > p_k$, then a graph of the left-hand side of the above equation immediately shows that there is a root at 0 and that there is one root in each gap between the ordered p_i's, which simply implies the inequalities in (35.11). If there are some equalities among the p_i's, inequalities in (35.11) still hold by a continuity argument and in this case some eigenvalues are values of the p_i's. The values of the λ_j, not equal to a p_i, could be found by applying Newton–Raphson iteration in each interval in (35.11); as pointed out by Watson (1996), this method would be computationally efficient when k is large. Further, with $\boldsymbol{v}_j$ as the eigenvector corresponding to λ_j, the spectral decomposition of $\boldsymbol{V}$ can be written as

$$\boldsymbol{V} = \text{Diag}(\boldsymbol{p}) - \boldsymbol{p}\boldsymbol{p}' = \sum_{j=1}^{k-1} \lambda_j \boldsymbol{v}_j \boldsymbol{v}_j'$$

so that the Moore-Penrose inverse of this matrix is

$$\boldsymbol{V}^{+} = (\text{Diag}(\boldsymbol{p}) - \boldsymbol{p}\boldsymbol{p}')^{+} = \sum_{j=1}^{k-1} \frac{1}{\lambda_j} \boldsymbol{v}_j \boldsymbol{v}_j' .$$

An alternate expression of this has been given by Tanabe and Sagae (1992); Watson (1996) has given an expression for $\boldsymbol{v}_j$.

The *conditional* joint distribution of $N_{a_1}, N_{a_2}, \ldots, N_{a_s}$, given the values $\{N_i = n_i\}$ of the remaining N's, is also multinomial with parameters

$$\left(n - t; \frac{p_{a_1}}{p_{a\bullet}}, \frac{p_{a_2}}{p_{a\bullet}}, \ldots, \frac{p_{a_s}}{p_{a\bullet}}\right),$$

where

$$t = \sum_{\substack{j \neq a_i \\ i = 1,2,\ldots,s}} n_j \qquad \text{and} \qquad p_{a\bullet} = \sum_{i=1}^{s} p_{a_i} .$$

It thus depends on the values of the remaining N_j's only through their sum t. It follows that

$$E[N_i \mid N_j = n_j] = (n - n_j)p_i/(1 - p_j) ; \tag{35.13}$$

similarly,

$$E\left[N_i \,\middle|\, \bigcap_{j=1}^{s}(N_{a_j} = n_{a_j})\right] = \left(n - \sum_{j=1}^{s} n_{a_j}\right) p_i \Big/ \left(1 - \sum_{j=1}^{s} p_{a_j}\right). \tag{35.14}$$

Thus, the regression of N_i on N_j is linear, and so is the multiple regression of N_i on $N_{a_1}, N_{a_2}, \ldots, N_{a_s}$. We have the corresponding conditional variances

$$\operatorname{var}(N_i \mid N_j = n_j) = (n - n_j)\frac{p_i}{1 - p_j}\left\{1 - \frac{p_i}{1 - p_j}\right\} \tag{35.15}$$

and

$$\operatorname{var}\left(N_i \,\middle|\, \bigcap_{j=1}^{s}(N_{a_j} = n_{a_j})\right)$$
$$= \left(n - \sum_{j=1}^{s} n_{a_j}\right)\frac{p_i}{1 - \sum_{j=1}^{s} p_{a_j}}\left\{1 - \frac{p_i}{1 - \sum_{j=1}^{s} p_{a_j}}\right\}. \tag{35.16}$$

Consequently, the regressions and multiple regressions are not homoscedastic.

We have already noted in Eq. (35.4) that the pgf of the multinomial distribution in (35.1) is $(p_1t_1 + p_2t_2 + \cdots + p_kt_k)^n$. The moment generating function (mgf) is then

$$\left(p_1e^{t_1} + p_2e^{t_2} + \cdots + p_ke^{t_k}\right)^n. \tag{35.17}$$

Hence, if $\boldsymbol{N} = (N_1, N_2, \ldots, N_k)'$ and $\boldsymbol{N}^* = (N_1^*, N_2^*, \ldots, N_k^*)'$ are mutually independent, having multinomial distributions with parameters $(n; p_1, p_2, \ldots, p_k)$ and $(n^*; p_1, p_2, \ldots, p_k)$ respectively, then

$$\boldsymbol{N} + \boldsymbol{N}^* = (N_1 + N_1^*, N_2 + N_2^*, \ldots, N_k + N_k^*)'$$

has a multinomial distribution with parameters $(n + n^*; p_1, p_2, \ldots, p_k)$. This property has been termed *reproducibility with respect to the n-parameter*. It clearly does not hold for cases when the p-parameters differ.

The convolution of m mutually independent multinomial distributions with parameters $(n_j; p_{1j}, p_{2j}, \ldots, p_{kj})$, $j = 1, 2, \ldots, m$—that is, the joint distribution of the sums $(N_{1\bullet}, N_{2\bullet}, \ldots, N_{k\bullet})'$, where $N_{i\bullet} = \sum_{j=1}^{m} N_{ij}$, of the mutually independent multinomial sets $(N_{1j}, N_{2j}, \ldots, N_{kj})'$, has probability generating function

$$\prod_{j=1}^{m}\left(\sum_{i=1}^{k} p_{ij}t_i\right)^{n_j}. \tag{35.18}$$

The probability mass function of this distribution is

$$\Pr\left[\bigcap_{i=1}^{k}(N_{i\bullet} = n_{i\bullet})\right] = \text{coefficient of } \textstyle\prod_{i=1}^{k} t_i^{n_{i\bullet}} \text{ in (35.18)}$$

$$= \sum_{j=1}^{m} \sum_{n_{ij}}^{*} \binom{n_j}{n_{1j}, n_{2j}, \ldots, n_{kj}} \prod_{i=1}^{k} p_{ij}^{n_{ij}}, \tag{35.19}$$

where the summation $\sum^*$ is over n_{ij}'s satisfying

$$\sum_{i=1}^{k} n_{ij} = n_j \qquad \text{and} \qquad \sum_{j=1}^{m} n_{ij} = n_{i\bullet} .$$

Note the distinction between n_j and $n_{i\bullet}$.

Shah (1975) derived a recurrence relationship satisfied by these probabilities. The expression is rather complex.

From (35.17), it follows that the cumulant generating function of the multinomial $(n; p_1, p_2, \ldots, p_k)$ distribution is

$$n \log \left(\sum_{i=1}^{k} p_i e^{t_i} \right) . \tag{35.20}$$

Guldberg's (1935) recurrence formula

$$\kappa_{r_1,\ldots,r_{i-1},r_i+1,r_{i+1},\ldots,r_k} = \frac{p_i}{p_k} \frac{\partial}{\partial(p_i/p_k)} [\kappa_{r_1,r_2,\ldots,r_k}] \tag{35.21}$$

may be derived from (35.20). Explicit formulas for several lower-order mixed cumulants were presented by Wishart (1949), including (in an obvious notation)

$$\begin{aligned} \kappa_{..21} &= -n\,p_i p_j (1 - 2p_i); \quad \kappa_{.111} = 2n\,p_i p_j p_g \,; \\ \kappa_{..31} &= -n\,p_i p_j \{1 - 6p_i(1 - p_i)\} \,; \\ \kappa_{..22} &= -n\,p_i p_j \{(1 - 2p_i)(1 - 2p_j) + 2p_i p_j\} \,; \\ \kappa_{..211.} &= 2n\,p_i p_j p_g (2 - 3p_i) \,; \quad \kappa_{.1111.} = -6n\,p_i p_j p_g p_h \,. \end{aligned} \tag{35.22}$$

[In each formula, the subscripts *i, j, g, h* correspond to the first, second, third, and fourth nonzero subscripts of the κ's, respectively.]

Khatri and Mitra (1968) have given the characteristic function of the multinomial distribution as

$$\begin{aligned} E\left[e^{it^T N}\right] &= E\left[e^{i(t_1 N_1 + t_2 N_2 + \cdots + t_k N_k)}\right] \\ &= n! \sum_{n_1,\ldots,n_k} \frac{(p_1 e^{it_1})^{n_1} \cdots (p_k e^{it_k})^{n_k}}{n_1! \cdots n_k!} \\ &= \left\{ \sum_{j=1}^{k} p_j\, e^{it_j} \right\}^n , \end{aligned}$$

where $i = \sqrt{-1}$.

By considering the entropy of a multinomial distribution defined by

$$H(\boldsymbol{p}) = -\sum_{\boldsymbol{x}} P(\boldsymbol{x}) \log P(\boldsymbol{x}) \,,$$

Mateev (1978) has shown that $\operatorname{argmax}_{\boldsymbol{p}} H(\boldsymbol{p}) = \frac{1}{k}\,\mathbf{1}' = (\frac{1}{k}, \ldots, \frac{1}{k})'$.

Sugg and Hutcheson (1972) developed computer algorithms to assist in the calculation of moments of linear combinations of functions of components of a multinomially distributed set of random variables. They provide computer results from a FORTRAN source code.

Price (1946) noted that the mode(s) of the distribution in (35.1) is(are) located, broadly speaking, near the expected value point $\{n_i = np_i, \; i = 1, 2, \ldots, k\}$. More precisely, Moran (1973) has shown [see also Feller (1957)] that the appropriate values of $\{n_i\}$ satisfy the inequalities

$$np_i - 1 \le n_i \le (n + k - 1)p_i, \qquad i = 1, 2, \ldots, k \,. \tag{35.23}$$

These inequalities, together with the condition $\sum_{i=1}^{k} n_i = n$, restrict the possible choices for modes to a relatively few points. If k is small, there may indeed be only one set of values of the n_i's satisfying conditions in (35.23) so that there is then a unique mode for the corresponding multinomial distribution.

Finucan (1964) has devised a procedure that will locate the possible modes in an efficient manner and that will also indicate possible configurations of joint modes. He has shown additionally that there can be only one *modal region*, in the sense that there is no local maximum remote from the set of equiprobable points, each giving the maximum value of $P(n_1, n_2, \ldots, n_k)$ in (35.1).

Olbrich (1965) has presented an interesting geometrical representation of a multinomial distribution (which did not receive sufficient attention in the literature during the last 30 years). By means of the transformation $M_j = \cos^{-1}\sqrt{(N_j/n)}$, $j = 1, 2, \ldots, k$, points $(M_1, M_2, \ldots, M_k)$ on a k-dimensional sphere of unit radius are obtained. The angle between the radii to two such points, corresponding to two independent multinomial random variables with parameters $(n_i; p_1, p_2, \ldots, p_k)$, $i = 1, 2$, respectively, has variance approximately equal to $(n_1^{-1} + n_2^{-1})$. [Note that this transformation is consequently approximately variance-stabilizing; see Chapter 1, Eq. (1.293).] Olbrich has, therefore, suggested that this transformed statistic may be used in developing a test for the equality of the values of $(p_1, p_2, \ldots, p_k)$ of two multinomial distributions.

Alam (1970) has shown that

$$\textbf{(i)}\ \Pr\left[\bigcap_{j=1}^{g}(N_{i_j} \ge c_{i_j})\right] \qquad (\text{for } g < k)$$

is a nondecreasing function of any one of $p_{i_1}, p_{i_2}, \ldots, p_{i_g}$; and

(ii) (in an obvious notation)

$$\Pr\left[\bigcap_{i=1}^{k}(N_i \geq c) \mid 1-(k-1)p, p, \ldots, p\right]$$
$$\leq \Pr\left[\bigcap_{i=1}^{k}(N_i \geq c) \mid p_1, p_2, \ldots, p_k\right]$$
$$\leq \Pr\left[\bigcap_{i=1}^{k}(N_i \geq c) \,\middle|\, \frac{1}{k}, \frac{1}{k}, \ldots, \frac{1}{k}\right], \qquad (35.24)$$

where $p \leq \min(p_1, p_2, \ldots, p_k)$.

Olkin (1972) has extended this result by considering two sets of p_i's, $\boldsymbol{p} = (p_1, p_2, \ldots, p_k)'$ and $\boldsymbol{p}^* = (p_1^*, p_2^*, \ldots, p_k^*)'$, with $p_{a_1} \leq p_{a_2} \leq \cdots \leq p_{a_k}$ and $p_{b_1}^* \leq p_{b_2}^* \leq \cdots \leq p_{b_k}^*$ and satisfying the conditions

$$\sum_{i=1}^{g} p_{a_i} \geq \sum_{i=1}^{g} p_{b_i}^* \qquad \text{for } g = 1, 2, \ldots, k-1 .$$

Olkin has specifically established that

$$\Pr\left[\bigcap_{i=1}^{k}(N_i \geq c) \mid \boldsymbol{p}\right] \leq \Pr\left[\bigcap_{i=1}^{k}(N_i \geq c) \mid \boldsymbol{p}^*\right] . \qquad (35.25)$$

Mallows (1968) has obtained the interesting inequality

$$\Pr\left[\bigcap_{i=1}^{k}(N_i \leq c_i)\right] \leq \prod_{i=1}^{k} \Pr\left[N_i \leq c_i\right] , \qquad (35.26)$$

which is valid for **any** set of values $c_1, c_2, \ldots, c_k$ and any parameter values $(n; p_1, p_2, \ldots, p_k)$. Jogdeo and Patil (1975) have shown that, in addition to the inequality in (35.26), the inequality

$$\Pr\left[\bigcap_{i=1}^{k}(N_i \geq c_i)\right] \leq \prod_{i=1}^{k} \Pr[N_i \geq c_i] \qquad (35.27)$$

also holds. The inequalities in (35.26) and (35.27) together establish the *negative dependence* among the components of a multinomially distributed set. This property also holds for the multivariate hypergeometric distributions (see Chapter 39).

By providing multivariate generalizations of the concept of a Schur convex-function, Rinott (1973) has obtained some majorization results and some general inequalities from which the above results follow as special cases. To be specific, let us consider the partial ordering $\boldsymbol{p} = (p_1, \ldots, p_k)' < \boldsymbol{p}^* = (p_1^*, \ldots, p_k^*)'$ (both with decreasingly rearranged components) if

$$\sum_{i=1}^{\ell} p_i \leq \sum_{i=1}^{\ell} p_i^* \qquad \text{for } \ell = 1, 2, \ldots, k\,.$$

Then, for the multinomial $(N; p_1, \ldots, p_k)$ distribution, let $\phi(n_1, \ldots, n_k)$ be a symmetric Schur function—that is, ϕ is symmetric and $\phi(\boldsymbol{n}) \leq \phi(\boldsymbol{n}^*)$ whenever $\boldsymbol{n} \prec \boldsymbol{n}^*$.

Define the function

$$\psi(\boldsymbol{p}) = E_{\boldsymbol{p}}[\phi(\boldsymbol{N})] = \sum_{\substack{0 \leq n_1, \ldots, n_k \leq N \\ \sum_{i=1}^{k} n_i = N}} \phi(\boldsymbol{n}) \binom{N}{n_1, \ldots, n_k} \prod_{i=1}^{k} p_i^{n_i}\,.$$

Then the function $\psi(\boldsymbol{p})$ is a Schur function.

Using this general result and the facts that $(\max_{1 \leq i \leq k} n_i)^\alpha$ and $-(\min_{1 \leq i \leq k} n_i)^\alpha$ are both symmetric Schur functions, the above result implies that

$$\psi_1(\boldsymbol{p}) = E_{\boldsymbol{p}}\left[\left(\max_{1 \leq i \leq k} N_i\right)^\alpha\right] \text{ and } \psi_2(\boldsymbol{p}) = -E_{\boldsymbol{p}}\left[\left(\min_{1 \leq i \leq k} N_i\right)^\alpha\right]$$

are both Schur functions. As a direct consequence, we readily observe that $E_{\boldsymbol{p}}\,[\max_{1 \leq i \leq k} N_i]$ is minimized for $\boldsymbol{p} = \left(\frac{1}{k}, \ldots, \frac{1}{k}\right)'$ since $\left(\frac{1}{k}, \ldots, \frac{1}{k}\right)' \prec (p_1, \ldots, p_k)'$ for any $\boldsymbol{p}$ with $\sum_{i=1}^{k} p_i = 1$. Furthermore, if we choose the functions

$$\phi_c(\boldsymbol{n}) = \begin{cases} 1 & \text{if} \quad \min_{1 \leq i \leq k} n_i \geq c \\ 0 & \text{if} \quad \min_{1 \leq i \leq k} n_i < c \end{cases}$$

and

$$\phi_c^*(\boldsymbol{n}) = \begin{cases} 1 & \text{if} \quad \min_{1 \leq i \leq k} n_i < c \\ 0 & \text{if} \quad \min_{1 \leq i \leq k} n_i \geq c \end{cases}$$

and observe that $-\phi_c(\boldsymbol{n})$ and $-\phi_c^*(\boldsymbol{n})$ (being monotone functions of $\min n_i$ and $\max n_i$, respectively) are symmetric Schur functions, we obtain the results that

$$\psi(\boldsymbol{p}) = E_{\boldsymbol{p}}[-\phi_c(\boldsymbol{N})] = -\Pr\left[\bigcap_{i=1}^{k} (N_i \geq c) \mid \boldsymbol{p}\right]$$

and

$$\psi^*(\boldsymbol{p}) = E_{\boldsymbol{p}}[-\phi_c^*(\boldsymbol{N})] = -\Pr\left[\bigcap_{i=1}^{k} (N_i < c) \mid \boldsymbol{p}\right]$$

are both Schur functions. Note that the above inequality is the same as Olkin's inequality in (35.25).

Chamberlain (1987) has presented a discussion of analysis when there are restrictions on certain moments. Note that if a crude moment (moment about zero) of a multinomial distribution is specified, there is a restriction on a polynomial in $p_1, p_2, \ldots, p_k$.

Manly (1974) studied a multinomial model in which the cell probabilities change sequentially, depending on the result of the previous trials. He supposed that if there were N_i individuals ($i = 1, 2, \ldots, k$) among those chosen in the first n_1 (say) trials, the probability that the next individual chosen will be from the ith cell is

$$p_i' = \beta_i N_i \Bigg/ \left(\sum_{j=1}^{k} \beta_j N_j \right) . \tag{35.28}$$

Stemming from the problem of selecting the most probable multinomial event discussed by Bechhofer, Elmaghraby, and Morse (1959), Kesten and Morse (1959) proved the following interesting property of the multinomial distribution: With $p_{[1]} \le p_{[2]} \le \cdots \le p_{[k]}$ denoting the ranked multinomial probabilities and $\phi_k = \phi(p_{[1]}, \ldots, p_{[k]})$ denoting the probability of correct selection of the procedure, ϕ_k is minimized among all configurations with $p_{[k]} \ge \theta p_{[i]}$, $i = 1, 2, \ldots, k-1$ (for any number $1 < \theta < \infty$), by the configuration $p_{[1]} = \cdots = p_{[k-1]} = p_{[k]}/\theta = 1/(\theta + k - 1)$. Thus, this configuration becomes the "least favorable" one; for more details, see Gupta and Panchapakesan (1979).

4 CALCULATION OF SUMS OF MULTINOMIAL PROBABILITIES

In Chapter 3 [see Eq. (3.37)], it has been noted that sums of binomial probabilities can be expressed as incomplete beta function ratios. Olkin and Sobel (1965) and Stoka (1966) have generalized this result, obtaining expressions for tail probabilities of multinomial distributions in terms of incomplete Dirichlet integrals. In particular,

$$\begin{aligned}\Pr\left[\bigcap_{i=1}^{k-1}(N_i > c_i)\right] &= \binom{n}{n-k-\sum_{i=1}^{k-1} c_i, c_1, c_2, \ldots, c_{k-1}} \\ &\quad \times \int_0^{p_1}\int_0^{p_2}\cdots\int_0^{p_{k-1}} \left(1 - \sum_{i=1}^{k-1} x_i\right)^{n-k-\sum_{i=1}^{k-1} c_i} \\ &\quad \times \prod_{i=1}^{k-1} x_i^{c_i}\, dx_{k-1}\cdots dx_2\, dx_1 \end{aligned} \tag{35.29}$$

with $\sum_{i=1}^{k-1} c_i \le n - k$.

The formula [Levin (1981)]

$$\begin{aligned} F_{\boldsymbol{N}}(\boldsymbol{n} \mid n) &= \Pr\left[\bigcap_{i=1}^{k}(N_i \le n_i) \mid n\right] \\ &= \frac{n!\, e^s}{s^n} \left\{ \prod_{i=1}^{k} \Pr[X_i \le n_i] \right\} \Pr[W = n] , \end{aligned} \tag{35.30}$$

where X_i has a Poisson (sp_i) distribution $(i = 1, 2, \ldots, k)$ and W is distributed as $Y_1 + Y_2 + \cdots + Y_k$ with the Y's being mutually independent and Y_i having a Poisson (sp_i) distribution truncated to the range $0, 1, \ldots, n_i$, can be used to evaluate the cumulative distribution function of $\max(N_1, N_2, \ldots, N_k) = M$ (say) by taking $n_1 = n_2 = \cdots = n_k = m$. Note that in (35.30), s is a parameter that may be chosen arbitrarily. Naturally, one will try to choose s so as to facilitate numerical calculations and for this reason it is called a *tuning parameter*.

Levin (1981) reported that C. L. Mallows informed him that formula (35.30) can be established by applying a version of Bayes' theorem to the genesis of a multinomial distribution as the conditional distribution of m mutually independent Poisson random variables, given their sum (see Section 1). With X_i's distributed as independent Poisson (sp_i) variables—as in (35.30)—we have using Bayes formula

$$\Pr\left[\bigcap_{i=1}^{k}(X_i \le n_i)\middle|\sum_{i=1}^{k} X_i = n\right]$$

$$= \frac{\Pr\left[\bigcap_{i=1}^{k}(X_i \le n_i)\right]}{\Pr\left[\sum_{i=1}^{k} X_i = n\right]} \Pr\left[\sum_{i=1}^{k} X_i = n\middle|\bigcap_{i=1}^{k}(X_i \le n_i)\right]. \tag{35.31}$$

[The conditional distribution of X_i, given $X_i \le n_i$, is of course truncated Poisson with range $0, 1, \ldots, n_i$.] Alternatively, as pointed out by Levin (1981), the formula can be obtained from the generating function identity

$$\sum_{n=0}^{\infty} F_N(\boldsymbol{n} \mid n) \frac{t^n}{n!} \equiv \prod_{i=1}^{k}\left\{1 + p_i t + \frac{1}{2!}(p_i t)^2 + \cdots + \frac{1}{n!}(p_i t)^{n_i}\right\}. \tag{35.32}$$

Mallows (1968) used the Stirling formula $a!e^a/a^a \simeq \sqrt{2\pi a}$ and a normal approximation to the distribution of W to evaluate $\Pr[W = n]$. For this purpose, Levin (1981) recommended using the four-term Edgeworth expansion (see Chapter 12)

$$\Pr[W = n] \simeq f\left(\frac{n - \sum_{i=1}^{k}\mu'_{1,i}}{\sqrt{\sum_{i=1}^{k}\mu_{2,i}}}\right)\sqrt{\sum_{i=1}^{k}\mu_{2,i}}\,, \tag{35.33}$$

where $\mu'_{1,i} = E[Y_i]$, $\mu_{r,i} = E[(Y_i - \mu'_{1,i})^r]$, and

$$f(x) = \frac{1}{\sqrt{2\pi}} e^{-x^2/2}\left\{1 + \frac{1}{6}(x^3 - 3x)\gamma_1 + \frac{1}{24}(x^4 - 6x^2 + 3)\gamma_2 + \frac{1}{72}(x^6 - 15x^4 + 45x^2 - 15)\gamma_1^2\right\} \tag{35.34}$$

with skewness and kurtosis (γ_1 and γ_2, respectively) given by

$$\gamma_1 = \left(\sum_{i=1}^{k}\mu_{3,i}\right) \Big/ \left(\sum_{i=1}^{k}\mu_{2,i}\right)^{3/2},$$

$$\gamma_2 = \left\{ \left(\sum_{i=1}^{k} \mu_{4,i} \right) \Big/ \left(\sum_{i=1}^{k} \mu_{2,i} \right)^2 \right\} - 3 .$$

This approximation seems to provide three- or four-decimal place accuracy.

For Y_i,

$$\mu'_{1,i} = sp_i \left\{ 1 - \frac{(sp_i)^{n_i}}{n_i!} \right\} \Big/ \left\{ \sum_{j=0}^{n_i} \frac{(sp_i)^j}{j!} \right\} , \tag{35.35}$$

and the factorial moments (see Chapter 34) can be evaluated from the recurrence relation

$$\mu_{(r+1),i} = sp_i\, \mu_{(r),i} - (sp_i)^{(r)}(sp_i - \mu'_{1,i}) , \tag{35.36}$$

where

$$\mu_{(r),i} = E[Y_i^{(r)}] = E[Y_i(Y_i - 1) \cdots (Y_i - r + 1)] \quad \text{[cf. (34.6)]}.$$

When n is large, taking $s = n$ appears to yield good results. Mallows (1968) considered an example with $n = 500$, $k = 50$, $p_i = 0.02$, and $n_i = m = 19$ for all $i = 1, 2, \ldots, k$. Using a modification of Bonferroni bounds, Mallows obtained the *Bonferroni–Mallows bounds*:

$$1 - \sum_{i=1}^{k} \Pr[N_i > m] \le \Pr[M \le m] \le \prod_{i=1}^{k} \Pr[N_i \le m], \tag{35.37}$$

which gave the inequalities

$$0.8437 \le \Pr[M \le 19] \le 0.8551$$

for the example considered. On the other hand, using (35.30) and (35.33), Levin (1981) obtained $\Pr[M \le 19] \simeq 0.8643$.

For the case $n = 12$, $k = 12$, $p_i = \frac{1}{12}$, and $m = 3$ [one of the sets considered by Barton and David (1959)], the normal approximation yields

$$\Pr[M \le 3] \simeq 0.8643$$

and the Edgeworth expansion (35.33) gives

$$\Pr[M \le 3] \simeq 0.8367$$

while the Bonferroni–Mallows inequalities (35.37) are

$$0.8340 \le \Pr[M \le 3] \le 0.8461$$

and the exact value (to four decimal places) is 0.8371.

Further discussion of approximations to the distribution of $M = \max(N_1, N_2, \ldots, N_k)$ in presented in Section 5. Freeman (1979) provided an enumerative method and a computer program for evaluating the distribution of M in the equiprobable case $p_i = 1/k, i = 1, 2, \ldots, k$.

Work of Glaz and Johnson (1984) on probabilities for multivariate distributions with dependence structures is also worth mentioning here in the present context.

Sobel, Uppuluri, and Frankowski (1977) considered the *incomplete Dirichlet integral of Type 1*, defined by the b-fold integral

$$I_p^{(b)}(r, n) = \frac{\Gamma(n+1)}{\{\Gamma(r)\}^b \Gamma(n+1-br)} \int_0^p \cdots \int_0^p \left(1 - \sum_{i=1}^b x_i\right)^{n-br} \prod_{i=1}^b x_i^{r-1} dx_i\,,$$

where $0 \le p \le 1/b, n \ge rb, b$ is an integer, and n, b, r are also positive. These authors have tabulated this incomplete Dirichlet integral of Type 1 quite extensively, and these tables can be used to determine certain multinomial cumulative probabilities. This is so as we observe that the above expression is nothing but the probability that the minimum frequency in a multinomial is at least r; here, it is supposed that there are n items and $b + 1$ cells, b of which have a common cell probability equal to p, and that the minimum frequency is taken over these b cells.

In a similar manner, Sobel, Uppuluri, and Frankowski (1985) have studied the following two forms, called *incomplete Dirichlet integrals of Type 2*:

$$C_a^{(b)}(r, m) = \frac{\Gamma(m+br)}{\{\Gamma(r)\}^b \Gamma(m)} \int_0^a \cdots \int_0^a \frac{(x_1 x_2 \cdots x_b)^{r-1}}{(1 + x_1 + \cdots + x_b)^{m+br}}\, dx_1 \cdots dx_b$$

and

$$D_a^{(b)}(r, m) = \frac{\Gamma(m+br)}{\{\Gamma(r)\}^b \Gamma(m)} \int_a^\infty \cdots \int_a^\infty \frac{(x_1 x_2 \cdots x_b)^{r-1}}{(1 + x_1 + \cdots + x_b)^{m+br}}\, dx_1 \cdots dx_b\,.$$

These authors have tabulated these integrals quite extensively, and these tables can in turn be used to determine certain negative multinomial cumulative probabilities.

5 APPROXIMATIONS

The most commonly used approximation connected with multinomial is the X^2-*approximation*. This can be obtained in a natural way by considering the limiting form of $P(n_1, n_2, \ldots, n_k)$ in (35.1) as n tends to ∞, with $p_1, p_2, \ldots, p_k$ remaining fixed; see, for example, Pearson (1900) and Johnson and Tetley (1950). Since (as mentioned earlier) N_i has a binomial (n, p_i) distribution, we have

$$\frac{N_i - np_i}{\sqrt{np_i(1-p_i)}} = O_p(1)\,;$$

see Chapter 3, Eq. (3.22). From (35.1), we have

$$P(n_1, n_2, \ldots, n_k) \doteq \left\{2\pi n \prod_{i=1}^{k} p_i\right\}^{-1/2} \exp\left\{-\frac{1}{2} \sum_{i=1}^{k} (n_i - np_i)^2/(np_i) - \frac{1}{2} \sum_{i=1}^{k} (n_i - np_i)/(np_i) + \frac{1}{6} \sum_{i=1}^{k} (n_i - np_i)^3/(np_i)^2\right\}. \tag{35.38}$$

Neglecting the terms in $n^{-1/2}$, we obtain from (35.38)

$$P(n_1, n_2, \ldots, n_k) \doteq \left\{2\pi n \prod_{i=1}^{k} p_i\right\}^{-1/2} \exp\left\{-\frac{1}{2} X^2\right\}, \tag{35.39}$$

where

$$X^2 = \sum_{i=1}^{k} \frac{(n_i - np_i)^2}{np_i} = \sum_{i=1}^{k} \frac{(\text{observed frequency} - \text{expected frequency})^2}{\text{expected frequency}} \tag{35.40}$$

—a familiar form, and a quantity often referred to as *chi-squared*.

There have been numerous numerical investigations of the accuracy of approximation in (35.39) [for example, Shanawany (1936), Lancaster (1961), Bennett (1962), Wise (1963, 1964), Lancaster and Brown (1965), Argentino, Morris, and Tolson (1967), and, from a more analytical standpoint, Struder (1966)], and it is not easy to give a simple summary of the results. The accuracy of the approximation increases as $n \times \min(p_1, p_2, \ldots, p_k)$ increases, and it decreases with increasing values of k. We note that, for $X^2 = \sum_{i=1}^{k} (N_i - np_i)^2/np_i$ (see Chapter 18, Sections 1 and 2),

$$E[X^2] = k - 1 = E[\chi_{k-1}^2] \tag{35.41}$$

and

$$\operatorname{var}(X^2) = 2(k-1)\left(1 - \frac{1}{n}\right) + \frac{1}{n}\left\{\left(\sum_{i=1}^{k} 1/p_i\right) - k^2\right\} = \left(1 - \frac{1}{n}\right)\operatorname{var}(\chi_{k-1}^2) + \frac{1}{n}\left\{\left(\sum_{i=1}^{k} 1/p_i\right) - k^2\right\} \tag{35.42}$$

where χ_{k-1}^2 denotes a chi-squared variable with $k - 1$ degrees of freedom (see Chapter 18). Note that in the equiprobable case ($p_1 = p_2 = \cdots = p_k = 1/k$), the last term on the right-hand side of (35.42) is zero so that

$$E[X^2] = E[\chi^2_{k-1}] \text{ and } \operatorname{var}(X^2) = \left(1 - \frac{1}{n}\right) \operatorname{var}(\chi^2_{k-1}) . \tag{35.43}$$

Rüst (1965) obtained an explicit but rather complicated expression for the moment generating function of X^2.

Wise (1964) suggested replacing X^2 in the exponent in (35.39) by

$$X'^2 = \sum_{i=1}^{k} \frac{(N_i - np_i)^2}{np_i + \frac{1}{2}} . \tag{35.44}$$

Pearson (1900) showed that X^2 is approximately distributed as χ^2_{k-1}. But, as mentioned in Section 2, Lancaster (1966) has noted that this limiting distributional result appears in the work of Bienaymé (1838). Hoel (1938) suggested an improvement on this chi-square approximation by expanding the moment generating function of the multinomial $(n; p_1, p_2, \ldots, p_k)$ distribution and expressing its leading terms as linear functions of moment generating functions of chi-square distributions. This leads to the formula

$$\Pr[X^2 \le c] \doteq \Pr[\chi^2_{k-1} \le c] + \frac{1}{n}(r_1 s_1 + r_2 s_2) , \tag{35.45}$$

where

$$s_1 = \frac{1}{8}\left\{\left(\sum_{i=1}^{k} 1/p_i\right) - (k^2 + 2k - 2)\right\} , \tag{35.46}$$

$$s_2 = \frac{1}{24}\left\{\left(\sum_{i=1}^{k} 1/p_i\right) - (3k^2 + 6k - 4)\right\} , \tag{35.47}$$

and, for k odd,

$$r_1 = \frac{c^{(k-1)/2} \, e^{-c/2}}{2 \cdot 4 \cdots (k-1)(k+1)} \{c - (k+1)\} , \tag{35.48}$$

while for k even

$$r_1 = \sqrt{\frac{2}{\pi}} \, \frac{c^{(k-1)/2} \, e^{-c/2}}{1 \cdot 3 \cdots (k-1)(k+1)} \{c - (k+1)\} , \tag{35.49}$$

and for all k (odd or even)

$$r_2 = \frac{c^2 - 2(k+3)c + (k+1)(k+3)}{(k+1)(k+3)\{c - (k+1)\}} \, r_1, \tag{35.50}$$

where r_1 is given by (35.48) or (35.49).

Vora (1950) derived bounds for $\Pr[X^2 \le c]$ in terms of cumulative probabilities for noncentral χ^2 distributions (see Chapter 29), of form

$$\tau_2 \Pr[\chi_{k-1}'^2(\delta_2^2) \le c_2] \le \Pr[X^2 \le c] \le \tau_1 \Pr[\chi_{k-1}'^2(\delta_1^2) \le c_1], \qquad (35.51)$$

where $\tau_1, \tau_2, \delta_1^2, \delta_2^2, c_1$ and c_2 depend on $c, n, p_1, p_2, \ldots, p_k$. The formulas for these constants are quite complicated and the bounds are also not very close. Vora (1950) has also derived bounds in terms of central χ^2 probabilities for $\Pr[X''^2 \le c]$, where

$$X''^2 = \sum_{i=1}^{k} \frac{(N_i + \frac{1}{2} - np_i)^2}{np_i}. \qquad (35.52)$$

It is important to bear in mind that there are *two* approximations to be considered: (i) the accuracy of the approximation in (35.39) for $P(n_1, n_2, \ldots, n_k)$, and (ii) the accuracy of the χ^2-approximation to the distribution of X^2.

Johnson (1960) suggested approximating the multinomial distribution in (35.1) by using a joint Dirichlet density function [see Chapter 40 of Johnson and Kotz (1972)] for the variables

$$Y_i = N_i/n, \qquad i = 1, 2, \ldots, k. \qquad (35.53)$$

Explicitly,

$$p_{Y_1,Y_2,\ldots,Y_k}(y_1, y_2, \ldots, y_k) = \Gamma(n-1) \prod_{i=1}^{k} \left\{ \frac{y_i^{(n-1)p_i - 1}}{\Gamma[(n-1)p_i]} \right\},$$

$$0 \le y_i, \quad \sum_{i=1}^{k} y_i = 1, \qquad (35.54)$$

which is to be interpreted as the joint density function of any $k - 1$ of the Y_i's, when the remaining Y_i is expressed in terms of these $k - 1$ Y_i's. This approximation provides the correct first- and second-order moments and product moments of the joint distribution of the Y_i's, and consequently also of the N_i's.

From the well-known properties of the Dirichlet distribution [see, for example, Johnson and Kotz (1972, Chapter 40)], the above-described approximation is equivalent to taking

$$Y_i = V_i \Big/ \sum_{j=1}^{k} V_j, \qquad (35.55)$$

where V_i's are mutually independent and are distributed as χ^2 with $2(n-1)p_i$ degrees of freedom, $i = 1, 2, \ldots, k$. This, in turn, implies that distributions of various functions of the Y_i's could be approximated by those of corresponding functions of independent χ^2's. Johnson and Young (1960) have used the Dirichlet approximation to obtain for the equiprobable case ($p_1 = p_2 = \cdots = p_k = 1/k$) approximations to the sampling distributions of $\max_{1\le i\le k} N_i$ and $(\max_{1\le i\le k} N_i)/(\min_{1\le i\le k} N_i)$; see also Young (1961) for additional details.

Tables of the distributions of $\max_{1\le i\le k} N_i$ and $\min_{1\le i\le k} N_i$ for $5 \le k \le 50$ and $10 \le n \le 200$ by Rappeport (in an unpublished work) have been reported by Mallows (1968); see also Pearson and Hartley (1973). Approximate percentage points for these extreme order statistics have been discussed by Kozelka (1956). Tables of the means and variances of $\max_{1\le i\le k} N_i$ and $\min_{1\le i\le k} N_i$ for the case

$$p_1 = p_2 = \cdots = p_{k-1} = \frac{1}{k-1+a} \qquad \text{and} \qquad p_k = \frac{a}{k-1+a}$$

where $k = 2(1)10$, $n = 2(1)15$, and $a = \frac{1}{5}, \frac{1}{4}, \frac{1}{3}, \frac{1}{2}, 1(1)5$ have been provided by Gupta and Nagel (1966, 1967).

For the case $k = 3$, with $p_1 = p_2 = p$ and $p_3 = 1 - 2p$, Greenwood and Glasgow (1950) proposed the approximations

$$E[\min(N_1, N_2)] \doteq np - \sqrt{\frac{np}{\pi}} \text{ and } E[\max(N_1, N_2)] \doteq np + \sqrt{\frac{np}{\pi}}. \tag{35.56}$$

These results clearly also apply to the general case when $p_1 = p_2$ [because the remaining $k - 2$ cells can be combined into one without affecting the distributions of $\min(N_1, N_2)$ or $\max(N_1, N_2)$].

Young (1962) proposed an approximation to the distribution of the sample range

$$W = \max_{1\le i\le k} N_i - \min_{1\le i\le k} N_i \tag{35.57}$$

when $p_1 = p_2 = \cdots = p_k = 1/k$, according to which

$$\Pr[W \le w] \doteq \Pr\left[W' \le w\sqrt{\frac{k}{n}}\right], \tag{35.58}$$

where W' is distributed as the range of k i.i.d. standard normal variables (see Chapter 13). Bennett and Nakamura (1968) gave values of the largest integer value of n for which $\Pr[W \ge w] \le \alpha$ for $\alpha = 0.01,\ 0.025,\ 0.05$, and

(i) $k = 2(1)4$; $w = 4(1)15$;
(ii) $k = 5$; $w = 3(1)13$;
(iii) $k = 6, 7$; $w = 3(1)10$;
(iv) $k = 8(1)10$; $w = 3(1)9$.

For the mean deviation

$$M = \frac{1}{k}\sum_{i=1}^{k}\left|N_i - \frac{n}{k}\right|, \tag{35.59}$$

Young (1967) proposed a similar approximation

$$\Pr[M \leq m] \doteq \Pr\left[M' \leq m\sqrt{\frac{k}{n}}\right], \tag{35.60}$$

where M' is distributed as the mean deviation of k i.i.d. standard normal variables (see Chapter 13).

Fisz (1964) showed that if the sequence of sets of random variables $\{a_{i,n}N_i + b_{i,n}\}$ with appropriate constants $\{a_{i,n}, b_{i,n}\}$, $(i = 1, 2, \ldots, k)$ has a proper $(k-1)$-dimensional limiting distribution as $n \to \infty$, then this must be (for some $\ell \leq k - 1$) the joint distribution of ℓ independent Poisson variables and $(k - 1 - \ell)$ multinormally distributed random variables, independent of the Poisson variables; see also Rjauba (1958). Khatri and Mitra (1968) constructed approximations based on this result (see below).

Approximations to sums of multinomial probabilities can be based on the identity

$$P(n_1, n_2, \ldots, n_k) = (e/n)^n\, n! \prod_{i=1}^{k} \left\{e^{-np_i}(np_i)^{n_i} \,/\, n_i!\right\}. \tag{35.61}$$

From (35.61), we see that

$$\Pr\left[\bigcap_{i=1}^{k}(N_i \leq c_i)\right] = (e/n)^n\, n! \sum_j{}^{*} \prod_{i=1}^{k} \Pr\left[N_i^+ = c_{ij}\right], \tag{35.62}$$

where N_i^+ has a Poisson distribution with mean np_i $(i = 1, 2, \ldots, k)$ and summation in $\sum_j^{*}$ is over all sets $\boldsymbol{c}_j = (c_{1j}, c_{2j}, \ldots, c_{kj})$ such that $c_{ij} \leq c_i$ and $\sum_{i=1}^{k} c_{ij} \leq n$. Approximations may then be applied to the Poisson probabilities.

Khatri and Mitra (1968) developed an interesting method for approximating the probability of the event

$$\bigcap_{i=1}^{s}(N_{a_i} \leq x_{a_i})$$

for any subset $(N_{a_1}, N_{a_2}, \ldots, N_{a_s})'$ of $\boldsymbol{N}$. Clearly, we need to take $s < k$ since, if $s = k$, then necessarily $\sum_{i=1}^{k} N_i = n$. As a matter of fact, we will take the subset $(N_1, N_2, \ldots, N_{k-1})'$ since any specific case can be reduced to this one by renumbering the cells and combining together all cells not in $(N_{a_1}, N_{a_2}, \ldots, N_{a_s})'$. Khatri and Mitra first note that $P(n_1, n_2, \ldots, n_k)$ may be written as

$$P(n_1, n_2, \ldots, n_k) = A \prod_{i=1}^{k-1} \left\{e^{-np_i}(np_i)^{n_i}/n_i!\right\} \tag{35.63}$$

with

$$A = \left\{1 - \frac{n(1 - p_k)}{n_k}\right\}^{n_k} \left\{\prod_{i=1}^{n-n_k-1} \left(1 - \frac{i}{n}\right)\right\} \exp\{n(1 - p_k)\}$$

and then approximate A. This leads to an approximation up to the order $O(n^{-4})$

$$\begin{aligned}
&\Pr\left[\bigcap_{i=1}^{k-1}(N_i \le x_i)\right] \\
&\doteq \left\{1 - \frac{1}{2n}\mathcal{D}^2 + \frac{1}{n^2}\left(\frac{1}{3}\mathcal{D}^3 + \frac{1}{8}\mathcal{D}^4\right)\right. \\
&\quad - \frac{1}{n^3}\left(\frac{1}{4}\mathcal{D}^4 + \frac{1}{6}\mathcal{D}^5 + \frac{1}{48}\mathcal{D}^6\right) \\
&\quad \left. + \frac{1}{n^4}\left(\frac{1}{5}\mathcal{D}^5 + \frac{13}{72}\mathcal{D}^6 + \frac{1}{24}\mathcal{D}^7 + \frac{1}{384}\mathcal{D}^8\right)\right\} \\
&\quad \cdot \prod_{i=1}^{k-1}\left\{\sum_{j=0}^{x_i} e^{-np_i}(np_i)^j/j!\right\}, \qquad (35.64)
\end{aligned}$$

where $\mathcal{D}$ is the operator $\sum_{i=1}^{k-1} np_i\nabla_{x_i}$ on the product of the sums, with the symbol ∇_{x_i} denoting the backward difference operator $\nabla_{x_i} f(x_i) \equiv f(x_i) - f(x_i - 1)$. Khatri and Mitra (1968) gave a numerical example for the case $k = 3$, $n = 10$, $p_1 = p_2 = 0.2$, $p_3 = 0.6$, $x_1 = 16$, and $x_2 = 24$ in which the first three successive approximants [corresponding to inclusion of higher powers of n^{-1} in (35.64)] to $\Pr[N_1 \le 16, N_2 \le 24]$ are 0.18641, 0.15516, and 0.15124, respectively. The exact value (to six decimal places) in this case is 0.150477.

Boardman and Kendell (1970) applied the method of Johnson (1960) [see Eq. (35.54)] to obtain approximations to complicated expressions for expected values of ratios and reciprocals of trinomial (case $k = 3$) random variables, truncated by omission of zero values. They have presented the approximate expression

$$E\left[N_1^{\alpha_1} N_2^{\alpha_2} N_3^{\alpha_3}\right] \doteq \frac{n^{\alpha}\,\Gamma(n-1)}{\Gamma(n-1+\alpha)} \prod_{i=1}^{3}\left\{\frac{\Gamma((n-1)p_i + \alpha_i)}{\Gamma((n-1)p_i)}\right\} \qquad (35.65)$$

when $\min_i\{(n-1)p_i + \alpha_i\} > 0$, where $\alpha = \sum_{i=1}^{3} \alpha_i$. In particular,

$$E[N_i/N_j] \doteq \{(n-1)p_i\}/\{(n-1)p_j - 1\}, \qquad (35.66)$$

$$E[(N_i/N_j)^2] \doteq \{(n-1)p_i\}^{[2]}/\{(n-1)p_j - 2\}^{[2]}, \qquad (35.67)$$

and

$$E[N_i/N_j^2] \doteq \{(n-1)(n-2)p_i\}/[n\{(n-1)p_j - 2\}^{[2]}]. \qquad (35.68)$$

The approximation

$$E[1/N_j] \doteq (n-2)\{(n-1)p_j - 1\}/n \qquad (35.69)$$

was obtained by Mendenhall and Lehmann (1960).

Boardman and Kendell (1970) tabulated exact and approximate values of $E[N_i^{\alpha_i} N_j^{\alpha_j}]$ for $(\alpha_1, \alpha_2) = (1, 1), (0, 1), (2, 2)$, and $(1, 2)$, $n = 10, 15, 25, 35, 50, 75$, $p_1 = 0.3(0.1)0.9$, $p_2 = 0.1(0.1)0.5$ with $p_1 + p_2 \leq 1$, and $p_3 = 1 - p_1 - p_2$. Approximations were found to be correct to two (three) significant figures for $np_1p_2 > 5\,(10)$ and $n \geq 20$.

6 ESTIMATION

Substantial attention has been paid to the estimation of multinomial probabilities in applied as well as theoretical literature in the past 30 years. Estimation (or inference) for the multinomial distribution seems to touch the most basic and profound aspects of modern statistical inference, especially the Bayesian approach, and goes to the heart of the foundation of statistics as the paper by Walley (1996) and the discussions following it indicate. We first consider situations in which n and k are known, and it is required to estimate the probabilities $p_1, p_2, \ldots, p_k$.

Given $N_1, N_2, \ldots, N_k$, the maximum likelihood estimators of $p_1, p_2, \ldots, p_k$ are the relative frequencies

$$\hat{p}_i = N_i/n, \qquad i = 1, 2, \ldots, k\,. \tag{35.70}$$

An elementary derivation, avoiding the use of calculus, has been provided by Davis and Jones (1992). They note that the geometric mean of the set of n numbers containing X_i numbers equal to p_i/X_i $(i = 1, 2, \ldots, k)$ is

$$\left[\prod_{i=1}^{k}(p_i/X_i)^{X_i}\right]^{1/n} \qquad \left(\text{with } \sum_{i=1}^{k} X_i = n\right), \tag{35.71}$$

while the arithmetic mean is

$$\frac{1}{n}\sum_{i=1}^{k} X_i(p_i/X_i) = \frac{1}{n}\sum_{i=1}^{k} p_i = \frac{1}{n}\,. \tag{35.72}$$

Since the geometric mean cannot exceed the arithmetic mean, it follows from (35.71) and (35.72) that

$$\prod_{i=1}^{k}(p_i/X_i)^{X_i} \leq (1/n)^n\,,$$

i.e.

$$\prod_{i=1}^{k} p_i^{X_i} \leq \prod_{i=1}^{k}(X_i/n)^{X_i}\,. \tag{35.73}$$

So the maximum value of $\prod_{i=1}^{k} p_i^{X_i}$ is attained by taking $p_i = X_i/n$, $i = 1, 2, \ldots, k$.

Boland and Proschan (1987) have established the Schur convexity of the maximum likelihood function for the multinomial distribution. Agresti (1990, Chapter 9) has presented a decomposition of the multinomial likelihood function into binomials.

From Eqs. (35.6) and (35.7), we readily have

$$E[\hat{p}_i] = p_i \qquad \text{and} \qquad \text{var}(\hat{p}_i) = p_i(1 - p_i)/n\,. \tag{35.74}$$

Quesenberry and Hurst (1964) obtained the following formula for boundaries of simultaneous confidence intervals for $p_1, p_2, \ldots, p_k$ with an approximate joint confidence coefficient $100(1 - \alpha)\%$. The limits for p_i are

$$\frac{1}{2(n + \chi^2_{k-1,1-\alpha})}\left[\chi^2_{k-1,1-\alpha} + 2N_i \pm \chi_{k-1,1-\alpha}\left\{\chi^2_{k-1,1-\alpha} + \frac{4}{n}N_i(n - N_i)\right\}^{1/2}\right]. \tag{35.75}$$

Goodman (1965) improved on these results (for $k > 2$), replacing $\chi^2_{k-1,1-\alpha}$ by $Z^2_{1-(\alpha/k)}$, where $\Phi(Z_\varepsilon) = \varepsilon$ and $\Phi(\cdot)$ is the standard normal cdf. He has also presented limits for a set of confidence intervals for $\{p_i - p_j\}$ with an approximate joint confidence coefficient $100(1 - \alpha)\%$. The limits for $p_i - p_j$ are

$$\hat{p}_i - \hat{p}_j \pm Z^2_{1-\alpha/\{k(k-1)\}}\left[\left\{\hat{p}_i + \hat{p}_j - \left(\hat{p}_i - \hat{p}_j\right)^2\right\}/n\right]^{1/2}. \tag{35.76}$$

Patel (1979) presented a uniformly most powerful unbiased conditional sign test for testing the equality of p_i and p_j.

Bromaghin (1993) discussed the determination of sample size needed to ensure that a set of confidence intervals for $p_1, p_2, \ldots, p_k$ of widths not exceeding $2d_1, 2d_2, \ldots, 2d_k$, respectively, each includes the true value of the appropriate estimated parameter with probability (simultaneous confidence coefficient) $100(1-\alpha)\%$.

The worst cases are those in which one of the p_i's exceeds 0.5. Based on consideration of such cases, Bromaghin (1993) recommended a minimum sample size of

$$1 + \left[\max_{1\le i\le k}\frac{Z^2_{1-(\alpha/2)}}{4\,d_i^2} - Z^2_{1-(\alpha/2)}\right] = 1 + \left[\left\{\frac{1}{4\,d^2_{\min}} - 1\right\}Z^2_{1-(\alpha/2)}\right], \tag{35.77}$$

where $d_{\min} = \min_{1\le i\le k} d_i$ and $[a]$ denotes the integer part of a. He also presented the following formula utilizing (suspected) values of the p_i's:

$$n \ge \max_{1\le i\le k}\left[p_i(1 - p_i) - 2\,d_i^2 + \left\{p_i^2(1 - p_i)^2 - d_i^2\{4p_i(1 - p_i) + 1\}\right\}^{1/2}\right] \times \frac{Z^2_{1-(\alpha/2)}}{2d_i^2}\,. \tag{35.78}$$

Thompson (1987) proposed the following formula, applicable when all $d_i = d$ and $p_i = \frac{1}{j}$ for j of the k categories ($j \le k$) and $p_i = 0$ for all other i:

$$n \ge 1 + \left[\max_{j=2,3,\ldots,k}\frac{1 - \frac{1}{j}}{j\,d^2}\,Z^2_{1-\alpha/(2j)}\right]. \tag{35.79}$$

The appropriate values of j are 2 for $\alpha < 0.0344$, 3 for $0.0344 < \alpha < 0.3466$, 4 for $0.3466 < \alpha < 0.6311$, 5 for $0.6311 < \alpha < 0.8934$, and 6 for $\alpha > 0.8934$.

Figure 35.1 [adapted from Bromaghin (1993)] compares, for the cases $\alpha = 0.05$ and $\alpha = 0.01$, $d_i = d = 0.10$, $k = 3(1)15$, the recommended minimum sample sizes under Bromaghin's and Thompson's procedures and under an earlier proposal by Tortora (1978) according to which

$$n \geq Z_{1-(\alpha/k)} \max_i \frac{p_i(1-p_i)}{d_i^2} \quad \text{if } p_i\text{'s are known,} \tag{35.80}$$

or for the 'worst' case ($p_i = 1/2$)

$$n \geq \frac{1}{4} Z_{1-(\alpha/k)} (\min_i d_i)^{-2} . \tag{35.81}$$

Bromaghin (1993) has remarked that

> Thompson's procedure is attractive in that recommended sample sizes are small relative to the other procedures, particularly as the number of multinomial categories increases. However, all intervals are of equal width and the confidence associated with individual intervals is variable. Therefore, if the simultaneous confidence level is the only property of concern, Thompson's procedure is the procedure of choice.

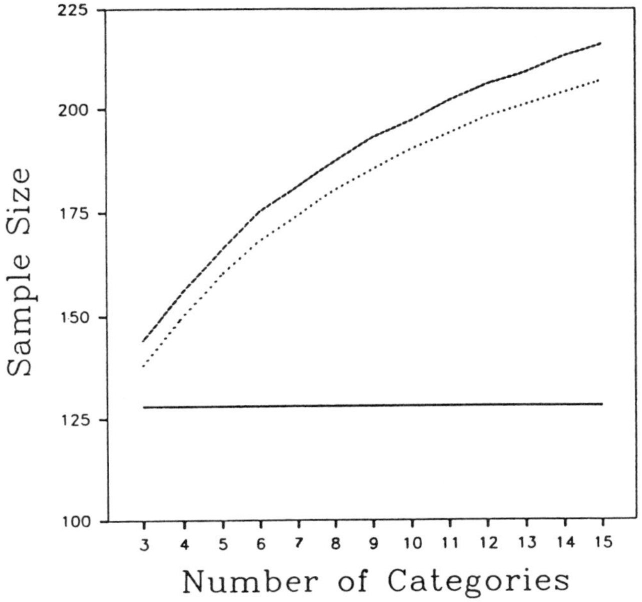

Figure 35.1. Recommended sample sizes for $\alpha = 0.05$, $d = 0.10$, and several values of k. - - - - -, Tortora; · · · · ·, Bromaghin; ———, Thompson.

Recently, Sison and Glaz (1993, 1995) presented a method of constructing simultaneous confidence intervals for $p_1, p_2, \ldots, p_k$ based on an application of a natural extension of Levin's formula (35.30) with $s = n$, namely

$$\Pr\left[\bigcap_{i=1}^{k}(b_i \le N_i \le a_i)\right]$$

$$\simeq \frac{n!e^n}{n^n}\left\{\prod_{i=1}^{k}\Pr[b_i \le X_i \le a_i]\right\} f\left(\frac{n - \sum_{i=1}^{k}\mu'_{1,i}}{\sqrt{\sum_{i=1}^{k}\mu_{2,i}}}\right)\sqrt{\sum_{i=1}^{k}\mu_{2,i}}\,, \tag{35.82}$$

where X_i's have Poisson (np_i) distributions $(i = 1, 2, \ldots, k)$, $f(x)$ is the Edgeworth expansion as defined in (35.34), and the moments $\mu'_{1,i}$, $\mu_{r,i}$ relate to mutually independent random variables Y_i $(i = 1, 2, \ldots, k)$ that have doubly truncated Poisson (np_i) distributions with ranges $b_i, b_i + 1, \ldots, a_i - 1, a_i$.

Sison and Glaz (1993, 1995) obtained sets of simultaneous confidence intervals of form

$$\hat{p}_i - \frac{c}{n} \le p_i \le \hat{p}_i + \frac{c+\gamma}{n}, \quad i = 1, 2, \ldots, k, \tag{35.83}$$

where $\hat{p}_i = N_i/n$ is the maximum likelihood estimator of p_i, and c and γ are chosen in such a manner that a specified $100(1-\alpha)\%$ simultaneous confidence coefficient is achieved.

These authors also described another method of constructing simultaneous confidence intervals for $p_1, p_2, \ldots, p_k$, with approximate simultaneous confidence coefficient $100(1-\alpha)\%$, based on evaluating

$$\Pr\left[\bigcap_{i=1}^{k}(b_i \le N_i \le a_i)\right] \simeq w_{1,m}\prod_{j=m+1}^{k}\frac{w_{j-m+1,j}}{w_{j-m+1,j-1}}\,, \tag{35.84}$$

where w_{ij} are truncated joint probabilities

$$w_{i,j} = \Pr\left[\bigcap_{h=i}^{j}(b_h \le N_h \le a_h)\right] \quad \text{for } i < j \tag{35.85}$$

[see also Glaz (1990)].

Sison and Glaz have compared their methods of construction with the others suggested earlier by Quesenberry and Hurst (1964), Goodman (1965), and Fitzpatrick and Scott (1987). Based on simulation results as well as on applications to data on personal crimes committed in New Orleans during a selected week in 1984 [Gelfand *et al.* (1992)], they reached the conclusion that their two procedures have the best properties among those mentioned above. The intervals suggested by Quesenberry and Hurst (1964) and Fitzpatrick and Scott (1987) are conservative involving rectangular regions with large volumes. Angers (1989) provided a table for conservative confidence intervals of the form $\hat{p} \pm c/\sqrt{n}$ for different choices of k and α. He used the fact that $\hat{p}_i$ $(i = 1, 2, \ldots, k)$ converge in distribution to a degenerate multivariate normal distribution.

The parameters $p_1, \ldots, p_k$ may possibly be functions of other parameters $\theta_1, \ldots, \theta_s$ (say), arising naturally in the formation of a model. For example [see Ryan (1992) and Catalano, Ryan, and Scharfstein (1994)], the cells of the multinomial distribution might represent successive stages of a disease, and the θ's are the probabilities of progressing from one stage to the next. Then,

$$p_1 = 1 - \theta_1,\ p_2 = \theta_1(1 - \theta_2), \ldots, p_{i+1} = \theta_1\theta_2 \cdots \theta_i(1 - \theta_{i+1}), \\ \ldots, p_k = \theta_1\theta_2 \cdots \theta_k\,, \tag{35.86}$$

and conversely

$$p_{i+1} = \left(1 - \sum_{j=1}^{i} p_j\right)(1 - \theta_{i+1})\,. \tag{35.87}$$

Such models may have wide applicability in survival analysis. In this case there is a one-to-one relationship between the p's and the θ's, and so estimation of the former follows in a straightforward manner from that of the latter. If $s < k - 1$, then application of maximum likelihood estimation must be based on replacement of each p by its expression in terms of the θ's. If $s > k - 1$, then in the absence of restrictions on the values of $\theta_1, \theta_2, \ldots, \theta_s$, it will not be possible to obtain separate estimates for each θ. Of course, estimators $\hat{p}_1, \hat{p}_2, \ldots, \hat{p}_k$ of the p's are estimators of the corresponding functions of the θ's.

If the probabilities p_i are differentiable functions $p_i(\theta_1, \theta_2, \ldots, \theta_s) \equiv p_i(\boldsymbol{\theta})$ of s $(< k - 1)$ parameters, the maximum likelihood estimators $\hat{\boldsymbol{\theta}} \equiv (\hat{\theta}_1, \hat{\theta}_2, \ldots, \hat{\theta}_s)$ satisfy the equations

$$\sum_{i=1}^{k} \frac{\hat{p}_i}{p_i(\boldsymbol{\theta})} \frac{\partial p_i(\boldsymbol{\theta})}{\partial \theta_j} = 0, \qquad j = 1, 2, \ldots, s\,. \tag{35.88}$$

For the case $s = 1$, Rao, Sinha, and Subramanyam (1981) have studied choice of a function $\xi(\cdot)$ (subject to some regularity conditions including continuity of derivatives up to and including fifth order) such that, if θ_1^* satisfies the equation

$$\sum_{i=1}^{k} \xi\left(\frac{\hat{p}_i}{p_i(\theta_1)}\right) \frac{\partial p_i(\theta_1)}{\partial \theta_1} = 0\,, \tag{35.89}$$

it then possesses second-order efficiency as an estimator of θ_1: a necessary and sufficient condition for this to be so is that $\xi'(1) \neq 0$ and $\xi''(1) = 0$. In addition, if $\xi'''(1) \neq 0$, then there exists a θ_1^{**} (a bias-adjusted modification of θ_1^*) with a smaller mean square error [to $o(n^{-1})$] than that of $\hat{\hat{\theta}}_1$ (a similar bias-adjusted modification of the maximum likelihood estimator $\hat{\theta}_1$). Incidentally, θ_1^{**} and $\hat{\hat{\theta}}_1$ have the same bias up to $O(n^{-2})$. The choice of

$$\xi(x) = x^{5/3} + x^{2/3} - 2x^{-1/3}$$

satisfies $\xi'(1) \neq 0$, $\xi''(1) = 0$, and $\xi'''(1) \neq 0$ and leads to the requirement that θ_1^* minimizes

$$\sum_{i=1}^{k} \left(\frac{\hat{p}_i}{p_i(\theta_1)} \right)^{2/3} \frac{\{\hat{p}_i - p_i(\theta_1)\}^2}{\hat{p}_i} \tag{35.90}$$

with respect to θ_1.

An example of this kind is provided in Barlow and Azen (1990) as a model for data from "a randomized clinical trial comparing two surgical methods for reattaching the retina." Initially the surgical method is applied R times. If success (S) is not achieved, then the surviving (F) group is split randomly into two subgroups. The first subgroup is treated a further ($T - R$) times by the same method; the other method is applied ($T - R$) times to the second subgroup. The results of the experiment on a total of n patients were modeled by a 10-cell multinomial $(n; p_{11}, p_{12}, \ldots, p_{15}, p_{21}, p_{22}, \ldots, p_{25})$ distribution according to the following scheme:

	First R Applications		Second $(T - R)$ Applications		
Cell (h)	Method	Result	Method	Result	p_{ih}
1	i	S	—	—	$1 - (1 - \pi_i)^R$
2	i	F	i	S	$(1 - \phi_i)(1 - \pi_i)^R\{1 - (1 - \pi_i)^{T-R}\}$
3	i	F	j	S	$\phi_i(1 - \pi_i)^R\{1 - (1 - \pi_j)^{T-R}\}$
4	i	F	i	F	$(1 - \phi_i)(1 - \pi_i)^T$
5	i	F	j	F	$\phi_i(1 - \pi_i)^R(1 - \pi_j)^{T-R}$

S—success, F—failure, $i = 1, 2,\ j = 3 - i$

π_i—probability that treatment i is successful on a single application

ϕ_i—probability that treatment i will be switched to j for second set of applications

In this case, we have $k = 5 \times 2 = 10$ and $s = 2 + 2 = 4$.

The maximum likelihood estimator of the entropy (Shannon measure of information)

$$H = -\sum_{i=1}^{k} p_i \log p_i \tag{35.91}$$

is

$$\hat{H} = -\sum_{i=1}^{k} \hat{p}_i \log \hat{p}_i\,. \tag{35.92}$$

Hutcheson and Shenton (1974) showed that

$$E[\hat{H}] = \log n - \sum_{j=1}^{n-1} \binom{n-1}{n-j} \log(n-j+1) \sum_{i=1}^{k} p_i^{n-j+1}(1-p_i)^{j-1} \tag{35.93}$$

and

$$\begin{aligned}\operatorname{var}(\hat{H}) &= \sum_{g=0}^{n-2} \binom{n-1}{g} \sum_{i=1}^{k} p_i^{n-g}(1-p_i)^g \\ &\quad \times \left[\sum_{j=g+1}^{n-1} \binom{n-1}{j} \sum_{i'=1}^{k} p_{i'}^{n-j}(1-p_{i'})^j \left\{ \log\left(\frac{n-g}{n-j}\right)\right\}^2 \right] \\ &\quad - \frac{n-1}{n} \sum_{g=0}^{n-3} \binom{n-2}{g} \left[\sum_{j=0}^{\gamma_g} \binom{n-g-2}{j} \sum_{i=1}^{k} \sum_{i'=1_{i\neq i'}}^{k} p_i^{n-g-j-1} p_{i'}^{j+1} \right. \\ &\quad \left. \times (1-p_i p_{i'})^g \left\{ \log\left(\frac{n-g-j-1}{j+1}\right)\right\}^2 \right], \end{aligned} \tag{35.94}$$

where $\gamma_g = \left[\frac{1}{2}(n-g-2)\right]$, the integer part of $\frac{1}{2}(n-g-2)$.

Cook, Kerridge, and Price (1974) proposed the "nearly unbiased" estimator

$$H^* = k - 1 + \frac{1}{n} \sum_{i=1}^{k} a(N_i), \tag{35.95}$$

where

$$a(N_i) = \begin{cases} 1 & \text{if } N_i = 0 \\ 0 & \text{if } N_i = n \\ \left(\sum_{j=N_i}^{n-1} 1/j\right) N_i & \text{otherwise.} \end{cases}$$

The bias of the estimator in (35.95) cannot exceed

$$k(1-p_{\max})^{n+1}/\{n(n+1)p_{\max}\}.$$

[Note that $\sum_{j=N_i}^{n-1} 1/j = \psi(n) - \psi(N_i)$, where $\psi(\cdot)$ is the digamma function defined in Chapter 1, Eq. (1.37).]

Sometimes, certain cells are combined together and frequencies in individual cells in such sets are not naturally recorded, and only the total frequency is shown. For example, if the first k' cells (out of k) are combined together, then only values of

$$\sum_{i=1}^{k'} N_i, N_{k'+1}, \ldots, N_k$$

are recorded. From these data, the maximum likelihood estimators

$$\hat{p}_j = N_j/n$$

can then be calculated only for $j = k' + 1, k' + 2, \ldots, k$. The only other information available is

$$\sum_{i=1}^{k'} \hat{p}_i = \left(\sum_{i=1}^{k'} N_i\right) \bigg/ n\,.$$

If, however, we have observations on *several independent* multinomial $(n_t; p_1, p_2, \ldots, p_k)$ $(t = 1, 2, \ldots, T)$ sets in which *different* groups of cells are combined together, it may be possible to pool the estimators from the T sets so as to obtain separate estimators for additional (possibly all) p_i's. As an example, if $k = 4$ and $T = 3$ with cells (1) and (2) combined together in set 1, (1), (3) and (4) combined together in set 2, and (1), (2) and (4) combined together in set 3, we have estimators of $p_1 + p_2$, p_3 and p_4 from set 1, $p_1 + p_3 + p_4$ and p_2 from set 2, and $p_1 + p_2 + p_4$ and p_3 from set 3. We thus have estimators of p_3 directly from sets 1 and 3, those of p_4 from set 1, those of p_2 from set 2, and then those of p_1 from sets 1, 2 and 3. This rather superficial "pooling" does not take into account the variances and covariances of the estimators, and therefore may not be the most efficient way of using the data for the purpose of estimation of parameters. It is presented merely as an illustration. Hocking and Oxspring (1971) provided an elegant, systematic, and detailed account of methods of utilizing these kinds of data for inferential purposes by means of matrix notation.

Alldredge and Armstrong (1974) discussed the maximum likelihood estimation as applied to the specific problem of estimating overlap sizes created by interlocking sampling frames in a multinomial setup. To be specific, let us consider a three-frame case with the frames denoted by A, B, and C and the domains denoted by a, b, c, ab, ac, bc, and abc. In this, a domain refers to the units in frame A only, ab domain refers to the units that are in both frames A and B, and so on. Then, with N_A, N_B, and N_C denoting the number of units in frames A, B, and C, respectively, and $N_a, N_b, \ldots, N_{abc}$ denoting the number of units in domains $a, b, \ldots, abc$, respectively, the likelihood function is

$$L = \alpha\, (N_a)^{n_{A\cdot a}} (N_b)^{n_{B\cdot b}} (N_c)^{n_{C\cdot c}} (N_{ab})^{n_{A\cdot ab} + n_{B\cdot ab}}$$
$$\times (N_{ac})^{n_{A\cdot ac} + n_{C\cdot ac}} (N_{bc})^{n_{B\cdot bc} + n_{C\cdot bc}} (N_{abc})^{n_{A\cdot abc} + n_{B\cdot abc} + n_{C\cdot abc}}\,.$$

In this equation, n_A, n_B, and n_C denote the number of units sampled from frames A, B, and C, respectively, and $n_{A\cdot a}, n_{A\cdot ab}, \ldots, n_{C\cdot abc}$ denote the number of units sampled from frames A, B, and C which are in domains $a, ab, \ldots, abc$, respectively. Then, the usual maximum likelihood estimation method will require the solution of a system of seven simultaneous nonlinear equations. However, Alldredge and Armstrong (1974) pointed out that this numerical problem can be simplified considerably by using the relationships $N_A = N_a + N_{ab} + N_{ac} + N_{abc}$, $N_B = N_b + N_{ab} + N_{bc} + N_{abc}$, and $N_C = N_c + N_{ac} + N_{bc} + N_{abc}$. Thus, the estimation problem can be reduced from a dimension of seven to a dimension of four. Alldredge and Armstrong (1974) also discussed the use of geometric programming [for details, see Duffin, Peterson, and Zener (1967)] for this numerical problem.

Oluyede (1994) considered estimation of the parameters $\boldsymbol{p}_j = (p_{1j}, p_{2j}, \ldots, p_{kj})'$, $j = 1, 2$, of two multinomial distributions with first parameters $n_1,\ n_2$ known, which are subject to the quasi-stochastic ordering

$$a_{i1} \geq a_{i2} \qquad (i = 1, 2, \ldots, k-1) \tag{35.96}$$

with strict inequality for at least one i, where

$$a_{ij} = \sum_{h=1}^{i} p_{hj} \Big/ \sum_{h=1}^{i+1} p_{hj} \qquad (i = 1, 2, \ldots, k-1; j = 1, 2)\,. \tag{35.97}$$

This ordering implies (but is not necessarily implied by) the standard stochastic ordering

$$\sum_{h=1}^{i} p_{h1} \geq \sum_{h=1}^{i} p_{h2} \qquad (i = 1, 2, \ldots, k-1) \tag{35.98}$$

with strict inequality for at least one i (see Chapter 33, p. 672).

Inverting (35.97), we obtain

$$\begin{aligned} p_{1j} &= \prod_{h=1}^{k-1} a_{hj} \\ p_{ij} &= (1 - a_{i-1,j}) \prod_{h=i}^{k-1} a_{hj} \quad (i = 2, 3, \ldots, k-1) \\ p_{kj} &= 1 - a_{k-1,j}\,. \end{aligned} \tag{35.99}$$

If $\boldsymbol{X}_j = (X_{1j}, X_{2j}, \ldots, X_{kj})'$ has a multinomial $(n_j; \boldsymbol{p}_j)$ distribution $(j = 1, 2)$, the likelihood function for $\boldsymbol{X}_j$ is

$$L_j = \frac{n_j!}{\prod_{i=1}^{k} X_{ij}!} \prod_{h=1}^{k-1} \left\{ a_{hj}^{\sum_{i=1}^{h} X_{ij}} (1 - a_{hj})^{X_{h+1,j}} \right\}, \ j = 1, 2. \tag{35.100}$$

(a) If the values of the p_{i2}'s, and so of the a_{i2}'s, are known, the MLEs of the a_{i1}'s are obtained by maximizing L_1 subject to the restriction (35.96). This leads to

$$\hat{a}_{i1} = \max\left\{ \frac{\sum_{h=1}^{i} X_{h1}}{\sum_{h=1}^{i+1} X_{h1}},\ a_{i2} \right\}, \quad i = 1, 2, \ldots, k-1\,. \tag{35.101}$$

The estimators $\hat{p}_{i1}$'s are, of course, the same functions of the $\hat{a}_{i1}$'s as the p_{i1}'s are of the a_{i1}'s as in Eq. (35.99).

(b) If the elements of $\boldsymbol{p}_2$ as well as those of $\boldsymbol{p}_1$ are unknown, the likelihood function to be maximized is

$$L_1 L_2 \propto \prod_{j=1}^{2} \prod_{i=1}^{k-1} \left\{ a_{ij}^{\sum_{h=1}^{i} X_{hj}} (1 - a_{ij})^{X_{i+1,j}} \right\} . \tag{35.102}$$

The MLEs of the a_{ij}'s are

$$\hat{a}_{ij} = \bar{a}_{ij} = \frac{\sum_{h=1}^{i} X_{hj}}{\sum_{h=1}^{i+1} X_{hj}} \qquad (h = 1, 2) \text{ if } \bar{a}_{i1} \geq \bar{a}_{i2} , \tag{35.103}$$

however,

$$\hat{a}_{i1} = \hat{a}_{i2} = \frac{\sum_{j=1}^{2} \sum_{h=1}^{i} X_{hj}}{\sum_{j=1}^{2} \sum_{h=1}^{i+1} X_{hj}} \qquad \text{if } \bar{a}_{i1} < \bar{a}_{i2} . \tag{35.104}$$

Chamberlain (1987) considered the derivation of bounds on asymptotic efficiency for a certain class of nonparametric models, but with the restriction that (at some point in the parameter space) the conditional expectation of a given function of the data and a parameter is zero, thus making the problem parametric. Chamberlain (1987) then derived bounds on asymptotic efficiency from the information matrix from a multinomial distribution. Brown and Muentz (1976) discussed reduced mean square estimation in the framework of contingency tables.

A popular *Bayesian approach* to the estimation of $\boldsymbol{p} \equiv (p_1, p_2, \ldots, p_k)'$ from observed values $\boldsymbol{n} \equiv (n_1, n_2, \ldots, n_k)'$ of $\boldsymbol{N} \equiv (N_1, N_2, \ldots, N_k)'$ is to assume a joint prior Dirichlet $(\alpha_1, \alpha_2, \ldots, \alpha_k)$ distribution [see, for example, Johnson and Kotz (1972, Chapter 40, Eq. (31))] for $\boldsymbol{p}$ with density function

$$\Gamma(\alpha_0) \prod_{i=1}^{k} \{p_i^{\alpha_i - 1} / \Gamma(\alpha_i)\} , \tag{35.105}$$

where $\alpha_0 = \sum_{i=1}^{k} \alpha_i$. The posterior distribution of $\boldsymbol{p}$, given $\boldsymbol{N} = \boldsymbol{n}$, is also Dirichlet $(\alpha_1 + n_1, \alpha_2 + n_2, \ldots, \alpha_k + n_k)$.

Fienberg and Holland (1973), Good (1967), and Leonard (1973, 1976) all investigated consequences of using a "two-stage" prior with $\boldsymbol{p}$ having a "second-stage" prior Dirichlet $(\alpha\xi_1, \alpha\xi_2, \ldots, \alpha\xi_k)$ distribution, where $\xi_1, \xi_2, \ldots, \xi_k$ are known prior means for $p_1, p_2, \ldots, p_k$ (with $\sum_{i=1}^{k} \xi_i = 1$, of course) and α has a "first-stage" prior density $\pi(\alpha)$. The posterior mean of p_i is

$$E[p_i \mid \boldsymbol{n}] = (1 - \lambda)\hat{p}_i + \lambda \xi_i \tag{35.106}$$

with

$$\begin{aligned} \lambda &= E\left[\frac{\alpha}{n + \alpha} \middle| \boldsymbol{n} \right] \\ &= \int_0^{\infty} \frac{\alpha}{n + \alpha} \, \pi(\alpha \mid \boldsymbol{n}) \, d\alpha , \end{aligned} \tag{35.107}$$

where $n = \sum_{i=1}^{k} n_i$ and

$$\pi(\alpha \mid \boldsymbol{n}) \propto \pi(\alpha) \prod_{i=1}^{k} \{\Gamma(\alpha \xi_i + n_i)\} / \Gamma(\alpha + n) .$$

Viana (1994) discussed recently the Bayesian small-sample estimation of multinomial probabilities by considering the apparent distribution and a matrix of misclassification probabilities.

Good (1965) suggested an alternative "frequentist" formula for λ as

$$\lambda^* = \min \left\{ \frac{k-1}{n \sum_{i=1}^{k} (\hat{p}_i - \xi_i)^2 / \xi_i} , 1 \right\} . \qquad (35.108)$$

This rectifies possible disadvantages arising from use of a value of λ, as per (35.107), that does not take account of discrepancies between values of p_i and ξ_i $(i = 1, 2, \ldots, k)$. Leonard (1976) modified λ^* in (35.108) to

$$\lambda^{**} = \begin{cases} \max \left\{ \frac{n\lambda^* - 1}{n-1} , 0 \right\} & \text{if} \quad \lambda^* \le 1 \\ \lambda^* & \text{if} \quad \lambda^* \ge 1 . \end{cases}$$

Some other "frequentist" alternatives were proposed by Sutherland, Holland, and Fienberg (1975) and Baranchik (1964). Leonard (1976) also provides explicit "Bayesian alternatives" to Good's formula. Good and Crook (1974) discussed a compromise between the Bayes and non-Bayes approaches.

Bhattacharya and Nandram (1996) have discussed the Bayesian inference for multinomial distributions under stochastic ordering. Specifically, they have considered g multinomial populations with probabilities for the ith population denoted by $\boldsymbol{p}_i = (p_{i1}, \ldots, p_{ik})'$ and counts denoted by $\boldsymbol{n}_i = (n_{i1}, \ldots, n_{ik})'$, and assumed the stochastic order restriction $\boldsymbol{N}_1 \overset{\text{st}}{\ge} \boldsymbol{N}_2 \overset{\text{st}}{\ge} \cdots \overset{\text{st}}{\ge} \boldsymbol{N}_g$; here, $\boldsymbol{N}_1 \overset{\text{st}}{\ge} \boldsymbol{N}_2$ if and only if

$$\sum_{i=1}^{s} p_{1i} \le \sum_{i=1}^{s} p_{2i} \qquad \text{for } i = 1, 2, \ldots, k-1 .$$

Then, by assigning (without any restriction on the p_{ij}) independent Dirichlet priors for $\boldsymbol{p}_i$ as

$$\Gamma \left(\sum_{j=1}^{k} \beta_{ij} \right) \prod_{j=1}^{k} \left\{ p_{ij}^{\beta_{ij}-1} \Big/ \Gamma(\beta_{ij}) \right\} ,$$

where β_{ij} are fixed quantities, Bhattacharya and Nandram (1996) have obtained the posterior distributions of $\boldsymbol{p} = (\boldsymbol{p}_1', \ldots, \boldsymbol{p}_g')'$ restricted to a subset $S^{(gk)}$ of the gk-Euclidean space ($R^{(gk)}$) and to $\bar{S}^{(gk)} = R^{(gk)} - S^{(gk)}$ as

$$\Pi_S(\boldsymbol{p} \mid \boldsymbol{n}') = \begin{cases} c(\boldsymbol{n}') \prod_{i=1}^{g} \Gamma \left(\sum_{j=1}^{k} n_{ij}' \right) \prod_{j=1}^{k} \left\{ p_{ij}^{n_{ij}'-1} \Big/ \Gamma(n_{ij}') \right\} & \text{if } \boldsymbol{p} \in S^{(gk)} \\ 0 & \text{otherwise} \end{cases}$$

and

$$\Pi_{\bar{S}}(\boldsymbol{p} \mid \boldsymbol{n}') = \begin{cases} \bar{c}(\boldsymbol{n}') \prod_{i=1}^{g} \Gamma\left(\sum_{j=1}^{k} n'_{ij}\right) \prod_{j=1}^{k} \left\{p_{ij}^{n'_{ij}-1} \Big/ \Gamma(n'_{ij})\right\} & \text{if } \boldsymbol{p} \in \bar{S}^{(gk)} \\ 0 & \text{otherwise,} \end{cases}$$

respectively. In the above equations,

$$n'_{ij} = n_{ij} + \beta_{ij}, \ \boldsymbol{n}'_i = (n'_{i1}, n'_{i2}, \ldots, n'_{ik})',$$

$$\boldsymbol{n}' = \{n'_{ij} : i = 1, \ldots, g;\ j = 1, \ldots, k\},$$

$$\frac{1}{c(\boldsymbol{n}')} = \int_{S^{(gk)}} \prod_{i=1}^{g} \Gamma\left(\sum_{j=1}^{k} n'_{ij}\right) \prod_{j=1}^{k} \left\{p_{ij}^{n'_{ij}-1} \Big/ \Gamma(n'_{ij})\right\} d\boldsymbol{p}$$

and

$$\frac{1}{\bar{c}(\boldsymbol{n}')} = \int_{\bar{S}^{(gk)}} \prod_{i=1}^{g} \Gamma\left(\sum_{j=1}^{k} n'_{ij}\right) \prod_{j=1}^{k} \left\{p_{ij}^{n'_{ij}-1} \Big/ \Gamma(n'_{ij})\right\} d\boldsymbol{p}.$$

Note that these expressions are valid for any restricted region $S^{(gk)} \subset R^{(gk)}$ and not necessarily just for the stochastic order restriction mentioned above.

Sequential estimation of $p_1, p_2, \ldots, p_k$ has been considered by several authors. By applying a general result of Wolfowitz (1947), Bhat and Kulkarni (1966) showed that, denoting by N the sample size ("number of trials") of a sequence of independent multinomial $(1; p_1, p_2, \ldots, p_k)$ sets and denoting the corresponding values of $N_1, N_2, \ldots, N_k$ by $N_1(N), N_2(N), \ldots, N_k(N)$, the variance of any unbiased estimator

$$T(N_1(N), N_2(N), \ldots, N_k(N))$$

of a differentiable function $g(p_1, p_2, \ldots, p_k)$ cannot be less than

$$\frac{1}{E[N]}\left\{\sum_{i=1}^{k} p_i \left(\frac{\partial g}{\partial p_i}\right)^2 - \left(\sum_{i=1}^{k} p_i \frac{\partial g}{\partial p_i}\right)^2\right\}, \tag{35.109}$$

with equality holding if and only if T is a linear function of the variables

$$\left\{\frac{N_i(N)}{p_i} - \frac{N_{i'}(N)}{p_{i'}}\right\}, \qquad i \neq i' = 1, 2, \ldots, k.$$

There is an infinite number of possible systems of sequential sampling. Among these are *inverse multinomial sampling*, in which sampling is terminated as soon as one or other of a specified subset of "classes" has occurred k times (see also Chapter 36).

In the special case of a *trinomial distribution* (i.e., multinomial with $k = 3$), a sequential method of estimation of the probabilities p_1, p_2, and p_3 was developed by

Muhamedanova and Suleimanova (1961). Their method is a natural generalization of that of Girshick, Mosteller, and Savage (1946) for the binomial distribution. It is based on the enumeration of the paths followed by the point (N_1, N_2, N_3), as N increases, which conclude at an observed termination point. (The termination points are points on the boundary of the continuation region, and this boundary defines the sampling procedure.) If the termination point is (N_1, N_2, N_3) and $K(N_1, N_2, N_3)$ is the number of possible paths from (0, 0, 0) ending at (N_1, N_2, N_3), while $K_i(N_1, N_2, N_3)$ is the number of such paths starting from (1, 0, 0) ($i = 1$), (0, 1, 0) ($i = 2$), or (0, 0, 1) ($i = 3$), then

$$K_i(N_1, N_2, N_3)/K(N_1, N_2, N_3) \tag{35.110}$$

is an unbiased estimator of p_i. This method can evidently be extended to values of k more than 3, but evaluation of K_i and K rapidly increases in complexity.

Lewontin and Prout (1956) described a rather unusual situation in which it is known that $p_1 = p_2 = \cdots = p_k = 1/k$ but k itself is unknown and must be estimated. The maximum likelihood estimator, $\hat{k}$, of k satisfies the equation

$$n\hat{k}^{-1} = \sum_{i=\hat{k}-K+1}^{\hat{k}} i^{-1}, \tag{35.111}$$

where K is the number of different 'classes' represented in a random sample of size n. Equation (35.111) is approximately equivalent to

$$n\hat{k}^{-1} \doteq \log\{\hat{k}\,/\,(\hat{k} - K + 1)\}. \tag{35.112}$$

For large values of n,

$$\text{var}(\hat{k}) \doteq k\left\{e^{n/k} - \left(1 + \frac{n}{k}\right)\right\}^{-1}. \tag{35.113}$$

We conclude this section with a brief discussion of minimax estimation of multinomial parameters.

For a binomial distribution with parameters (n, p) the *minimax estimator* of p, based on a single observed value N, with risk function of form

$$R(\hat{p}; p) = a(\hat{p} - p)^2, \tag{35.114}$$

where $\hat{p}$ is an estimator of p, is [Hodges and Lehmann (1950)]

$$p^* = \left(N + \frac{1}{2}\sqrt{n}\right)\Big/(n + \sqrt{n}), \tag{35.115}$$

that is,

$$\min_{\hat{p}} \max_{0\le p\le 1} E[(\hat{p} - p)^2] = \max_{0\le p\le 1} E[(p^* - p)^2]. \tag{35.116}$$

For a multinomial distribution with parameters $(n; p_1, p_2, \ldots, p_k) \equiv (n; \boldsymbol{p})$, a natural generalization of (35.114) is the risk function

$$R(\hat{\boldsymbol{p}}; \boldsymbol{p}) = \sum_{i=1}^{k} a_i(\hat{p}_i - p_i)^2 \tag{35.117}$$

with $a_i \geq 0$. For convenience we suppose that the a's are ordered so that $a_1 \geq a_2 \geq \cdots \geq a_k \geq 0$, with $a_2 > 0$. The special case $a_1 = a_2 = \cdots = a_k = 1/k$ yields the minimax estimators, based on an observed set $(N_1, N_2, \ldots, N_k)'$,

$$p_i^* = \left(N_i + \frac{1}{k}\sqrt{n}\right) \Big/ (n + \sqrt{n}), \qquad i = 1, 2, \ldots, k \tag{35.118}$$

[Steinhaus (1957)]. For the general case, Trybula (1962) obtained the minimax estimators

$$p_i^* = (N_i + c_i\sqrt{n})/(n + \sqrt{n}), \qquad i = 1, 2, \ldots, k, \tag{35.119}$$

where

$$c_i = \begin{cases} \frac{1}{2}\left\{1 - \frac{m-2}{a_i}\sum_{j=1}^{m}\frac{1}{a_j}\right\} & \text{if } i \leq m \\ 0 & \text{if } i > m\,, \end{cases} \tag{35.120}$$

and m is the largest integer such that

$$\sum_{j=1}^{m}\frac{1}{a_j} > \frac{m-2}{a_m} \qquad \text{and} \qquad a_m > 0\,. \tag{35.121}$$

Note that the inequality (35.121) is certainly satisfied when $m = 2$, and so m always exists and cannot be less than 2. For the case $a_1 = a_2 = \cdots = a_k > 0$ we find $m = k$ and $a_i = 1/k$, as in Steinhaus (1957).

Later, Trybula (1985) considered risk function of form

$$\begin{aligned} R(\hat{\boldsymbol{p}}; \boldsymbol{p}) &= (\hat{\boldsymbol{p}} - \boldsymbol{p})'\boldsymbol{A}(\hat{\boldsymbol{p}} - \boldsymbol{p}) \\ &= \sum_{i=1}^{k}\sum_{j=1}^{k} a_{ij}(\hat{p}_i - p_i)(\hat{p}_j - p_j)\,, \end{aligned} \tag{35.122}$$

where $\boldsymbol{A}$ is a non-negative definite $k \times k$ matrix. He showed that there is always a set of numbers $c_1, c_2, \ldots, c_k$ satisfying $c_i \geq 0$, $\sum_{i=1}^{k} c_i = 1$, such that $\boldsymbol{p}^* = (p_1^*, \ldots, p_k^*)'$ with

$$p_i^* = (N_i + c_i\sqrt{n})/(n + \sqrt{n}), \qquad i = 1, 2, \ldots, k\,, \tag{35.123}$$

is a minimax set of estimators. Wilczyński (1985) showed that appropriate values of the c_i's are those that maximize

$$\text{diag}(\boldsymbol{A})\boldsymbol{c}' - \boldsymbol{c}'\boldsymbol{A}\boldsymbol{c}\,, \tag{35.124}$$

subject, of course, to $c_i \geq 0$ and $\sum_{i=1}^{k} c_i = 1$.

He (1990) has shown that the minimax estimator of $\boldsymbol{p} = (p_1, \ldots, p_k)'$ (when n is fixed) given by

$$\begin{array}{ll} \left(\dfrac{1}{k}, \ldots, \dfrac{1}{k}\right)' & \text{if} \quad n = 0 \\ \left(\dfrac{N_1 + \sqrt{n}/k}{n + \sqrt{n}}, \ldots, \dfrac{N_k + \sqrt{n}/k}{n + \sqrt{n}}\right)' & \text{if} \quad n > 0 \end{array}$$

is not minimax for the squared error loss function

$$L_1(\boldsymbol{a}, \boldsymbol{p}) = \sum_{i=1}^{k} (a_i - p_i)^2$$

when there exists an ancillary statistic—that is, a statistic N whose distribution F does not depend on $\boldsymbol{p}$. In this case, one can think of $\boldsymbol{N}$ as being obtained by the following two-stage experiment; observe first a random quantity N with distribution F; given $N = n$, observe a quantity $\boldsymbol{N}$ with multinomial distribution $P(\boldsymbol{p} \mid n)$. He (1990), however, has shown that this minimax estimator of $\boldsymbol{p}$ for fixed sample size n is still minimax for relative squared error loss function

$$L_2(\boldsymbol{a}, \boldsymbol{p}) = \sum_{i=1}^{k} (a_i - p_i)^2 / p_i \,.$$

Chan, Eichenauer-Hermann, and Lehn (1989) and Brockmann and Eichenauer-Hermann (1992) studied Γ-minimax estimation of the parameters of multinomial distribution. A Γ-minimax estimator minimizes the maximum of the Bayes risk with respect to the elements of a subset Γ of prior distributions of a specific form. In particular, Brockmann and Eichenauer-Hermann (1992) determined the Γ-minimax estimators of the parameters of a multinomial distribution under arbitrary squared-error loss assuming that the available vague prior information can be described by a class of priors whose vector of the first moments belongs to a suitable convex and compact set.

Botha (1992) has discussed the adaptive estimation of multinomial probabilities with applications in the analysis of contingency tables.

7 CHARACTERIZATIONS

Bol'shev (1965) provided one of the earliest characterizations of the multinomial distribution by showing that the independent, non-negative, integer valued random variables $N_1, N_2, \ldots, N_k$ have nondegenerate Poisson distributions if and only if the conditional distribution of these variables for a fixed sum $N_1 + N_2 + \cdots + N_k = n$ is a nondegenerate multinomial distribution. He also presented a number of examples where this characterization can be proved useful in simplifying testing of hypotheses problems.

Given two mutually independent random variables N' and N'' which take on only non-negative integer values with $\Pr[N' = 0]$ and $\Pr[N'' = 0]$ both nonzero, Patil and Seshadri (1964) showed that if the distribution of N', given $N' + N''$, is hypergeometric with parameters n', n'' and $N' + N''$, then the distributions of N' and N'' are both binomial with parameters (n', p) and (n'', p), respectively (they share the same p). This result was extended by Janardan (1974) to the case of multivariate hypergeometric distributions (see Chapter 39) and multinomial distributions. In turn, Panaretos (1983) extended Janardan's characterization result by removing the assumption of independence.

Shanbhag and Basawa (1974) established that if $\boldsymbol{N}$ and $\boldsymbol{N}^*$ are mutually independent $k \times 1$ random vectors and $\boldsymbol{N} + \boldsymbol{N}^*$ has a multinomial distribution, then $\boldsymbol{N}$ and $\boldsymbol{N}^*$ both have multinomial distributions with the same $\boldsymbol{p}$ as $\boldsymbol{N} + \boldsymbol{N}^*$.

Griffiths (1974), however, showed that the supplementary condition that the sum of any subset of components of $\boldsymbol{N}$ has a binomial distribution is not sufficient to characterize the distribution of $\boldsymbol{N}$. An additional condition is required. Griffiths showed that the additional condition, "the canonical variables of any two disjoint subsets of the variables are the orthogonal polynomials in the sums of the variables of the two subsets," does suffice to establish the characterization. He showed the need for an additional condition for the characterization by means of the counterexample of a trivariate distribution with probability generating function

$$\left\{\frac{1}{3}(t_1 + t_2 + t_3)\right\}^{12} + c\, g(t_2, t_3) g(t_1, t_3) g(t_1, t_2)\,, \tag{35.125}$$

where $g(a, b) = (a-1)(b-1)(a-b)(a+b-2)$ and c, is sufficiently small for (35.125) to be a proper probability generating function (i.e., to ensure that all coefficients are non-negative—they will add to 1 in any case, as can be seen by setting $t_1 = t_2 = t_3$). Clearly, (35.125) does not correspond to a multinomial distribution, but does satisfy the condition that the sum of any subset has a binomial distribution (as can be seen by setting any one or two of the t_i's equal to one and the other(s) equal to t, say).

A characterization based on a multivariate splitting model has been derived by Rao and Srivastava (1979).

A conjecture of Neyman (1963), reformulated by Ghosh, Sinha, and Sinha (1977), regarding a characterization of multinomial distributions is that within the class of power series distributions (see Chapter 38), the multinomial distributions are characterized by the following properties:

(i) The regression of N_i on the remaining variables is a linear function of the sum of the remaining variables.

(ii) The distribution of $N_1 + N_2 + \cdots + N_k$ is of the power series type.

Ghosh, Sinha, and Sinha (1977) showed the conjecture to be affirmative under the conditions $a(0, 0, \ldots, 0) > 0$ and $a(1, 0, \ldots, 0) + \cdots + a(0, 0, \ldots, 1) > 0$ [see Eq.

(38.1) for a definition of these terms]. Sinha and Gerig (1985) provided a counterexample in order to prove that, without these conditions, the above claims turn out to be false. Furthermore, under the following conditions, Sinha and Gerig (1985) proved that there exist integers $r \geq 1, \ell_1, \ell_2, \ldots, \ell_k \geq 0$ and a random variable $\boldsymbol{M} = (M_1, \ldots, M_k)'$ having a joint multinomial distribution such that $N_i = rM_i + \ell_i$ $(1 \leq i \leq k)$ with probability one:

(a) The conditional distribution of N_i given $\{N_j = n_j \ (1 \leq j \leq k,\ j \neq i)\}$ is nondegenerate for at least one set of values of the n_j's and the regression is linear and, moreover, depends on the n_j's only through $\sum_{\substack{j=1 \\ j \neq i}}^{k} n_j$. Also, $\sum_{\substack{j=1 \\ j \neq i}}^{k} n_j$ assumes at least three distinct values $(1 \leq i \leq k)$.

(b) The distribution of $N_\bullet = N_1 + N_2 + \cdots + N_k$ is of the power series type.

Sinha and Gerig then stated that this result is the correct form of the Neyman conjecture regarding the multinomial distributions.

Dinh, Nguyen, and Wang (1996) presented some characterizations of the joint multinomial distribution of two discrete random vectors by assuming multinomial distributions for the conditionals. For example, by taking $\boldsymbol{X} = (X_1, \ldots, X_k)'$ and $\boldsymbol{Y} = (Y_1, \ldots, Y_\ell)'$ as two discrete random vectors with components taking values on the set of non-negative integers and assuming $\boldsymbol{X} \mid \boldsymbol{Y} = \boldsymbol{y} = (y_1, \ldots, y_\ell)'$ and $\boldsymbol{Y} \mid \boldsymbol{X} = \boldsymbol{x} = (x_1, \ldots, x_k)'$ to be distributed as multinomial $\left(n - \sum_{i=1}^{\ell} y_i;\ p_1, \ldots, p_\ell\right)$ and multinomial $\left(n - \sum_{i=1}^{k} x_i;\ p_1^*, \ldots, p_k^*\right)$ (where $\sum_{i=1}^{\ell} y_i \leq n$ and $\sum_{i=1}^{k} x_i \leq n$), these authors have shown that $\boldsymbol{X}$ and $\boldsymbol{Y}$ have a joint multinomial distribution.

8 RANDOM NUMBER GENERATION

Brown and Bromberg (1984) proposed a two-stage simulation procedure based on Bol'shev's (1965) characterization described in the beginning of the last section. Their simulational algorithm is as follows:

Stage 1: k independent Poisson variables $N_1, N_2, \ldots, N_k$ are generated with mean $\theta_i = n' p_i$, where n' is less than n and depends on n and k.

Stage 2: If $\sum_{i=1}^{k} N_i > n$, the sample is rejected;
if $\sum_{i=1}^{k} N_i = n$, the sample is used;
if $\sum_{i=1}^{k} N_i < n$, the sample is expanded by the addition of $n - \sum_{i=1}^{k} N_i$ individuals, sampling from a multinomial $(1; p_1, p_2, \ldots, p_k)$ population.

Of course, one could directly obtain the whole sample by taking just n samples from multinomial $(1; p_1, p_2, \ldots, p_k)$, but Brown and Bromberg's method results in a saving in the expected number of samples needed. Quite often, both n and k are large, in which case Brown and Bromberg suggested using their procedure to obtain samples of sizes ${}_1n, {}_2n, \ldots, {}_cn$ with $\sum_{i=1}^{c} {}_in = n$ and then combining these samples. If n/c is an integer, one would usually take ${}_in = n/c$ for all i.

Kemp and Kemp (1987) improved Brown and Bromberg's method by using a chain of binomial random variables. They choose

N_1 as a binomial (n, p_1) variable,

then N_2 as a binomial $\left(n - N_1, \frac{p_2}{1-p_1}\right)$ variable,

then N_3 as a binomial $\left(n - N_1 - N_2, \frac{p_3}{1-p_1-p_2}\right)$ variable,

and so on.

Generally, N_r is a binomial $\left(n - \sum_{i=1}^{r-1} N_i, \frac{p_r}{1-\sum_{i=1}^{r-1} p_i}\right)$ variable. If n and k are large, this requires a considerable amount of binomial sampling. For this reason, Kemp (1986) proposed a very efficient algorithm for generating binomial variables, the so-called *fast variable-parameter* algorithm; see Chapter 1, Section C3 and Chapter 3, Section 7.

Davis (1993) commented that "although the Brown–Bromberg algorithm is clearly the preferred generator of multinomial random variates in terms of speed, this advantage is gained at the expense of requiring a complex set-up phase and a large number of storage locations." He compared it with several other methods including Kemp and Kemp's modification, which Davis termed the *conditional distribution method*, and also with the following two methods:

(i) The *directed method* [Dagpunar (1988)], also known as the *ball-in-urn method* [Devroye (1986)] and the *naive method* [Ho, Gentle, and Kennedy (1979)], which first generates the cumulative probabilities

$$p_{\le i} = \sum_{j=1}^{i} p_j, \qquad i = 1, 2, \ldots, k\,.$$

The initial values of $\boldsymbol{N} = (N_1, N_2, \ldots, N_k)'$ are set to zero. Then, mutually independent random variables $U_1, U_2, \ldots, U_n$, each having a standard uniform (0, 1) distribution (see Chapter 26), are generated. Next, for $i = 1, 2, \ldots, n$, the ith component of $\boldsymbol{N}$ is incremented by 1 if

$$p_{\le i-1} < U_i \le p_{\le i} \qquad (\text{with } p_{\le 0} = 0)\,.$$

Generation time may be reduced (if the p_i's are not all equal) by arranging the p_i's in descending order of magnitude.

(ii) The *alias method*, which is a general method for generating random variates from any discrete distribution; see Chapter 1, Section C2 for more details. In the present situation, it is simply applied by choosing one of the k categories (with probabilities $p_1, p_2, \ldots, p_k$) on n successive occasions and summing the results.

Davis (1993), on the basis of several applications, concluded that Kemp and Kemp's conditional distribution method "is a good all-purpose algorithm for multinomial random variate generation, since it provides a nice balance between the competing criteria of execution speed and simplicity." This is especially true for large n, but Dagpunar (1988) is of the opinion that, for small values of n, the method is "not competitive"; see also Devroye (1986). For a discussion on sampling from general multivariate discrete distributions, one may refer to Loukas and Kemp (1983).

Lyons and Hutcheson (1996) have presented an algorithm for generating ordered multinomial frequencies.

9 APPLICATIONS OF MULTINOMIAL DISTRIBUTIONS

These distributions are employed in so many diverse fields of statistical analysis that an exhaustive catalogue would simply be very lengthy and unwieldy. They are used in the same situations as those in which a binomial distribution might be used, when there are multiple categories of events instead of a simple dichotomy.

An important classical field of application is in the kinetic theory of physics. "Particles" (e.g., molecules) are considered to occupy cells in phase space. That is, each particle is assigned to a "cell" in a six-dimensional space (three dimensions for position and three for velocity). The "cells" are formed by a fine subdivision of space. Each allocation of n particles among the k cells available constitutes a "microstate." If the particles are indistinguishable, each configuration corresponds to a "macrostate." The *thermodynamic probability* of a macrostate is proportional to the number of ways it can be realized by different "microstates." Thus, for the macrostate consisting of n_i particles in state i ($i = 1, 2, \ldots, k$), the probability is

$$\frac{n!}{\prod_{i=1}^k n_i!}\left\{\sum_{n_1}\cdots\sum_{n_k}\frac{n!}{\prod_{i=1}^k n_i!}\right\}^{-1} = \frac{n!}{\prod_{i=1}^k n_i!}\,k^{-n}. \tag{35.126}$$

This corresponds to "Maxwell–Boltzmann" statistical thermodynamics.

In general, multinomial distributions may be applicable whenever data obtained by random sampling are grouped in a finite number of groups. Provided that successive observations are independent (e.g., effects of finite sample size are negligible), conditions for a multinomial distribution are satisfied. Therefore, one may apply a multinomial distribution to many problems requiring estimation of a population distribution, since we are, in effect, estimating the "cell" probabilities of such a distribution. (In fact approximations, such as the use of continuous distributions, are often introduced.)

Another situation where multinomial distributions are commonly used is in the analysis of contingency tables [e.g., two-, three-, or M-way tables representing the joint incidence of two-, three-, or M-factors each at a number of different levels].

Among some recent typical applications in accounting, bounds based on Dirichlet mixtures of multinomial distributions have been used by Tsui, Matsumara, and Tsui

(1985) in construction of a Bayesian model for the grouped distribution of "percentage overstatement in account totals" using "dollar-unit" sampling [see Fienberg, Neter, and Leitch (1978) and Neter, Leitch, and Fienberg (1978)]. It is assumed that auditors' prior knowledge of proportions likely to fall in the different groups can be adequately represented by a Dirichlet distribution. A major aim of their analysis was estimation of a "multinomial" error bound on the overstatement.

Other recent applications of multinomial distributions include such diverse areas as formulation of connection acceptance control in an asynchronous transfer model telecommunications network [Marzo *et al.* (1994)], clinical trials [Barlow and Azen (1990)], time-series analysis [Harvey *et al.* (1989)], cytometry [Van Putten *et al.* (1993)], sickness and accident analysis [Lundberg (1940)], genetics [Fisher (1936)], medical epidemiology [Froggatt (1970)], cross-validation [Stone (1974)], social research [Thomas and Lohaus (1993)], and photon-counting [Koellner (1991)], to mention just a few examples. Sinoquet and Bonhomme (1991) introduced multinomial distributions in a model for the effect of interception of sunlight by a canopy with two or more species, to allow for various combinations of numbers of interceptions by each species and their different efficacies.

Itoh (1988) described a coding strategy to encode in high resolution in which a multinomial distribution with unknown order for the smaller alphabet and a uniform distribution within each of the subalphabets have been assumed. Terza and Wilson (1990) combined a flexible generalized Poisson model with the multinomial distribution to jointly predict households' choices among types and frequency of trips. Al-Hussaini (1989) generalized a detector with multinomial input for on–off communication systems to generalized multinomial detectors that include binary antipodal signals with arbitrary shapes; see also the paper by Beaulieu (1991) and the comments by Al-Hussaini. Kelleher and Masterson (1992) used the multinomial distribution while modeling equations for condensation biosynthesis using stable isotopes and radioisotopes. Ryan (1992) proposed a method for estimating an exposure level at which the overall risk of any adverse effect is acceptably low, and the method is based on a "continuation ratio formulation" of a multinomial distribution with an additional scale parameter to account for overdispersion. Catalano, Ryan, and Scharfstein (1994) used this method to provide a model for risk while jointly modeling fetal death and malformation. Kanoh and Li (1990) proposed a method in which the qualitative survey data are regarded as a sample from a multinomial distribution whose parameters are time-variant functions of inflation expectations, and then they applied it to Japanese data from the Business and Investment Survey of Incorporated Enterprises. Attias and Petitjean (1993) used the distribution in the statistical analysis of topological neighborhoods and multivariate representations of a large chemical file. Radons *et al.* (1994) discussed the use of multinomial distribution in the analysis, classification, and coding of multielectrode data. They also compared its performance with that of hidden Markov models. Applications in clinical trials were discussed by Shah (1986), among others. Walley (1996) discussed applications of the "imprecise Dirichlet model" in the analysis of medical data; see also Walley (1991).

10 INCOMPLETE AND MODIFIED MULTINOMIAL DISTRIBUTIONS

Suppose that $N_{11}, N_{12}, \ldots, N_{1k}$ have a joint multinomial $(n_1; p_1, p_2, \ldots, p_k)$ distribution. Suppose further that $N_{21}, N_{22}, \ldots, N_{2k'}$ $(k' < k)$ have a joint multinomial $\left(n_2; \frac{p_1}{\sum_{i=1}^{k'} p_i}, \frac{p_2}{\sum_{i=1}^{k'} p_i}, \ldots, \frac{p_{k'}}{\sum_{i=1}^{k'} p_i}\right)$ distribution, independent of $N_{11}, N_{12}, \ldots, N_{1k}$. This corresponds to a situation in which the data represented by $N_{11}, N_{12}, \ldots, N_{1k}$ are supplemented by a further "incomplete" experiment in which cells $k' + 1, k' + 2, \ldots, k$ are not observed.

Asano (1965) demonstrated that the maximum likelihood estimators of the probabilities p_i are

$$\hat{p}_i = \begin{cases} (N_{1i} + N_{2i}) \Big/ \left[n_1 \left(1 + \frac{n_2}{\sum_{j=1}^{k'} N_{1j}} \right) \right] & \text{if } i \le k' \\ N_{1i}/n_1 & \text{if } i > k'. \end{cases} \tag{35.127}$$

These formulas apply to the "simple truncated" sampling described above. They are special cases of more general results established by Batschelet (1960) and Geppert (1961), corresponding to "compound truncation," in which there are R sets $(N_{ra_{r_1}}, N_{ra_{r_2}}, \ldots, N_{ra_{r_{k_r}}})$ with cells $a_{r_1}, a_{r_2}, \ldots, a_{r_{k_r}}$ only observed in the rth set $(r = 1, 2, \ldots, R)$, and $\sum_{i=1}^{k_r} N_{ra_{r_i}} = n_r$. Asano (1965) proved that for the p_i's to be estimable, a necessary and sufficient condition is that

(i) each p_i appears in at least one set $\{a_{r_j}\}$ and

(ii) every set $\{a_{r_j}\}$ has at least one member in common with at least one other set $\{a_{r'_j}\}$, $r \neq r'$.

A modification of the multinomial distribution results in the so-called *multinomial class size distribution* with probability mass function [Wilks (1962, p. 152) and Patil and Joshi (1968)]

$$\begin{aligned} &P(x_0, x_1, \ldots, x_k) \\ &\quad = \frac{(m+1)!}{x_0! x_1! \cdots x_k!} \frac{k!}{(0!)^{x_0}(1!)^{x_1} \cdots (k!)^{x_k}} \left(\frac{1}{m+1} \right)^k, \\ &\qquad x_i = 0, 1, \ldots, m+1, \ i = 0, 1, \ldots, k, \ \sum_{i=0}^{k} x_i = m+1, \ \sum_{i=1}^{k} i x_i = k. \end{aligned} \tag{35.128}$$

The distribution (35.128) has the following interesting genesis. If $\boldsymbol{Y} = (Y_1, \ldots, Y_m)'$ has a multinomial distribution with parameters $(k; p_1, p_2, \ldots, p_m)$ where $p_i = \frac{1}{m+1}$, $i = 1, 2, \ldots, m$, then the random variables $X_0, X_1, \ldots, X_k$ denoting the number of components among $Y_0, Y_1, \ldots, Y_m$ $(Y_0 = k - \sum_{i=1}^{m} Y_i)$ which are $0, 1, 2, \ldots, k$, respectively, have their joint distribution as in (35.128). Evidently,

$$\mu_i = (m+1)\,\frac{k}{i}\left(\frac{1}{m+1}\right)^i \left(\frac{m}{m+1}\right)^{k-i} .$$

11 TRUNCATED MULTINOMIAL DISTRIBUTIONS

Let us consider the multinomial distribution in (35.1) and assume that the ith variable is doubly truncated (i.e., N_i is restricted to the range $b+1, b+2, \ldots, c$). The probability mass function of the resulting truncated distribution is [Gerstenkorn (1977)]

$$P^T(n_1, n_2, \ldots, n_k) = \frac{\prod_{j=1}^k \left(p_j^{n_j}/n_j!\right)}{\sum_{n_i=b+1}^{c} \frac{1}{n_i!(n-n_i)!}\, p_i^{n_i}(1-p_i)^{n-n_i}} ,$$

$$\sum_{j=1}^k n_j = n,\ 0 \le n_j \le n \text{ for } j \ne i,\ b+1 \le n_i \le c,\ \sum_{j=1}^k p_j = 1 . \qquad (35.129)$$

Similarly, the probability mass function of the truncated multinomial distribution (with the variables $N_1, N_2, \ldots, N_\ell$ all doubly truncated, where N_i is restricted to the range $b_i+1, b_i+2, \ldots, c_i$ for $i = 1, 2, \ldots, \ell$) is

$$P^T(n_1, n_2, \ldots, n_k) = \frac{\prod_{j=1}^k (p_j^{n_j}/n_j!)}{\sum_{n_1=b_1+1}^{c_1} \cdots \sum_{n_\ell=b_\ell+1}^{c_\ell} \left\{\prod_{i=1}^\ell \frac{p_i^{n_i}}{n_i!}\right\} \frac{(1-\sum_{i=1}^\ell p_i)^{n-\sum_{i=1}^\ell n_i}}{(n-\sum_{i=1}^\ell n_i)!}} , \qquad (35.130)$$

where $\sum_{j=1}^k n_j = n$, $\sum_{j=1}^k p_j = 1$, $b_i + 1 \le n_i \le c_i$ for $i = 1, 2, \ldots, \ell$, and $0 \le n_i \le n$ for $i = \ell+1, \ell+2, \ldots, k$.

The rth descending factorial moment of the truncated (single variable) multinomial distribution in (35.129) is

$$\begin{aligned}
\mu'_{(\boldsymbol{r})T}(\boldsymbol{N}) &= \sum_{n_1=0}^{n} \cdots \sum_{n_{i-1}=0}^{n} \sum_{n_i=b+1}^{c} \sum_{n_{i+1}=0}^{n} \cdots \sum_{n_k=0}^{n} \left\{\prod_{j=1}^k n_j^{(r_j)}\right\} P^T(n_1, n_2, \ldots, n_k) \\
&= \frac{1}{\sum_{n_i=b+1}^{c} \binom{n}{n_i} p_i^{n_i}(1-p_i)^{n-n_i}} \sum_{n_1=0}^{n} \cdots \sum_{n_{i-1}=0}^{n} \sum_{n_i=b+1}^{c} \\
&\quad \times \sum_{n_{i+1}=0}^{n} \cdots \sum_{n_k=0}^{n} \prod_{j=1}^{k} \left\{n_j^{(r_j)}\, p_j^{n_j}/n_j!\right\}
\end{aligned}$$

$$= n^{(r)} \left\{ \prod_{j=1}^{k} p_j^{r_j} \right\} \frac{\sum_{n_i=b+1-r_i}^{c-r_i} \binom{n-r}{n_i} p_i^{n_i} (1-p_i)^{n-r-n_i}}{\sum_{n_i=b+1}^{c} \binom{n}{n_i} p_i^{n_i} (1-p_i)^{n-n_i}}, \tag{35.131}$$

where $r = \sum_{j=1}^{k} r_j$. In particular, (35.131) yields the following formulas:

$$\begin{aligned} E[N_j] &= np_j \frac{\sum_{n_i=b+1}^{c} \binom{n-1}{n_i} p_i^{n_i} (1-p_i)^{n-1-n_i}}{\sum_{n_i=b+1}^{c} \binom{n}{n_i} p_i^{n_i} (1-p_i)^{n-n_i}}, \quad j \neq i \\ &= np_i \frac{\sum_{n_i=b}^{c-1} \binom{n-1}{n_i} p_i^{n_i} (1-p_i)^{n-1-n_i}}{\sum_{n_i=b+1}^{c} \binom{n}{n_i} p_i^{n_i} (1-p_i)^{n-n_i}}, \quad j = i; \end{aligned} \tag{35.132}$$

$$\begin{aligned} E[N_j^2] &= n(n-1)p_j^2 \frac{\sum_{n_i=b+1}^{c} \binom{n-2}{n_i} p_i^{n_i} (1-p_i)^{n-2-n_i}}{\sum_{n_i=b+1}^{c} \binom{n}{n_i} p_i^{n_i} (1-p_i)^{n-n_i}}, \quad j \neq i \\ &= n(n-1)p_i^2 \frac{\sum_{n_i=b-1}^{c-2} \binom{n-2}{n_i} p_i^{n_i} (1-p_i)^{n-2-n_i}}{\sum_{n_i=b+1}^{c} \binom{n}{n_i} p_i^{n_i} (1-p_i)^{n-n_i}}, \quad j = i; \end{aligned} \tag{35.133}$$

and

$$\begin{aligned} &E[N_j N_{j'}] \\ &= n(n-1)p_j p_{j'} \frac{\sum_{n_i=b+1}^{c} \binom{n-2}{n_i} p_i^{n_i} (1-p_i)^{n-2-n_i}}{\sum_{n_i=b+1}^{c} \binom{n}{n_i} p_i^{n_i} (1-p_i)^{n-n_i}}, \quad j, j' \neq i \\ &= n(n-1)p_i p_{j'} \frac{\sum_{n_i=b}^{c-1} \binom{n-2}{n_i} p_i^{n_i} (1-p_i)^{n-2-n_i}}{\sum_{n_i=b+1}^{c} \binom{n}{n_i} p_i^{n_i} (1-p_i)^{n-n_i}}, \quad j = i,\ j' \neq i. \end{aligned} \tag{35.134}$$

Expressions for the variances, covariances and correlation coefficients can be derived easily from (35.132)–(35.134).

The corresponding results for the left-truncated case (i.e., when N_i is restricted to the range $b+1, b+2, \ldots, n$) can be deduced from Eqs. (35.129) and (35.131)–(35.134) simply by setting $c = n$.

For the general truncated multinomial distribution (with the variables $N_1, N_2, \ldots, N_\ell$ all doubly truncated) in (35.130), we similarly obtain the following rather cumbersome expression for the $\boldsymbol{r}$th descending factorial moment

$$\begin{aligned} &\mu'_{(\boldsymbol{r})T}(\boldsymbol{N}) \\ &= \sum_{n_1=b+1}^{c_1} \cdots \sum_{n_\ell=b_\ell+1}^{c_\ell} \sum_{n_{\ell+1}=0}^{n} \cdots \sum_{n_k=0}^{n} \left\{ \prod_{j=1}^{k} n_j^{(r_j)} \right\} P^T(n_1, n_2, \ldots, n_k) \\ &= \frac{1}{\sum_{n_1=b_1+1}^{c_1} \cdots \sum_{n_\ell=b_\ell+1}^{c_\ell} \{\prod_{i=1}^{\ell} \frac{p_i^{n_i}}{n_i!}\} \frac{(1-\sum_{i=1}^{\ell} p_i)^{n-\sum_{i=1}^{\ell} n_i}}{(n-\sum_{i=1}^{\ell} n_i)!}} \end{aligned}$$

$$\times \sum_{n_1=b_1+1}^{c_1} \cdots \sum_{n_\ell=b_\ell+1}^{c_\ell} \sum_{n_{\ell+1}=0}^{n} \cdots \sum_{n_k=0}^{n} \prod_{j=1}^{k} \left\{ n_j^{(r_j)} p_j^{n_j} / n_j! \right\}$$

$$= n^{(r)} \left\{ \prod_{j=1}^{k} p_j^{r_j} \right\} \left[\sum_{n_1=b_1+1-r_1}^{c_1-r_1} \cdots \sum_{n_\ell=b_\ell+1-r_\ell}^{c_\ell-r_\ell} \binom{n-r}{n_1, \ldots, n_\ell, n-r-\Sigma_{i=1}^{\ell} n_i} \right.$$

$$\left. \times p_1^{n_1} p_2^{n_2} \cdots p_\ell^{n_\ell} \left(1 - \sum_{i=1}^{\ell} p_i\right)^{n-r-\Sigma_{i=1}^{\ell} n_i} \right]$$

$$\times \left[\sum_{n_1=b_1+1}^{c_1} \cdots \sum_{n_\ell=b_\ell+1}^{c_\ell} \binom{n}{n_1, \ldots, n_\ell, n-\Sigma_{i=1}^{\ell} n_i} \right.$$

$$\left. \times p_1^{n_1} \cdots p_\ell^{n_\ell} \left(1 - \sum_{i=1}^{\ell} p_i\right)^{n-\Sigma_{i=1}^{\ell} n_i} \right]^{-1}, \tag{35.135}$$

where, as before, $r = \sum_{i=1}^{k} r_i$. In particular, (35.135) yields the following formulas[1]:

$$E[N_j] = np_j \left[\sum_{n_1=b_1+1}^{c_1} \cdots \sum_{n_{j-1}=b_{j-1}+1}^{c_{j-1}} \sum_{n_j=b_j}^{c_j-1} \right.$$

$$\times \sum_{n_{j+1}=b_{j+1}+1}^{c_{j+1}} \cdots \sum_{n_\ell=b_\ell+1}^{c_\ell} \binom{n-1}{n_1, \ldots, n_\ell, n-1-\Sigma_{i=1}^{\ell} n_i}$$

$$\left. \times p_1^{n_1} \cdots p_\ell^{n_\ell} \left(1 - \sum_{i=1}^{\ell} p_i\right)^{n-1-\Sigma_{i=1}^{\ell} n_i} \right]$$

$$\times \left[\sum_{n_1=b_1+1}^{c_1} \cdots \sum_{n_\ell=b_\ell+1}^{c_\ell} \binom{n}{n_1, \ldots, n_\ell, n-\Sigma_{i=1}^{\ell} n_i} \right.$$

$$\left. \times p_1^{n_1} \cdots p_\ell^{n_\ell} \left(1 - \sum_{i=1}^{\ell} p_i\right)^{n-\Sigma_{i=1}^{\ell} n_i} \right]^{-1},$$

$$j = 1, 2, \ldots, \ell,$$

$$= np_j \left[\sum_{n_1=b_1+1}^{c_1} \cdots \sum_{n_\ell=b_\ell+1}^{c_\ell} \binom{n-1}{n_1, \ldots, n_\ell, n-1-\Sigma_{i=1}^{\ell} n_i} \right.$$

$$\left. \times p_1^{n_1} p_2^{n_2} \cdots p_\ell^{n_\ell} \left(1 - \sum_{i=1}^{\ell} p_i\right)^{n-1-\Sigma_{i=1}^{\ell} n_i} \right]$$

[1] Though these are cumbersome formulas, we have presented them here for the convenience of the readers.

$$\times \left[\sum_{n_1=b_1+1}^{c_1} \cdots \sum_{n_\ell=b_\ell+1}^{c_\ell} \binom{n}{n_1, \ldots, n_\ell, n - \Sigma_{i=1}^{\ell} n_i} \times p_1^{n_1} \cdots p_\ell^{n_\ell} \left(1 - \sum_{i=1}^{\ell} p_i\right)^{n - \Sigma_{i=1}^{\ell} n_i} \right]^{-1},$$

$$j = \ell + 1, \ell + 2, \ldots, k; \tag{35.136}$$

$$E[N_j^2] = n(n-1)p_j^2 \left[\sum_{n_1=b_1+1}^{c_1} \cdots \sum_{n_{j-1}=b_{j-1}+1}^{c_{j-1}} \sum_{n_j=b_j-1}^{c_j-2} \times \sum_{n_{j+1}=b_{j+1}+1}^{c_{j+1}} \cdots \sum_{n_\ell=b_\ell+1}^{c_\ell} \binom{n-2}{n_1, \ldots, n_\ell, n-2-\Sigma_{i=1}^{\ell} n_i} \times p_1^{n_1} \cdots p_\ell^{n_\ell} \left(1 - \sum_{i=1}^{\ell} p_i\right)^{n-2-\Sigma_{i=1}^{\ell} n_i} \right]$$

$$\times \left[\sum_{n_1=b_1+1}^{c_1} \cdots \sum_{n_\ell=b_\ell+1}^{c_\ell} \binom{n}{n_1, \ldots, n_\ell, n - \Sigma_{i=1}^{\ell} n_i} \times p_1^{n_1} \cdots p_\ell^{n_\ell} \left(1 - \sum_{i=1}^{\ell} p_i\right)^{n - \Sigma_{i=1}^{\ell} n_i} \right]^{-1},$$

$$j = 1, 2, \ldots, \ell,$$

$$= n(n-1)p_j^2 \left[\sum_{n_1=b_1+1}^{c_1} \cdots \sum_{n_\ell=b_\ell+1}^{c_\ell} \binom{n-2}{n_1, \ldots, n_\ell, n-2-\Sigma_{i=1}^{\ell} n_i} \times p_1^{n_1} \cdots p_\ell^{n_\ell} \left(1 - \sum_{i=1}^{\ell} p_i\right)^{n-2-\Sigma_{i=1}^{\ell} n_i} \right]$$

$$\times \left[\sum_{n_1=b_1+1}^{c_1} \cdots \sum_{n_\ell=b_\ell+1}^{c_\ell} \binom{n}{n_1, \ldots, n_\ell, n - \Sigma_{i=1}^{\ell} n_i} \times p_1^{n_1} \cdots p_\ell^{n_\ell} \left(1 - \sum_{i=1}^{\ell} p_i\right)^{n - \Sigma_{i=1}^{\ell} n_i} \right]^{-1},$$

$$j = \ell + 1, \ell + 2, \ldots, k; \tag{35.137}$$

and

$$E[N_jN_{j'}] = n(n-1)p_jp_{j'}\Bigg[\sum_{n_1=b_1+1}^{c_1}\cdots\sum_{n_{j-1}=b_{j-1}+1}^{c_{j-1}}\sum_{n_j=b_j}^{c_j-1}$$

$$\times\sum_{n_{j+1}=b_{j+1}+1}^{c_{j+1}}\cdots\sum_{n_{j'-1}=b_{j'-1}+1}^{c_{j'-1}}\sum_{n_{j'}=b_{j'}}^{c_{j'}-1}$$

$$\times\sum_{n_{j'+1}=b_{j'+1}+1}^{c_{j'+1}}\cdots\sum_{n_\ell=b_\ell+1}^{c_\ell}\binom{n-2}{n_1,\ldots,n_\ell,n-2-\Sigma_{i=1}^{\ell}n_i}$$

$$\times p_1^{n_1}\cdots p_\ell^{n_\ell}\left(1-\sum_{i=1}^{\ell}p_i\right)^{n-2-\Sigma_{i=1}^{\ell}n_i}\Bigg]$$

$$\times\Bigg[\sum_{n_1=b_1+1}^{c_1}\cdots\sum_{n_\ell=b_\ell+1}^{c_\ell}\binom{n}{n_1,\ldots,n_\ell,n-\Sigma_{i=1}^{\ell}n_i}$$

$$\times p_1^{n_1}\cdots p_\ell^{n_\ell}\left(1-\sum_{i=1}^{\ell}p_i\right)^{n-\Sigma_{i=1}^{\ell}n_i}\Bigg]^{-1},$$

$$1\le j<j'\le\ell,$$

$$= n(n-1)p_jp_{j'}\Bigg[\sum_{n_1=b_1+1}^{c_1}\cdots\sum_{n_{j-1}=b_{j-1}+1}^{c_{j-1}}\sum_{n_j=b_j}^{c_j-1}$$

$$\times\sum_{n_{j+1}=b_{j+1}+1}^{c_{j+1}}\cdots\sum_{n_\ell=b_\ell+1}^{c_\ell}\binom{n-2}{n_1,\ldots,n_\ell,n-2-\Sigma_{i=1}^{\ell}n_i}$$

$$\times p_1^{n_1}\cdots p_\ell^{n_\ell}\left(1-\sum_{i=1}^{\ell}p_i\right)^{n-2-\Sigma_{i=1}^{\ell}n_i}\Bigg]$$

$$\times\Bigg[\sum_{n_1=b_1+1}^{c_1}\cdots\sum_{n_\ell=b_\ell+1}^{c_\ell}\binom{n}{n_1,\ldots,n_\ell,n-\Sigma_{i=1}^{\ell}n_i}$$

$$\times p_1^{n_1}\cdots p_\ell^{n_\ell}\left(1-\sum_{i=1}^{\ell}p_i\right)^{n-\Sigma_{i=1}^{\ell}n_i}\Bigg]^{-1},$$

$$1\le j\le\ell<j'\le k,$$

$$= n(n-1)p_jp_{j'}\Bigg[\sum_{n_1=b_1+1}^{c_1}\cdots\sum_{n_\ell=b_\ell+1}^{c_\ell}\binom{n-2}{n_1,\ldots,n_\ell,n-2-\Sigma_{i=1}^{\ell}n_i}$$

$$\times\, p_1^{n_1} \cdots p_\ell^{n_\ell} \left(1 - \sum_{i=1}^{\ell} p_i\right)^{n-2-\Sigma_{i=1}^{\ell} n_i} \Bigg]$$

$$\times \left[\sum_{n_1=b_1+1}^{c_1} \cdots \sum_{n_\ell=b_\ell+1}^{c_\ell} \binom{n}{n_1, \ldots, n_\ell, n - \Sigma_{i=1}^{\ell} n_i} \right.$$

$$\left. \times\, p_1^{n_1} \cdots p_\ell^{n_\ell} \left(1 - \sum_{i=1}^{\ell} p_i\right)^{n-\Sigma_{i=1}^{\ell} n_i} \right]^{-1},$$

$$\ell + 1 \le j < j' \le k\,. \quad (35.138)$$

Expression for the variances, covariances, and correlation coefficients can easily be derived from (35.136)–(35.138).

The corresponding results for the left-truncated case (i.e., when N_i is restricted to the range $b_i + 1, b_i + 2, \ldots, n$ for $i = 1, 2, \ldots, \ell$) can be deduced from Eqs. (35.130) and (35.135)–(35.138) simply by setting $c_1 = c_2 = \cdots = c_\ell = n$.

12 MIXTURES OF MULTINOMIALS

Morel and Nagaraj (1993) proposed a finite mixture of multinomial random variables to model categorical data exhibiting overdispersion. Suppose that $\boldsymbol{N}^* = (\boldsymbol{N}', n - \mathbf{1}'_{k-1}\boldsymbol{N})'$ is a k-dimensional multinomial random variable with probability vector $\boldsymbol{p}^* = (\boldsymbol{p}', 1 - \mathbf{1}'_{k-1}\boldsymbol{p})' = (p_1, p_2, \ldots, p_{k-1}, p_k)'$, where $\mathbf{1}_{k-1}$ denotes a $(k-1) \times 1$ column vector of ones. In this discussion, the last component of both vectors $\boldsymbol{N}^*$ and $\boldsymbol{p}^*$ will be left out. Let $\boldsymbol{Y}, \boldsymbol{Y}_1^0, \ldots, \boldsymbol{Y}_n^0$ be independent and identically distributed as multinomial with parameters $n = 1$ and $\boldsymbol{p}$. Let $U_1, U_2, \ldots, U_n$ be independent and identically distributed Uniform (0, 1) variables and let ρ $(0 < \rho < 1)$ be a real number, $0 < \rho < 1$. For $i = 1, 2, \ldots, n$, define

$$\boldsymbol{Y}_i = \boldsymbol{Y}\, I(U_i \le \rho) + \boldsymbol{Y}_i^0\, I(U_i > \rho)\,, \quad (35.139)$$

where $I(\cdot)$ denotes the indicator function, and

$$\boldsymbol{T} = \sum_{i=1}^{n} \boldsymbol{Y}_i\,. \quad (35.140)$$

Note that

$$E[\boldsymbol{T}] = n\boldsymbol{p} \qquad \text{and} \qquad \text{var}(\boldsymbol{T}) = \{1 + \rho^2(n-1)\} n\boldsymbol{V}(\boldsymbol{p})\,, \quad (35.141)$$

where $\boldsymbol{V}(\boldsymbol{p}) = ((\text{diag}(\boldsymbol{p}) - \boldsymbol{p}\boldsymbol{p}'))$. Thus, the distribution of $\boldsymbol{T}$ has extra variation relative to the multinomial distribution with parameters $(n; \boldsymbol{p})$.

The random variable $\boldsymbol{T}$ can be represented as

$$\boldsymbol{T} = \boldsymbol{Y}X + (\boldsymbol{N}|X)\,, \tag{35.142}$$

where X is distributed as Binomial(n, ρ), $\boldsymbol{Y}$ is distributed as multinomial with parameters 1 and $\boldsymbol{p}$, X and $\boldsymbol{Y}$ are independent, and $\boldsymbol{N}|X$ is distributed as multinomial with parameters $n - X$ (if $X < n$) and $\boldsymbol{p}$. Equation (35.142) shows the clustering mechanism: a random number X of "extra counts" is added to a category selected randomly by $\boldsymbol{Y}$. Note that the distribution of $\boldsymbol{T}$ is a *finite mixture of multinomials*. This facilitates the evaluation of the estimators, but (35.142) is more easily interpreted. The distribution of $\boldsymbol{T}$ is a finite mixture of multinomial random variables, because the moment generating function of $\boldsymbol{T}$ can be computed by conditioning on X as

$$\begin{aligned}
M_{\boldsymbol{T}}(\boldsymbol{z}) &= E_X\left[E[e^{\boldsymbol{z}'\boldsymbol{Y}X+\boldsymbol{z}'(\boldsymbol{N}|X)} \mid X]\right] \\
&= \sum_{i=1}^{k-1} p_i \left\{(1-\rho)p_k + \rho\, e^{z_i} + \sum_{j=1}^{k-1}(1-\rho)p_j\, e^{z_j}\right\}^n \\
&\qquad + p_k \left\{(1-\rho)p_k + \rho + \sum_{j=1}^{k-1}(1-\rho)p_j\, e^{z_j}\right\}^n \\
&= \sum_{i=1}^{k} p_i\, M_{\boldsymbol{N}_i}(\boldsymbol{z})\,,
\end{aligned} \tag{35.143}$$

where $\boldsymbol{N}_i$ is distributed as multinomial$(n; (1-\rho)\boldsymbol{p} + \rho\,\boldsymbol{e}_i)$ for $i = 1, 2, \ldots, k-1$, $\boldsymbol{N}_k$ is distributed as multinomial$(n; (1-\rho)\boldsymbol{p})$, and $\boldsymbol{e}_i$ is the ith column of the $(k-1)\times(k-1)$ identity matrix.

Let $\boldsymbol{t}$ be a realization of $\boldsymbol{T}$. Then by (35.143),

$$\Pr[\boldsymbol{T} = \boldsymbol{t}] = \sum_{i=1}^{k} p_i \Pr[\boldsymbol{N}_i = \boldsymbol{t}]\,. \tag{35.144}$$

If $\boldsymbol{T}$ is distributed according to (35.143), we denote it as mixture-multinomial $(n; \boldsymbol{p}, \rho)$. The weights in the finite mixture (35.143) are part of the parameters of the multinomials being weighted. Let $\boldsymbol{t}_1, \boldsymbol{t}_2, \ldots, \boldsymbol{t}_m$ be observed values of independent random variables $\boldsymbol{T}_1, \boldsymbol{T}_2, \ldots, \boldsymbol{T}_m$, where $\boldsymbol{T}_j$ is distributed as mixture-multinomial$(n_j; \boldsymbol{p}, \rho)$. Let $\boldsymbol{\theta} = (p_1, \ldots, p_{k-1}, \rho)'$ be the k-dimensional vector of unknown parameters, where $p_i > 0$, $0 < \sum_{i=1}^{k-1} p_i < 1$, and $0 < \rho < 1$. Let $L(\boldsymbol{\theta}; \boldsymbol{t}_j)$ denote $\Pr[\boldsymbol{T}_j = \boldsymbol{t}_j] = \sum_{i=1}^{k} p_i \Pr[\boldsymbol{N}_{ji} = \boldsymbol{t}_j]$, where $\boldsymbol{t}_j = (t_{j1}, \ldots, t_{j,k-1})'$, the observed vector of counts for the jth cluster, $t_{ji} \geq 0$, and $t_{jk} = n_j - \sum_{i=1}^{k-1} t_{ji}$. Let $\nabla_{\boldsymbol{\theta}} \ln L(\boldsymbol{\theta}; \boldsymbol{t}_j)$ represent the column vector of partial derivatives of the log-likelihood $\ln L(\boldsymbol{\theta}; \boldsymbol{t}_j)$ with respect to the components of $\boldsymbol{\theta}$. Then,

$$\nabla'_{\boldsymbol{\theta}} \ln L(\boldsymbol{\theta}; \boldsymbol{t}_j) = \left(\frac{\partial}{\partial p_1}, \ldots, \frac{\partial}{\partial p_{k-1}}, \frac{\partial}{\partial \rho}\right) \ln L(\boldsymbol{\theta}; \boldsymbol{t}_j), \tag{35.145}$$

where

$$\frac{\partial}{\partial p_i} \ln L(\boldsymbol{\theta}; \boldsymbol{t}_j) = \frac{1}{\Pr[\boldsymbol{T}_j = \boldsymbol{t}_j]} \{a_i(\boldsymbol{\theta}; \boldsymbol{t}_j) - a_k(\boldsymbol{\theta}; \boldsymbol{t}_j)\} \tag{35.146}$$

with

$$a_\ell(\boldsymbol{\theta}; \boldsymbol{t}_j) = \frac{t_{j\ell}}{p_\ell} \Pr[\boldsymbol{T}_j = \boldsymbol{t}_j] + \Pr[\boldsymbol{N}_{j\ell} = \boldsymbol{t}_j] \left\{1 - \frac{\rho t_{j\ell}}{(1-\rho)p_\ell + \rho}\right\}, \tag{35.147}$$

and

$$\frac{\partial}{\partial \rho} \ln L(\boldsymbol{\theta}; \boldsymbol{t}_j) = \frac{1}{(1-\rho)\Pr[\boldsymbol{T}_j = \boldsymbol{t}_j]} \sum_{i=1}^{k} p_i \Pr[\boldsymbol{N}_{ji} = \boldsymbol{t}_j] \times \left[\frac{t_{ji} - n_j\{(1-\rho)p_i + \rho\}}{(1-\rho)p_i + \rho}\right]. \tag{35.148}$$

The vector of scores is $\sum_{j=1}^{n} \nabla_{\boldsymbol{\theta}} \ln L(\boldsymbol{\theta}; \boldsymbol{t}_j)$, and the solution to the likelihood equations can be obtained iteratively by using Fisher's scoring method.

Morel and Nagaraj (1993) have shown that if $\hat{\boldsymbol{\theta}} = (\hat{\boldsymbol{p}}', \hat{\rho})'$ represents the maximum likelihood estimator of $\boldsymbol{\theta} = (\boldsymbol{p}', \rho)'$ and

$$\boldsymbol{A}_{\boldsymbol{\theta\theta}n} = \begin{pmatrix} \boldsymbol{A}_{11\boldsymbol{\theta\theta}n} & \boldsymbol{A}_{12\boldsymbol{\theta\theta}n} \\ \boldsymbol{A}_{21\boldsymbol{\theta\theta}n} & \boldsymbol{A}_{22\boldsymbol{\theta\theta}n} \end{pmatrix} \tag{35.149}$$

denotes the Fisher information $(k \times k)$ matrix partitioned in such a manner that $\boldsymbol{A}_{11\boldsymbol{\theta\theta}n}$ is $(k-1) \times (k-1)$ and $\boldsymbol{A}_{22\boldsymbol{\theta\theta}n}$ is 1×1, then asymptotically as $n \to \infty$,

(i) $\boldsymbol{A}_{11\boldsymbol{\theta\theta}n} \sim (n\boldsymbol{A}_{11} + \boldsymbol{A}_{\boldsymbol{pp}})$, where

$$\boldsymbol{A}_{11} = (1-\rho)^2\{\text{diag}(\beta_1, \ldots, \beta_{k-1}) + \beta_k \, \mathbf{1}_{k-1} \, \mathbf{1}'_{k-1}\},$$

$$\beta_i = \frac{p_i}{(1-\rho)p_i + \rho} + \frac{1-p_i}{(1-\rho)p_i}, \qquad i = 1, 2, \ldots, k,$$

$$\boldsymbol{A}_{\boldsymbol{pp}} = \left\{\text{diag}\left(\frac{1}{p_1}, \ldots, \frac{1}{p_{k-1}}\right) + \frac{1}{p_k} \mathbf{1}_{k-1} \, \mathbf{1}'_{k-1}\right\},$$

where $\mathbf{1}_{k-1} \, \mathbf{1}'_{k-1}$ is a $(k-1) \times (k-1)$ matrix of ones;

(ii) $\boldsymbol{A}_{12\boldsymbol{\theta\theta}n} \sim n\,\boldsymbol{A}_{12}$, where $\boldsymbol{A}_{12}$ is a vector given by

$$\boldsymbol{A}_{12} = (1-\rho)\{(\gamma_1, \ldots, \gamma_{k-1})' - \gamma_k \, \mathbf{1}_{k-1}\},$$

$$\gamma_i = \frac{p_i(1-p_i)}{(1-\rho)p_i + \rho} + \frac{p_i}{1-\rho}, \qquad i = 1, 2, \ldots, k;$$

and

(iii) $\boldsymbol{A}_{22\boldsymbol{\theta\theta}n} \sim n\boldsymbol{A}_{22}$, where

$$\boldsymbol{A}_{22} = \frac{1}{1-\rho} \sum_{i=1}^{k} \frac{p_i(1-p_i)}{(1-\rho)p_i + \rho} .$$

(In the above formulas, $A \sim B$ means that the difference $A - B$ goes to 0.) The above expressions are useful in computing estimated standard errors of the maximum likelihood estimators when n (the cluster size) is large. Morel and Nagaraj (1993), in fact, claim that this approximation is satisfactory even for moderate values of n.

13 COMPOUND MULTINOMIAL AND MARKOV MULTINOMIAL DISTRIBUTIONS

13.1 Compound Multinomial

Compound multivariate distributions are formed in the same way as compound univariate distributions, by assigning distributions to some (or all) of the parameters of a multivariate distribution. Let us consider

$$\text{Multinomial}(n; p_1, \ldots, p_k) \bigwedge_{p_1,\ldots,p_k} \text{Dirichlet}(\alpha_1, \ldots, \alpha_k) . \tag{35.150}$$

The probability mass function of this compound distribution is

$$\begin{aligned} P(n_1, \ldots, n_k) &= \frac{n!}{\prod_{i=1}^k n_i!} E\left[\prod_{i=1}^{k} p_i^{n_i}\right] \\ &= \frac{n!\Gamma(\sum_{i=1}^k \alpha_i)}{\{\prod_{i=1}^k n_i!\}\Gamma(n + \sum_{i=1}^k \alpha_i)} \prod_{i=1}^{k} \left\{\frac{\Gamma(n_i + \alpha_i)}{\Gamma(\alpha_i)}\right\} \\ &= \frac{n!}{(\sum_{i=1}^k \alpha_i)^{[n]}} \prod_{i=1}^{k} \left\{\frac{\alpha_i^{[n_i]}}{n_i!}\right\}, \quad n_i \geq 0, \ \sum_{i=1}^{k} n_i = n . \end{aligned} \tag{35.151}$$

This simple distribution is a natural generalization of the Binomial$(n, p)\bigwedge_p$ Beta(α, β) distribution discussed by Johnson, Kotz, and Kemp (1992). In fact, the marginal distributions of (35.151) are of this form. The distribution (35.151) is called the *Dirichlet (or Beta)-compound multinomial distribution*. It was derived using the compounding argument as above by Mosimann (1962) and was also studied by Ishii and Hayakawa (1960), who referred to it as the *multivariate binomial-beta distribution*. An alternative form of this compound multinomial distribution [Hoadley (1969)] is

$$\frac{n!\,\Gamma(\alpha_\bullet)}{\Gamma(n+\alpha_\bullet)} \prod_{i=1}^{k} \left\{\frac{\Gamma(n_i + \alpha_i)}{n_i!\,\Gamma(\alpha_i)}\right\}, \quad \sum_{i=1}^{k} n_i = n, \ \alpha_\bullet = \sum_{i=1}^{k} \alpha_i , \tag{35.152}$$

and is denoted by CMtn$(n; \boldsymbol{\alpha})$.

For the distribution (35.151), the marginal distribution of N_i is

$$\text{Binomial}(n, p_i) \bigwedge_{p_i} \text{Beta}(\alpha_i,\ \alpha_\bullet - \alpha_i) \tag{35.153}$$

which has the probability mass function

$$\begin{aligned}
\Pr[N_i = x] &= \binom{n}{x} \frac{1}{B(\alpha_i, \alpha_\bullet - \alpha_i)} \int_0^1 p_i^{x+\alpha_i-1}(1-p_i)^{n-x+\alpha_\bullet-\alpha_i-1} dp_i \\
&= \binom{n}{x} \frac{\Gamma(\alpha_\bullet)}{\Gamma(\alpha_i)\Gamma(\alpha_\bullet - \alpha_i)} \frac{\Gamma(x+\alpha_i)\Gamma(n-x+\alpha_\bullet-\alpha_i)}{\Gamma(n+\alpha_\bullet)} \\
&= \binom{x+\alpha_i-1}{x}\binom{n-x+\alpha_\bullet-\alpha_i-1}{n-x} \Big/ \binom{n+\alpha_\bullet-1}{n},
\end{aligned}$$

$$\text{where } \alpha_\bullet = \sum_{i=1}^{k} \alpha_i\,. \tag{35.154}$$

For integer values of α_i's, (35.154) is the negative hypergeometric distribution (see Chapter 6) with parameters $n + \alpha_\bullet - 1,\ \alpha_i - 1,\ n$. Therefore, the distribution (35.151) is sometimes called a *negative multivariate hypergeometric distribution* [Cheng Ping (1964)].

The joint distribution of the subset $N_{a_1}, N_{a_2}, \ldots, N_{a_s},\ n - \sum_{i=1}^{s} N_{a_i}$ is also of the form (35.151) with parameters $(n; \alpha_{a_1}, \ldots, \alpha_{a_s}, \alpha_\bullet - \alpha_{a_\bullet})$, where $\alpha_\bullet = \sum_{i=1}^{k} \alpha_i$ and $\alpha_{a_\bullet} = \sum_{i=1}^{s} \alpha_{a_i}$. This can be established easily by the following arguments. If $N_1, N_2, \ldots, N_k$ have the joint multinomial distribution (35.1), then (as noted earlier) the subset $N_{a_1}, \ldots, N_{a_s},\ n - \sum_{i=1}^{s} N_{a_i}$ jointly have a multinomial distribution with parameters $(n; p_{a_1}, \ldots, p_{a_s}, 1 - \sum_{i=1}^{s} p_{a_i})$. Also, if $p_1, p_2, \ldots, p_k$ jointly have a Dirichlet distribution with parameters $\alpha_1, \alpha_2, \ldots, \alpha_k$, then the subset $p_{a_1}, p_{a_2}, \ldots, p_{a_s}$, $1 - \sum_{i=1}^{s} p_{a_i}$ jointly have a Dirichlet distribution with parameters $\alpha_{a_1}, \alpha_{a_2}, \ldots,$ $\alpha_{a_s}, \alpha_\bullet - \alpha_{a_\bullet}$. A similar argument also leads to the result that the conditional distribution of the remaining N_i's, given $n_{a_1}, n_{a_2}, \ldots, n_{a_s}$, is of the form (35.151) with parameters $n - \sum_{i=1}^{s} n_{a_i}, \{\alpha_i,\ i \neq a_1, \ldots, a_s\}$.

The ***r***th descending factorial moment of the Dirichlet-compound multinomial distribution (35.151) is

$$\mu'_{(\boldsymbol{r})} = E\left[\prod_{i=1}^{k} N_i^{(r_i)}\right] = n^{(r_\bullet)} \left\{\prod_{i=1}^{k} \alpha_i^{[r_i]}\right\} \Big/ \alpha_\bullet^{[r_\bullet]}\,; \tag{35.155}$$

[see Chapter 34, Eqs. (34.10) and (34.11)]. Using the relationship between the descending factorial moments and the raw moments (see Chapter 34), Hoadley (1969) derived rather complicated explicit expressions for the raw moments (moments about zero). From (35.155) we obtain, in particular,

$$E[N_i] = np'_i, \quad \text{var}(N_i) = \left(\frac{n+\alpha_\bullet}{1+\alpha_\bullet}\right) np'_i(1-p'_i), \tag{35.156}$$

$$\operatorname{cov}(N_i, N_j) = -\left(\frac{n + \alpha_\bullet}{1 + \alpha_\bullet}\right) np_i' p_j', \tag{35.157}$$

and, hence,

$$\operatorname{corr}(N_i, N_j) = -\sqrt{\frac{p_i' p_j'}{(1 - p_i')(1 - p_j')}}, \tag{35.158}$$

where $\alpha_\bullet = \sum_{i=1}^k \alpha_i$ as before, and $p_i' = \alpha_i / \alpha_\bullet$. Note that $\sum_{i=1}^k p_i' = 1$. From (35.156) and (35.157), we observe that the variance–covariance matrix of the Dirichlet-compound multinomial distribution (35.151) is

$$\left(\frac{n + \alpha_\bullet}{1 + \alpha_\bullet}\right) \times \text{Variance–covariance matrix of Multinomial}(n; p_1', \dots, p_k'). \tag{35.159}$$

The multiple regression of N_i on $N_{a_1}, N_{a_2}, \dots, N_{a_s}$ $(i \neq a_1, a_2, \dots, a_s)$ is

$$E\left[N_i \mid \bigcap_{j=1}^s (N_{a_j} = n_{a_j})\right] = \left(n - \sum_{j=1}^s n_{a_j}\right) \alpha_i \Big/ (\alpha_\bullet - \alpha_{a_\bullet}) \tag{35.160}$$

which is linear. Hoadley (1969) has in fact shown that if $\boldsymbol{N} \rightsquigarrow \text{CMtn}(n; \boldsymbol{a})$, then $\{(N_1, \dots, N_t)' \mid (N_{t+1}, \dots, N_k)' = (n_{t+1}, \dots, n_k)'\} \rightsquigarrow \text{CMtn}(n - \sum_{i=t+1}^k n_i;\ a_1, \dots, a_t)$. Similarly, if $T_1, T_2, \dots, T_r$ are disjoint nonempty subsets of $\{1, 2, \dots, k\}$ for which $\bigcup_{i=1}^r T_i = \{1, 2, \dots, k\}$ and $\boldsymbol{N}_i$ is the vector of variables whose components are the a_j's for which $j \in T_i$, then $(N_{1\bullet}, N_{2\bullet}, \dots, N_{r\bullet})' \rightsquigarrow \text{CMtn}(n; a_{1\bullet}, a_{2\bullet}, \dots, a_{r\bullet})$; further, conditional on $(N_{1\bullet}, N_{2\bullet}, \dots, N_{r\bullet})' = (n_{1\bullet}, \dots, n_{r\bullet})'$, $\boldsymbol{N}_1, \dots, \boldsymbol{N}_r$ are mutually independent and distributed as $\text{CMtn}(n_{1\bullet}; \boldsymbol{a}_1), \dots, \text{CMtn}(n_{r\bullet}; \boldsymbol{a}_r)$, respectively.

The distribution (35.151) depends on the parameters $n, \alpha_1, \dots, \alpha_k$. Mosimann (1963) discussed the problem of estimating the parameters α_i (n being assumed to be known) given independent sets of observed values $(n_{1t}, n_{2t}, \dots, n_{kt})$ with $\sum_{i=1}^k n_{it} = n$ $(t = 1, 2, \dots, m)$. If the method of moments is used, then we obtain from (35.156)

$$\bar{n}_i = n\, \tilde{\alpha}_i \Big/ \sum_{j=1}^k \tilde{\alpha}_j, \qquad i = 1, 2, \dots, k, \tag{35.161}$$

where $\bar{n}_i = \frac{1}{m} \sum_{t=1}^m n_{it}$. From the k equations in (35.161), the ratios of the $\tilde{\alpha}_i$'s one to another are determined. In order to get an equation from which $\sum_{i=1}^k \tilde{\alpha}_i$ can be obtained, one can make use of (35.159) and equate the determinant of the sample variance–covariance matrix (sample generalized variance) to $\left(\frac{n+\sum_{i=1}^k \tilde{\alpha}_i}{1+\sum_{i=1}^k \tilde{\alpha}_i}\right)^{k-1}$ times the variance–covariance matrix of a multinomial $(n; \frac{\bar{n}_1}{n}, \dots, \frac{\bar{n}_k}{n})$ [only $k - 1$ of the variables being used, to avoid having singular matrices.] From the resulting equation, $\sum_{i=1}^k \tilde{\alpha}_i$ can be determined, and then the individual values $\tilde{\alpha}_1, \tilde{\alpha}_2, \dots, \tilde{\alpha}_k$ can all be calculated from (35.161). Numerical examples, applied to pollen counts, have been given by Mosimann (1963).

13.2 Markov Multinomial

Wang and Yang (1995) recently discussed a *Markov multinomial distribution* which is an extension of a generalized form of binomial distribution given by Edwards (1960). Let the sequence of random variables $X_1, X_2, \ldots$ form a Markov chain with state space $S = 0, 1, 2, \ldots, k$, where $k \geq 2$. Let its stationary transition probability matrix $\boldsymbol{P}$ be

$$\begin{array}{cc} & X_{i+1} \\ & \begin{array}{cccc} 0 & 1 & \cdots & k \end{array} \\ X_i \begin{array}{c} 0 \\ 1 \\ \vdots \\ k \end{array} & \begin{pmatrix} \Pi + (1-\Pi)p_0 & (1-\Pi)p_1 & \cdots & (1-\Pi)p_k \\ (1-\Pi)p_0 & \Pi + (1-\Pi)p_1 & \cdots & (1-\Pi)p_k \\ \vdots & \vdots & & \vdots \\ (1-\Pi)p_0 & (1-\Pi)p_1 & \cdots & \Pi + (1-\Pi)p_k \end{pmatrix} \end{array}$$

and let the initial probabilities be

$$\Pr[X_1 = j] = p_j \geq 0 \qquad \text{for } j \in S\,,$$

where $p_0 + p_1 + \cdots + p_k = 1$ and $0 \leq \Pi \leq 1$. Let $Y_i(j)$ be the indicator variable of the event $\{X_i = j\}$ and write

$$\boldsymbol{Y}_i = (Y_i(0), Y_i(1), \ldots, Y_i(k))'$$

so that $\sum_{i=0}^{k} Y_i(j) = 1$. Further, let vector of category frequencies be

$$\boldsymbol{Z}_n = (Z_n(0), Z_n(1), \ldots, Z_n(k))'\,,$$

where $Z_n(j) = \sum_{i=1}^{n} Y_i(j)$ for $j = 0, 1, \ldots, k$, so that $\sum_{j=0}^{k} Z_n(j) = n$ for all n. The joint probability generating function of $(\boldsymbol{Z}_n)$, having this Markov multinomial distribution, is

$$G_n(\boldsymbol{t}) = E\left[\prod_{j=0}^{k} t_j^{Z_n(j)}\right] = (1, t_1, \ldots, t_k)\boldsymbol{Q}^{n-1}(\boldsymbol{t})\boldsymbol{p}\,,$$

where $\boldsymbol{t} = (1, t_1, \ldots, t_k)$ $(0 \leq t_j \leq 1)$,

$$\boldsymbol{Q}(\boldsymbol{t}) = \begin{bmatrix} \Pi + (1-\Pi)p_0 & (1-\Pi)p_0 t_1 & \cdots & (1-\Pi)p_0 t_k \\ (1-\Pi)p_1 & \Pi + (1-\Pi)p_1 t_1 & \cdots & (1-\Pi)p_1 t_k \\ \vdots & \vdots & & \vdots \\ (1-\Pi)p_k & (1-\Pi)p_k t_1 & \cdots & \Pi + (1-\Pi)p_k t_k \end{bmatrix},$$

and $\boldsymbol{p} = (p_0, p_1, \ldots, p_k)'$; for further details, see Wang and Yang (1995).

BIBLIOGRAPHY

Agresti, A. (1990). *Categorical Data Analysis*, New York: John Wiley & Sons.

Alam, K. (1970). Monotonicity properties of the multinomial distribution, *Annals of Mathematical Statistics,* **41,** 315–317.

Al-Hussaini, E. K. (1989). Generalized multinomial detectors for communication signals, *IEEE Transactions on Communications*, **37**, 1099–1102.

Alldredge, J. R., and Armstrong, D. W. (1974). Maximum likelihood estimation for the multinomial distribution using geometric programming, *Technometrics*, **16**, 585–587.

Angers, C. (1989). Note on quick simultaneous confidence intervals for multinomial proportions, *The American Statistician*, **43**, 91.

Argentino, P. D., Morris, R. A., and Tolson, R. H. (1967). *The X^2-statistics and the goodness of fit test*, Report X-551-67-519, Goddard Space Flight Center, Greenbelt, MD.

Asano, C. (1965). On estimating multinomial probabilities by pooling incomplete samples, *Annals of the Institute of Statistical Mathematics,* **17,** 1–13.

Attias, R., and Petitjean, M. (1993). Statistical analysis of atom topological neighborhoods and multivariate representations of a large chemical file, *Journal of Chemical Information and Computing Science*, **33**, 649–656.

Baranchik, A. J. (1964). Multiple regression and estimation of the mean of a multi-variant normal distribution, *Technical Report No. 51,* Department of Statistics, Stanford University, Stanford, CA.

Barlow, W., and Azen, S. (1990). The effect of therapeutic treatment crossovers on the power of clinical trials, *Controlled Clinical Trials*, **11**, 314–326.

Barton, D. E., and David, F. N. (1959). Combinatorial extreme value distributions, *Mathematika*, **6**, 63–76.

Batschelet, E. (1960). Über eine Kontingenztafel mit fehlenden Daten, *Biometrische Zeitschrift,* **2,** 236–243.

Beaulieu, N. C. (1991). On the generalized multinomial distribution, optimal multinomial detectors and generalized weighted partial decision detectors, *IEEE Transactions on Communications*, **39**, 193–194. (Comment by E. K. Al-Hussaini, p. 195).

Bechhofer, R. E., Elmaghraby, S., and Morse, N. (1959). A single-sample multiple-decision procedure for selecting the multinomial event which has the highest probability, *Annals of Mathematical Statistics*, **30**, 102–119.

Bennett, B. M., and Nakamura, E. (1968). Percentage points of the range from a symmetric multinomial distribution, *Biometrika,* **55,** 377–379.

Bennett, R. W. (1962). Size of the χ^2 test in the multinomial distribution, *Australian Journal of Statistics,* **4,** 86–88.

Bernoulli, J. (1713). *Ars Conjectandi*, Basilea: Thurnisius.

Bhat, B. R., and Kulkarni, N. V. (1966). On efficient multinomial estimation, *Journal of the Royal Statistical Society, Series B,* **28,** 45–52.

Bhattacharya, B., and Nandram, B. (1996). Bayesian inference for multinomial populations under stochastic ordering, *Journal of Statistical Computation and Simulation*, **54**, 145–163.

Bienaymé, I. J. (1838). Mémoire sur la probabilité des résultats moyens des observations; demonstration directe de la régle de Laplace, *Mémoires de l'Académie de Sciences de l'Institut de France, Paris, Series Étrangers*, **5**, 513–558.

Boardman, T. J., and Kendell, R. J. (1970). Tables of exact and approximate values of certain ratios and reciprocals of multinomial random variables useful in life testing, *Technometrics*, **12,** 901–908.

Boland, P. J., and Proschan, F. (1987). Schur convexity of the maximum likelihood function for the multivariate hypergeometric and multinomial distributions, *Statistics & Probability Letters*, **5**, 317–322.

Bol'shev, L. N. (1965). On a characterization of the Poisson distribution, *Teoriya Veroyatnostei i ee Primeneniya*, **10**, 446–456.

Botha, L. M. (1992). Adaptive estimation of multinomial probabilities with application in the analysis of contingency tables, M.Sc. thesis, Potchefstroom University for CHE (see also *South African Statistical Journal*, **27**, 221, for abstract).

Brockmann, M., and Eichenauer-Hermann, J. (1992). Gamma-minimax estimation of multinomial probabilities, *Zastosowania Matematyki*, **21**, 427–435.

Bromaghin, J. E. (1993). Sample size determination for interval estimation of multinomial probability, *The American Statistician,* **47,** 203–206.

Brown, C. C., and Muentz, L. R. (1976). Reduced mean square error estimation in contingency tables, *Journal of the American Statistical Association*, **71**, 176–182.

Brown, M., and Bromberg, J. (1984). An efficient two-stage procedure for generating random variables for the multinomial distribution, *The American Statistician,* **38,** 216–219.

Catalano, P., Ryan, L., and Scharfstein, D. (1994). Modeling fetal death and malformation in developmental toxicity studies, *Risk Analysis*, **14**, 629–637.

Chamberlain, G. (1987). Asymptotic efficiency in estimation with conditional moment restrictions, *Journal of Econometrics*, **34**, 305–324.

Chan, L., Eichenauer-Hermann, J., and Lehn, J. (1989). Gamma-minimax estimators for the parameters of a multinomial distribution, *Zastosowania Matematyki*, **20**, 561–564.

Cheng Ping (1964). Minimax estimates of parameters of distributions belonging to the exponential family, *Acta Mathematica Sinica*, **5**, 277–299.

Cook, G. W., Kerridge, D. F., and Price, J. D. (1974). Estimation of functions of a binomial parameter, *Sankhyā, Series A,* **36,** 443–448.

Dagpunar, J. (1988). *Principles of Random Number Generation*, Oxford, England: Clarendon Press.

Davis, C. S. (1993). The computer generation of multinomial random variates, *Computational Statistics & Data Analysis*, **16**, 205–217.

Davis, C. S., and Jones, H. A. (1992). Maximum likelihood estimation for the multinomial distribution, *Teaching Statistics*, **14**(3), 9–11.

de Moivre, A. (1730). *Miscellanea Analytica de Seriebus et Quadratis*, London: Tomson and Watts.

de Montmort, P. R. (1708). *Essay d'Analyse sur les Jeux de Hazard*, Paris: Quillan (published anonymously). Second augmented edition, Quillan, Paris (1913) (published anonymously). Reprinted by Chelsea, New York (1980).

Devroye, L. (1986). *Non-Uniform Random Variate Generation*, New York: Springer-Verlag.

Dinh, K. T., Nguyen, T. T., and Wang, Y. (1996). Characterizations of multinomial distributions based on conditional distributions, *International Journal of Mathematics and Mathematical Sciences*, **19**, 595–602.

Duffin, R. J., Peterson, E. L., and Zener, C. M. (1967). *Geometric Programming*, New York: John Wiley & Sons.

Edwards, A. W. F. (1960). The meaning of binomial distribution, *Nature*, **186**, 1074.

Feller, W. (1957). *An Introduction to Probability Theory and Its Applications*, New York: John Wiley & Sons.

Feller, W. (1968). *An Introduction to Probability Theory and Its Applications*, third edition, New York: John Wiley & Sons.

Fienberg, S. E., and Holland, P. W. (1973). Simultaneous estimation of multinomial cell probabilities, *Journal of the American Statistical Association,* **68,** 683–689.

Fienberg, S. E., Neter, J., and Leitch, R. A. (1978). The total overstatement error in accounting populations, *Journal of the American Statistical Association*, **73**, 297–302.

Finucan, H. M. (1964). The mode of a multinomial distribution, *Biometrika,* **51,** 513–517.

Fisher, R. A. (1936). Has Mendel's work been rediscovered? *Annals of Science*, **1**, 115–137.

Fisz, M. (1964). The limiting distributions of the multinomial distribution, *Studia Mathematica,* **14,** 272–275.

Fitzpatrick, S., and Scott, A. (1987). Quick simultaneous confidence intervals for multinomial proportions, *Journal of the American Statistical Association*, **82**, 875–878.

Freeman, P. R. (1979). Exact distribution of the largest multinomial frequency, *Applied Statistics*, **28**, 333–336.

Froggatt, P. (1970). Application of discrete distribution theory to the study of noncommunicable events in medical epidemiology, in *Random Counts in Scientific Work,* Vol. 2 (Ed. G. P. Patil), pp. 15–40, Dordrecht: Reidel.

Gelfand, A. E., Glaz, J., Kuo, L., and Lee, T. M. (1992). Inference for the maximum cell probability under multinomial sampling, *Naval Research Logistics*, **39**, 97–114.

Geppert, M. P. (1961). Erwartungstreue plausibelste Schätzen aus dreieckig gestützten Kontingenztafeln, *Biometrische Zeitschrift,* **3,** 55–67.

Gerstenkorn, T. (1977). Multivariate doubly truncated discrete distributions, *Acta Universitatis Lodziensis*, 1–115.

Ghosh, J. K., Sinha, B. K., and Sinha, B. K. (1977). Multivariate power series distributions and Neyman's properties for multinomials, *Journal of Multivariate Analysis*, **7**, 397–408.

Girshick, M. A., Mosteller, F., and Savage, L. J. (1946). Unbiased estimates for certain binomial sampling problems with applications, *Annals of Mathematical Statistics,* **17,** 13–23.

Glaz, J. (1990). A comparison of Bonferroni-type and product-type inequalities in the presence of dependence, in *Topics in Statistical Independence* (Eds. H. W. Block, A. R. Sampson, and T. H. Savits), pp. 223–235, Hayward, CA: Institute of Mathematical Statistics.

Glaz, J., and Johnson, B. (1984). Probability for multivariate distributions with dependence structures, *Journal of the American Statistical Association*, **79**, 436–440.

Good, I. J. (1965). *The Estimation of Probabilities,* Cambridge, MA: MIT Press.

Good, I. J. (1967). A Bayesian significance test for multinomial distributions, *Journal of the Royal Statistical Society, Series B,* **29,** 399–431.

Good, I. J. (1976). On the application of symmetric Dirichlet distributions and their mixtures to contingency tables, *Annals of Statistics*, **4**, 1159–1189.

Good, I. J., and Crook, J. F. (1974). The Bayes/non-Bayes compromise and the multinomial distribution, *Journal of the American Statistical Association*, **69**, 711–720.

Goodman, L. A. (1965). On simultaneous confidence intervals for multinomial proportions, *Technometrics,* **7,** 247–254.

Greenwood, R. E., and Glasgow, M. O. (1950). Distribution of maximum and minimum frequencies in a sample drawn from a multinomial distribution, *Annals of Mathematical Statistics,* **21,** 416–424.

Griffiths, R. C. (1974). A characterization of the multinomial distribution, *Australian Journal of Statistics,* **16,** 53–56.

Guldberg, S. (1935). Recurrence formulae for the semi-invariants of some discontinuous frequency functions of n variables, *Skandinavisk Aktuarietidskrift,* **18,** 270–278.

Gupta, S. S., and Nagel, K. (1966). On selection and ranking procedures and order statistics from the multinomial distribution, *Mimeo Series No. 70,* Department of Statistics, Purdue University, West Lafayette, IN.

Gupta, S. S., and Nagel, K. (1967). On selection and ranking procedures and order statistics from the multinomial distribution, *Sankhyā, Series B*, **29**, 1–34.

Gupta, S. S., and Panchapakesan, S. (1979). *Multiple Decision Procedures: Theory and Methodology of Selecting and Ranking Populations*, New York: John Wiley & Sons.

Hald, A. (1990). *A History of Probability and Statistics and Their Applications before 1750*, New York: John Wiley & Sons.

Harvey, A. C., Fernandes, C., Wecker, W. E., and Winkler, R. L. (1989). Time series models for count or qualitative observations, *Journal of Business and Economics*, **7**, 407–422.

He, K. (1990). An ancillary paradox in the estimation of multinomial probabilities, *Journal of the American Statistical Association*, **85**, 824–828.

Ho, F. C. M., Gentle, J. E., and Kennedy, W. J. (1979). Generation of random variates from the multinomial distribution, *Proceedings of the American Statistical Association Statistical Computing Section*, pp. 336–339.

Hoadley, B. (1969). The compound multinomial distribution, *Journal of the American Statistical Association*, **64**, 216–229.

Hocking, R. R., and Oxspring, M. H. (1971). Maximum likelihood estimation with incomplete multinomial data, *Journal of the American Statistical Association,* **66,** 65–70.

Hodges, J. L., and Lehmann, E. L. (1950). Some problems in minimax estimation, *Annals of Mathematical Statistics*, **21**, 182–197.

Hoel, P. G. (1938). On the chi-square distribution for small samples, *Annals of Mathematical Statistics*, **9**, 158–165.

Holst, L. (1979). On testing fit and estimating the parameters in complete or incomplete Poisson or multinomial samples, Unpublished manuscript, Department of Mathematics, Uppsala University, Uppsala, Sweden.

Huschens, S. (1990). On upper bounds for the characteristic values of the covariance matrix for multinomial, Dirichlet and multivariate hypergeometric distributions, *Statistische Hefte*, **31**, 155–159.

Hutcheson, K., and Shenton, L. R. (1974). Some moments of an estimate of Shannon's measure of information, *Communications in Statistics,* **3,** 89–94.

Ishii, G., and Hayakawa, R. (1960). On the compound binomial distribution, *Annals of the Institute of Statistical Mathematics*, **12**, 69–80. Errata, **12**, 208.

Itoh, S. (1988). Universal modeling and coding for continuous source, *IEEE International Symposium on Information Theory*, Kobe, Japan. (Abstracts, Paper No. 25).

Janardan, K. G. (1974). Characterization of certain discrete distributions, in *Statistical Distributions in Scientific Work—3* (Eds. G. P. Patil, S. Kotz, and J. K. Ord), pp. 359–364, Dordrecht: Reidel.

Jogdeo, K., and Patil, G. P. (1975). Probability inequalities for certain multivariate discrete distributions, *Sankhyā, Series B,* **37,** 158–164.

Johnson, N. L. (1960). An approximation to the multinomial distribution: some properties and applications, *Biometrika,* **47,** 93–102.

Johnson, N. L., and Kotz, S. (1972). *Distributions in Statistics—Continuous Multivariate Distributions*, New York: John Wiley & Sons.

Johnson, N. L., Kotz, S., and Kemp, A. W. (1992). *Univariate Discrete Distributions*, second edition, New York: John Wiley & Sons.

Johnson, N. L., and Tetley, H. (1950). *Statistics: An Intermediate Text Book,* Vol. II, London: Cambridge University Press.

Johnson, N. L., and Young, D. H. (1960). Some applications of two approximations to the multinomial distribution, *Biometrika,* **47,** 463–469.

Kanoh, S., and Li, Z. D. (1990). A method of exploring the mechanism of inflationary expectations based on qualitative survey data, *Journal of Business and Economics*, **8**, 395–403.

Kelleher, J. K., and Masterson, J. M. (1992). Model equations for condensation, biosynthesis using stable isotopes and radio isotopes, *American Journal of Physiology*, **262**, 118–125.

Kemp, C. D. (1986). A modal method for generating binomial variables, *Communications in Statistics—Theory and Methods,* **15,** 805–815.

Kemp, C. D., and Kemp, A. W. (1987). Rapid generation of frequency tables, *Applied Statistics,* **36,** 277–282.

Kesten, H., and Morse, N. (1959). A property of the multinomial distribution, *Annals of Mathematical Statistics*, **30**, 120–127.

Khatri, C. G., and Mitra, S. K. (1968). Some identities and approximations concerning positive and negative multinomial distributions, *Technical Report 1/68,* Indian Statistical Institute, Calcutta, India.

Koellner, M. (1991). Best estimates of exponential and decay parameters and the design of single-photon-counting experiments, *Proceedings of 7th International Symposium on Ultrafast Processes in Spectroscopy*, Bayreuth, Germany.

Kozelka, R. M. (1956). Approximate upper percentage points for extreme values in multinomial sampling, *Annals of Mathematical Statistics*, **27**, 507–512.

Lancaster, H. O. (1961). Significance tests in discrete distributions, *Journal of the American Statistical Association*, **56**, 223–224. (Errata, **57**, 919.)

Lancaster, H. O. (1966). Forerunners of the Pearson χ^2, *Australian Journal of Statistics*, **8**, 117–126.

Lancaster, H. O., and Brown, T. A. I. (1965). Size of the χ^2 test in the symmetrical multinomials, *Australian Journal of Statistics,* **7,** 40–44.

Leonard, T. (1973). A Bayesian method for histograms, *Biometrika*, **60**, 297–308.

√ Leonard, T. (1976). Some alternative approaches to multi-parameter estimation, *Biometrika,* **63,** 69–75.

Levin, B. (1981). A representation for the multinomial cumulative distribution function, *Annals of Statistics*, **9**, 1123–1126.

Lewontin, R. C., and Prout, T. (1956). Estimation of the number of different classes in a population, *Biometrics,* **12,** 211–223.

Loukas, S., and Kemp, C. D. (1983). On computer sampling from trivariate and multivariate discrete distributions, *Journal of Statistical Computation and Simulation*, **17**, 113–123.

Lundberg, O. (1940). *On Random Processes and Their Application to Sickness and Accident Statistics*, Uppsala, Sweden: Almqvist and Wiksell.

Lyons, N. I., and Hutcheson, K. (1996). Algorithm AS 303. Generation of ordered multinomial frequencies, *Applied Statistics*, **45**, 387–393.

Mallows, C. L. (1968). An inequality involving multinomial probabilities, *Biometrika*, **55**, 422–424.

Manly, B. F. J. (1974). A model for certain types of selection expression, *Biometrics*, **30**, 281–294.

Marzo, J. L., Fabregat, R., Domingo, J., and Sole, J. (1994). Fast calculation of the CAC algorithm using the multinomial distribution function, *Tenth U.K. Teletraffic Symposium on Performance Engineering in Telecommunications*, London, U.K.

Mateev, P. (1978). On the entropy of the multinomial distribution, *Theory of Probability and Its Applications*, **23**, 188–190.

Mendenhall, W., and Lehmann, E. H. (1960). An approximation to the negative moments of the positive binomial useful in life testing, *Technometrics*, **2**, 233–239.

Moran, P. A. P. (1973). A central limit theorem for exchangeable variates with geometric applications, *Journal of Applied Probability*, **10**, 837–846.

Morel, J. G., and Nagaraj, N. K. (1993). A finite mixture distribution for modelling multinomial extra variables, *Biometrika*, **80**, 363–371.

Mosimann, J. E. (1962). On the compound multinomial distribution, the multivariate β-distribution and correlations among proportions, *Biometrika*, **49**, 65–82.

Mosimann, J. E. (1963). On the compound negative multinomial distribution and correlations among inversely sampled pollen counts, *Biometrika*, **50**, 47–54.

Muhamedanova, R., and Suleimanova, M. (1961). Unbiased estimation for selection problems in the case of multinomial distributions, *Proceedings of the Institute of Mathematics, AN UzSSR*, **22**, 121–129 (in Russian).

Neter, J., Leitch, R. A., and Fienberg, S. E. (1978). Dollar-unit sampling: Multinomial bounds for total overstatement and understatement errors, *The Accounting Review*, **53**, 77–93.

Neyman, J. (1963). Certain chance mechanisms involving discrete distributions (Inaugural Address), *Proceedings of the International Symposium on Discrete Distributions, Montreal*, in *Classical and Contagious Discrete Distributions* (Ed. G. P. Patil), pp. 4–14, Calcutta: Calcutta Statistical Publishing Society.

Olbrich, E. (1965). Geometrical interpretation of the multinomial distribution and some conclusions, *Biometrische Zeitschrift*, **7**, 96–101.

Olkin, I. (1972). Monotonicity properties of Dirichlet integrals with applications to the multinomial distribution and the analysis of variance test, *Biometrika*, **59**, 303–307.

Olkin, I., and Sobel, M. (1965). Integral expressions for tail probabilities of the multinomial and the negative multinomial distributions, *Biometrika*, **52**, 167–179.

Oluyede, B. (1994). Estimation and testing of multinomial parameters under order restrictions, *Proceedings of the ASA Section on Quality and Productivity*, pp. 276–280.

Panaretos, J. (1983). An elementary characterization of the multinomial and the multivariate hypergeometric distributions, in *Stability Problems for Stochastic Models* (Eds. V. V. Kalashnikov and V. M. Zolotarev), Lecture Notes in Mathematics, No. 982, pp. 156–164, Berlin: Springer-Verlag.

Patel, S. R. (1979). Optimum test for equality of two parameters in multinomial distributions, *Pure and Applied Mathematical Sciences*, **10**, 101–103.

Patil, G. P., and Bildikar, S. (1967). Multivariate logarithmic series distribution as a probability model in population and community ecology and some of its statistical properties, *Journal of the American Statistical Association*, **62**, 655–674.

Patil, G. P., and Joshi, S. W. (1968). *A Dictionary and Bibliography of Discrete Distributions*, Edinburgh: Oliver and Boyd.

Patil, G. P., and Seshadri, V. (1964). Characterization theorems for some univariate probability distributions, *Journal of the Royal Statistical Society, Series B*, **26,** 286–292.

Pearson, E. S., and Hartley, H. O. (Eds.) (1973). *Biometrika Tables*, third edition, London: Cambridge University Press.

Pearson, K. (1900). On the criterion that a given system of deviations from the probable in the case of a correlated system of variables is such that it can be reasonably be supposed to have arisen from random sampling, *Philosophical Magazine, 5th Series,* **50,** 157–175.

Price, G. B. (1946). Distributions derived from the multinomial expansion, *American Mathematical Monthly*, **53**, 59–74.

Quesenberry, C. P., and Hurst, D. C. (1964). Large sample simultaneous confidence intervals for multinomial proportions, *Technometrics*, **6,** 191–195.

Radons, G., Becker, J. D., Dülfer, B., and Krüger, J. (1994). Analysis, classification, and coding of multielectrode spike trains with hidden Markov models, *Biological Cybernetics*, **71**, 359–373.

Rao, C. R. (1973). *Linear Statistical Inference and Its Applications*, second edition, New York: John Wiley & Sons.

Rao, C. R., Sinha, B. V., and Subramanyam, K. (1981). Third order efficiency of the maximum likelihood estimator in the multinomial distribution, *Technical Report 81-21,* Department of Mathematics and Statistics, University of Pittsburgh, Pittsburgh, PA.

Rao, C. R., and Srivastava, R. C. (1979). Some characterizations based on multivariate splitting model, *Sankhyā, Series A*, **41**, 121–128.

Rinott, Y. (1973). Multivariate majorization and rearrangement inequalities with some applications to probability and statistics, *Israel Journal of Mathematics*, **15**, 60–77.

Rjauba, B. (1958). Asymptotic laws for the trinomial distribution, *Vilnius Universitet, Mosklo Akademijos Darbai, Series P,* **25,** 17–32 (in Russian).

Ronning, G. (1982). Characteristic values and triangular factorization of the covariance matrix for multinomial, Dirichlet and multivariate hypergeometric distributions, and some related results, *Statistische Hefte*, **23**, 152–176.

Rüst, H. (1965). Die Momente der Testgrosschen χ^2- tests, *Zeitschrift für Wahrscheinlichkeitsrechnung und Verwandte Gebiete,* **4,** 222–231.

Ryan, L. (1992). Quantitative risk assessment for developmental toxicity, *Biometrics*, **48**, 163–174.

Seber, G. A. F. (1982). *The Estimation of Animal Abundance*, second edition, New York: Macmillan.

Shah, B. K. (1975). Application of multinomial distribution to clinical trial, *Gujarat Statistical Review,* **2**(2), 19–23.

Shanawany, M. R. El. (1936). An illustration of the accuracy of the χ^2-approximation, *Biometrika,* **28,** 179–187.

Shanbhag, D. N., and Basawa, I. V. (1974). On a characterization property of the multinomial distribution, *Trabajos de Estadística y de Investigaciones Operativas,* **25,** 109–112.

Sinha, B. K., and Gerig, T. M. (1985). On Neyman's conjecture: A characterization of the multinomials, *Journal of Multivariate Analysis*, **16**, 440–450.

Sinoquet, H., and Bonhomme, R. (1991). A theoretical analysis of radiation interception in a two-species canopy, *Mathematical Biosciences*, **105**, 23–31.

Sison, C. P., and Glaz, J. (1993). Simultaneous confidence intervals and sample size determination for multinomial proportions, *Technical Report No. 93-05*, Department of Statistics, University of Connecticut.

Sison, C. P., and Glaz, J. (1995). Simultaneous confidence intervals and sample size determination for multinomial proportions, *Journal of the American Statistical Association*, **90**, 366–369.

Sobel, M., Uppuluri, V. R. R., and Frankowski, K. (1977). Dirichlet distribution—Type 1, *Selected Tables in Mathematical Statistics*—**4**, Providence, Rhode Island: American Mathematical Society.

Sobel, M., Uppuluri, V. R. R., and Frankowski, K. (1985). Dirichlet integrals of Type 2 and their applications, *Selected Tables in Mathematical Statistics*—**9**, Providence, Rhode Island: American Mathematical Society.

Steinhaus, H. (1957). The problem of estimation, *Annals of Mathematical Statistics*, **28**, 633–648.

Stoka, M. I. (1966). Asupra functiei de repartite a unei repartitii multinomiale, *Studii si Cercetari Matematice,* **18,** 1281–1285 (in Rumanian).

Stone, M. E. (1974). Cross-validation and multinomial prediction, *Biometrika*, **61**, 509–515.

Struder, H. (1966). Prüfung der exakten χ^2 Verteilung durch die stetige χ^2 Verteilung, *Metrika,* **11,** 55–78.

Sugg, M. N., and Hutcheson, K. (1972). Computational algorithm for higher order moments of multinomial type distributions, *Journal of Statistical Computation and Simulation,* **1,** 149–155.

Sutherland, M., Holland, P. W., and Fienberg, S. E. (1975). Combining Bayes and frequency approaches to estimate a multinomial parameter, in *Studies in Bayesian Econometrics and Statistics* (Eds. S. E. Fienberg and A. Zellner), pp. 585–617, Amsterdam: North-Holland.

Tanabe, K., and Sagae, M. (1992). An exact Cholesky decomposition and the generalized inverse of the variance–covariance matrix of the multinomial distribution, with applications, *Journal of the Royal Statistical Society, Series B*, **54**, 211–219.

Terza, J. V., and Wilson, P. W. (1990). Analyzing frequencies of several types of events: A multivariate multinomial–Poisson approach, *Review of Economics and Statistics*, **72**, 108–115.

Thomas, H., and Lohaus, A. (1993). Modeling growth and individual differences in spatial tasks, *Monographs on Social Research and Child Development*, **58**, 1–19.

Thompson, S. K. (1987). Sample size for estimating multinomial proportions, *The American Statistician,* **41,** 42–46.

Tortora, R. D. (1978). A note on sample size estimation for multinomial populations, *The American Statistician*, **32**, 100–102.

Tracy, D. S., and Doss, D. C. (1980). A characterization of multivector multinomial and negative multinomial distributions, in *Multivariate Statistical Analysis* (Ed. R. P. Gupta), pp. 281–289, Amsterdam: North-Holland.

Trybula, S. (1958). Some problems in simultaneous minimax estimation, *Annals of Mathematical Statistics*, **29**, 245–253.

Trybula, S. (1962). On the minimax estimation of the parameters in a multinomial distribution, *Selected Translations in Mathematical Statistics and Probability*, pp. 225–238, Providence, Rhode Island: American Mathematical Society.

Trybula, S. (1985). Some investigations in minimax estimation, *Dissertationes Mathematicae*, **240**.

Trybula, S. (1991). Systematic estimation of parameters of multinomial and multivariate hypergeometric distribution, *Statistics*, **22**, 59–67.

Tsui, K.-W., Matsumara, E. M., and Tsui, K.-L. (1985). Multinomial-Dirichlet bounds for dollar-unit sampling in auditing, *The Accounting Review*, **60**, 76–96.

Van Putten, W. L., Kortboyer, J., Bolhuis, R. L., and Gratana, J. W. (1993). Three-marker phenotypic analysis of lymphocytes based on two-color immunofluorescence using a multinomial model for flow cytometric counts and maximum likelihood estimation, *Cytometry*, **14**, 179–187.

Viana, M. A. G. (1994). Bayesian small-sample estimation of misclassified multinomial data, *Biometrics*, **50**, 237–243.

Vora, S. A. (1950). *Bounds on the distribution of chi-square,* Ph.D. dissertation, Department of Statistics, University of North Carolina, Chapel Hill, NC.

Walley, P. (1991). *Statistical Reasoning with Imprecise Probabilities*, London: Chapman and Hall.

Walley, P. (1996). Inferences from multinomial data: Learning about a bag of marbles (with discussion), *Journal of the Royal Statistical Society, Series B*, **58**, 3–57.

Wang, Y. H., and Yang, Z. (1995). On a Markov multinomial distribution, *The Mathematical Scientist*, **20**, 40–49.

Watson, G. S. (1996). Spectral decomposition of the covariance matrix of a multinomial, *Journal of the Royal Statistical Society, Series B*, **58**, 289–291.

Wilczyński, M. (1985). Minimax estimation for the multinomial and multivariate hypergeometric distributions, *Sankhyā, Series A*, **47**, 128–132.

Wilks, S. S. (1962). *Mathematical Statistics*, New York: John Wiley & Sons.

Wise, M. E. (1963). Multinomial probabilities and the χ^2 and X^2 distributions, *Biometrika,* **50,** 145–154.

Wise, M. E. (1964). A complete multinomial distribution compared with the X^2 distribution and an improvement on it, *Biometrika,* **51,** 277–281.

Wishart, J. (1949). Cumulants of multivariate multinomial distributions, *Biometrika*, **36**, 47–58.

Wolfowitz, J. (1947). The efficiency of sequential estimation and Wald's equation for sequential processes, *Annals of Mathematical Statistics,* **18,** 215–230.

Young, D. H. (1961). Quota fulfilment using unrestricted random sampling, *Biometrika*, **48**, 333–342.

Young, D. H. (1962). Two alternatives to the standard χ^2 test of the hypothesis of equal cell frequencies, *Biometrika,* **49,** 107–110.

Young, D. H. (1967). A note on the first two moments of the mean deviation of the symmetrical multinomial distribution, *Biometrika,* **54,** 312–314.

CHAPTER 36

Negative Multinomial and Other Multinomial-Related Distributions

1 DEFINITION

Recall that the binomial distribution has the probability generating function $(q + pt)^n$ with $q+p = 1$ ($p > 0$ and $n > 0$) [cf. Chapter 3, Eq. (3.2)], but the negative binomial distribution has the probability generating function $(Q-Pt)^{-N}$ with $Q-P = 1$ ($P > 0$ and $N > 0$) [cf. Chapter 5, Eq. (5.2)]. Note that while it is possible to define *two* random variables (X and $n - X$) in the binomial case, there is only a *single* random variable for the negative binomial.

When considering the construction of k-variate distributions analogous to the univariate negative binomial distributions, it is therefore natural to expect the k-variate distributions to be related to multinomial distributions with $k + 1$ (and not k) variables. Since the probability generating function of such a multinomial distribution is of the form $\left(\sum_{i=1}^{k+1} p_i t_i\right)^n$ (with $\sum_{i=1}^{k+1} p_i = 1$), we (by analogy with the univariate case) define a *k-variate negative multinomial distribution* to correspond to the probability generating function

$$\left(Q - \sum_{i=1}^{k} P_i t_i\right)^{-n} \tag{36.1}$$

with $P_i > 0$ for all i, $n > 0$ and $Q - \sum_{i=1}^{k} P_i = 1$.

From (36.1), we obtain the joint probability mass function

$$\Pr\left[\bigcap_{i=1}^{k}(N_i = n_i)\right] = P(n_1, n_2, \ldots, n_k)$$

$$= \frac{\Gamma\left(n + \sum_{i=1}^{k} n_i\right)}{\Gamma(n)\prod_{i=1}^{k} n_i!} Q^{-n} \prod_{i=1}^{k}\left(\frac{P_i}{Q}\right)^{n_i}, \qquad n_i \geq 0\,. \tag{36.2}$$

This is termed a *negative multinomial* distribution with parameters $n, P_1, P_2, \ldots, P_k$ (since $Q = 1 + \sum_{i=1}^{k} P_i$, it need not be mentioned though sometimes it is also included).

There is clearly some similarity in formal expression between the negative multinomial and multinomial distributions, just as there is between the negative binomial and binomial distributions. Perhaps this will be more easily visible if (36.2) is expressed in terms of new parameters $p_0 = Q^{-1}$, $p_i = P_i/Q$ $(i = 1, 2, \ldots, k)$ as

$$P(n_1, n_2, \ldots, n_k) = \binom{n + \sum_{i=1}^{k} n_i - 1}{n_1, n_2, \ldots, n_k, n - 1} p_0^n \prod_{i=1}^{k} p_i^{n_i} . \tag{36.3}$$

However, note that n can be fractional for negative multinomial distributions.

An alternative representation is

$$P(n_1, n_2, \ldots, n_k) = \frac{\Gamma(n + \sum_{i=1}^{k} n_i)}{n_1! n_2! \cdots n_k! \, \Gamma(n)} p_0^n \prod_{i=1}^{k} p_i^{n_i} ,$$

$$n_i = 0, 1, 2, \ldots, \quad i = 1, 2, \ldots, k, \tag{36.4}$$

where $0 < p_i < 1$ for $i = 0, 1, 2, \ldots, k$ and $\sum_{i=0}^{k} p_i = 1$ $(k > 0)$.

Olkin and Sobel (1965) showed that

$$\Pr\left[\bigcap_{i=1}^{k} (N_i \le n_i)\right] = \int_{m_k}^{\infty} \int_{m_{k-1}}^{\infty} \cdots \int_{m_1}^{\infty} f_{\boldsymbol{n}}(\boldsymbol{u}) \, du_1 \, du_2 \ldots du_k \tag{36.5}$$

and

$$\Pr\left[\bigcap_{i=1}^{k} (N_i > n_i)\right] = \int_{0}^{m_k} \int_{0}^{m_{k-1}} \cdots \int_{0}^{m_1} f_{\boldsymbol{n}}(\boldsymbol{u}) \, du_1 \, du_2 \ldots du_k , \tag{36.6}$$

where

$$f_{\boldsymbol{n}}(\boldsymbol{u}) = \frac{\Gamma\left(n + \sum_{i=1}^{k} n_i + k\right)}{\Gamma(n) \prod_{i=1}^{k} \Gamma(n_i + 1)} \frac{\prod_{i=1}^{k} u_i^{n_i}}{\left(1 + \sum_{i=1}^{k} u_i\right)^{n + \sum_{i=1}^{k} n_i + k}} , \quad u_i > 0 , \tag{36.7}$$

and $m_i = p_i/p_0$ $(i = 1, 2, \ldots, k)$.

Joshi (1975) provided an alternative (and more elementary) proof of these formulas. See also Khatri and Mitra (1968), who presented similar expressions for sums of negative multinomial probabilities similar to those described for multinomial probabilities in Chapter 35.

2 HISTORICAL REMARKS AND APPLICATIONS

Recall that (Chapter 5, Section 4) the negative binomial distribution is a compound Poisson distribution. In fact, the distribution

$$\text{Poisson}(\Theta) \bigwedge_{\Theta} \text{Gamma}(\alpha, \beta) \tag{36.8}$$

is a negative binomial distribution. If

$$p_\Theta(\theta) = \frac{1}{\Gamma(\alpha)} \beta^\alpha \theta^{\alpha-1} e^{-\beta\theta}, \qquad \theta \geq 0,\ \alpha > 0,\ \beta > 0, \tag{36.9}$$

then the probability generating function of (36.8) is

$$\frac{1}{\Gamma(\alpha)} \beta^\alpha \int_0^\infty \theta^{\alpha-1} e^{-\theta(\beta+1-t)}\, d\theta$$
$$= \beta^\alpha(\beta + 1 - t)^{-\alpha} = (1 + \beta^{-1} - \beta^{-1}t)^{-\alpha}. \tag{36.10}$$

Suppose we now consider the joint distribution of k random variables $N_1, N_2, \ldots, N_k$ which correspond to the same Θ. For a specific value of Θ, the joint probability generating function is given by

$$\prod_{i=1}^{k} e^{-\Theta(1-t_i)} = e^{-\Theta(k-\sum_{i=1}^k t_i)}. \tag{36.11}$$

Then, the joint probability generating function of $N_1, N_2, \ldots, N_k$, averaged over the distribution of Θ, is

$$\frac{1}{\Gamma(\alpha)} \beta^\alpha \int_0^\infty \theta^{\alpha-1} e^{-\theta(\beta+k-\sum_{i=1}^k t_i)} d\theta = \left(1 + k\beta^{-1} - \beta^{-1}\sum_{i=1}^{k} t_i\right)^{-\alpha}, \tag{36.12}$$

which clearly corresponds to a special form of the negative multinomial distribution [compare with (36.1)] with $P_1 = P_2 = \cdots = P_k = \beta^{-1}$. If the values of Θ for successive N's are not equal, but are in known ratios, the more general form of the negative multinomial distribution in (36.3) is obtained.

This model has been applied [see, in particular, Bates and Neyman (1952)] to represent the joint distribution of numbers of accidents suffered by the same individual in k separate periods of time. There are many other physical situations where this kind of model is appropriate; Sibuya, Yoshimura, and Shimizu (1964) presented seven examples [also see Patil (1968)]. Neyman (1963) presented a lucid account of the historical development of the negative multinomial distribution, with particular reference to its applications in models representing accident proneness data and also to spatial distribution of galaxies, military aviation, and the theory of epidemics. Sinoquet and Bonhomme (1991) used the negative multinomial distribution in the analysis of radiation interception in a two-species plant canopy. Guo (1995) discussed negative multinomial regression models for clustered event counts.

In connection with studies on industrial absenteeism, Arbous and Sichel (1954) introduced a *symmetric bivariate negative binomial distribution* with probability mass function

$$\Pr\left[\bigcap_{i=1}^{2}(X_i = x_i)\right] = \left(\frac{\theta}{\theta + 2\phi}\right)^\theta \frac{(\theta - 1 + x_1 + x_2)!}{(\theta - 1)!x_1!x_2!} \left(\frac{\phi}{\theta + 2\phi}\right)^{x_1+x_2},$$
$$x_1, x_2, = 0, 1, \ldots;\ \theta, \phi > 0, \tag{36.13}$$

which is a special case of Lundberg's (1940) bivariate negative binomial distributions. The regression of X_2 on X_1 is

$$E[X_2 \mid X_1 = x_1] = \frac{\phi}{\theta + \phi}(\theta + x_1) = \rho(\theta + x_1) \tag{36.14}$$

where $\rho = \phi/(\theta + \phi)$, and the array variance is

$$\operatorname{var}(X_2 \mid X_1 = x_1) = \frac{\theta(\theta + 2\phi)}{(\theta + \phi)^2}(\theta + x_1) = \rho(1 + \rho)(\theta + x_1)\,. \tag{36.15}$$

There are similar expressions for $E[X_1 \mid X_2 = x_2]$ and $\operatorname{var}(X_1 \mid X_2 = x_2)$. The regression functions of X_2 on X_1, and of X_1 on X_2 have the same slope, $\rho = \phi/(\theta+\phi)$. Note also that for all x_1,

$$\frac{\operatorname{var}(X_2 \mid X_1 = x_1)}{E[X_2 \mid X_1 = x_1]} = 1 + \rho\,. \tag{36.16}$$

Froggatt (1970) discussed applications of this distribution in the fields of accident frequency, doctor–patient consultations in general practice, and also one-day industrial absence. Rao *et al.* (1973) suggested these distributions as possible bases for a model of the joint distribution of numbers of boys and girls in a family. They supposed that, conditional on the total number (N) of children in a family, the numbers of boys (X_1) and girls (X_2) each have binomial ($N, \frac{1}{2}$) distributions. The joint probability mass function of X_1 and X_2 is

$$\Pr\left[\bigcap_{i=1}^{2}(X_i = x_i)\right] = \Pr[N = x_1 + x_2]\,\frac{(x_1 + x_2)!}{x_1!x_2!}\left(\frac{1}{2}\right)^{x_1+x_2}. \tag{36.17}$$

Among the special cases considered by Rao *et al.* (1973), the one they described as the "most realistic" uses the negative binomial distribution with

$$\Pr[N = n] = \binom{\theta + n - 1}{n}\omega^\theta(1 - \omega)^n, \quad n = 0, 1, \ldots, \quad 0 < \omega < 1. \tag{36.18}$$

Inserting this expression (with $n = x_1 + x_2$) in (36.17) leads to

$$\Pr\left[\bigcap_{i=1}^{2}(X_i = x_i)\right] = \binom{\theta - 1 + x_1 + x_2}{x_1 + x_2}\binom{x_1 + x_2}{x_1}\omega^\theta\left\{\frac{1}{2}(1 - \omega)\right\}^{x_1+x_2}. \tag{36.19}$$

Setting $\omega = \theta/(\theta + 2\phi)$, we arrive at (36.13). Reference may also be made to Sinha (1985).

Basu and Dhar (1995) have introduced a bivariate geometric distribution as the discrete analog of the bivariate exponential distribution of Marshall and Olkin (1967), and then extended it to define a multivariate geometric distribution. The joint probability mass function of their bivariate geometric distribution is

$$P(x_1, x_2) = \begin{cases} (p_1p_{12})^{x_1-1}p_2^{x_2-1}q_2(1 - p_1p_{12}), & x_2 < x_1 \\ p_1^{x_1-1}(p_2p_{12})^{x_2-1}q_1(1 - p_2p_{12}), & x_1 < x_2 \\ (p_1p_2p_{12})^{x_1-1}\{1 - p_2p_{12} - p_1p_{12} + p_1p_2p_{12}\}, & x_1 = x_2, \end{cases}$$

for $x_1, x_2 \geq 1$, $0 < p_1, p_2 < 1$, $0 < p_{12} \leq 1$, and $q_i = 1 - p_i$. The corresponding probability generating function is

$$\begin{aligned} &p_2q_1p_{12}(1 - p_2p_{12})t_1t_2^2(1 - p_2p_{12}t_2)^{-1}(1 - p_1p_2p_{12}t_1t_2)^{-1} \\ &\quad + p_1q_2p_{12}(1 - p_1p_{12})t_1^2t_2(1 - p_1p_{12}t_1)^{-1}(1 - p_1p_2p_{12}t_1t_2)^{-1} \\ &\quad + t_1t_2(1 - p_1p_{12} - p_2p_{12} + p_1p_2p_{12})(1 - p_1p_2p_{12}t_1t_2)^{-1}. \end{aligned}$$

This probability generating function can also be obtained from the generalized geometric distribution discussed earlier by Hawkes (1972). Basu and Dhar (1995) have also illustrated an application of this distribution in reliability theory. Marshall and Olkin (1995) similarly started with $\boldsymbol{Y} = (Y_1, Y_2, \ldots, Y_k)'$ having a multivariate exponential distribution and discussed the class of multivariate geometric distributions arising as the distribution of $\boldsymbol{X} = ([Y_1] + 1, [Y_2] + 1, \ldots, [Y_k] + 1)'$, where $[z]$ is the greatest integer $\leq z$.

Another bivariate geometric distribution that has been discussed in literature is the one with joint probability mass function

$$P(x_1, x_2) = \binom{x_1 + x_2}{x_1} p_1^{x_1}p_2^{x_2}p_0, \qquad x_2, x_2 = 0, 1, 2, \ldots,$$

where $0 < p_0, p_1, p_2 < 1$ and $p_0 + p_1 + p_2 = 1$. For this distribution, Phatak and Sreehari (1981) and Nagaraja (1983) have established some characterizations by generalizing appropriately the corresponding characteristic properties of the univariate geometric distribution (see Chapter 5, Section 9). For example, Nagaraja (1983) has shown that

$$\begin{aligned} \Pr[X_1 = x_1, X_2 = x_2] &= c_1 \Pr[X_1 = x_1 - 1, X_2 = x_2] \\ &\quad + c_2 \Pr[X_1 = x_1, X_2 = x_2 - 1], \\ &\qquad x_1, x_2 = 0, 1, 2, \ldots, (x_1, x_2) \neq (0, 0), \end{aligned}$$

where $c_1, c_2 > 0$ and $c_1 + c_2 < 1$, iff $(X_1, X_2)'$ has the above given bivariate geometric distribution.

Bivariate negative binomial distributions have been discussed in detail by Kocherlakota and Kocherlakota (1992). The bivariate negative binomial distribution with probability generating function

$$\{1 + \gamma_0 + \gamma_1 - \beta_2 - (\gamma_0 - \beta_2)t_1 - (\gamma_1 - \beta_2)t_2 - \beta_2t_1\}^{-n}$$

was introduced by Subrahmaniam and Subrahmaniam (1973), who also discussed the estimation of its parameters. Papageorgiou and Loukas (1988) derived conditional even point estimates of the parameters under the assumption that n is known. A similar discussion has been provided earlier by Papageorgiou and Loukas (1987) concerning the *bivariate negative binomial–Poisson distribution* with joint probability generating function

$$\left(Q - Pt_1\, e^{\lambda(t_2-1)}\right)^{-n},$$

where n, λ, and P are all positive, and $Q - P = 1$.

3 PROPERTIES

Due to the fact that the marginal distributions of $N_1, N_2, \ldots, N_k$ are all negative binomial, this distribution is also sometimes called a *multivariate negative binomial distribution*. The name *negative multinomial distribution* seems to be more accurately descriptive, however.

A negative multinomial distribution has a number of properties relating to conditional distributions, which are quite similar to properties of multinomial distributions. Setting $t_j = 1\ (j \neq a_1, a_2, \ldots, a_s)$ in (36.1), we obtain the joint probability generating function of $N_{a_1}, N_{a_2}, \ldots, N_{a_s}$ as

$$\left(Q - \sum_{i \neq a_j} P_i - \sum_{i=1}^{s} P_{a_i} t_{a_i}\right)^{-n}. \tag{36.20}$$

Thus, the joint distribution of $N_{a_1}, N_{a_2}, \ldots, N_{a_s}$ is a negative multinomial ($n; P_{a_1}, P_{a_2}, \ldots, P_{a_s}$) distribution. The probability mass function is [see (36.2)]

$$P(n_{a_1}, n_{a_2}, \ldots, n_{a_s}) = \frac{\Gamma\left(n + \sum_{i=1}^{s} n_{a_i}\right)}{\Gamma(n) \prod_{i=1}^{s} n_{a_i}!}\, Q'^{-n} \prod_{i=1}^{s} \left(\frac{P_{a_i}}{Q'}\right)^{n_{a_i}}, \tag{36.21}$$

where $Q' = Q - \sum_{i \neq a_j} P_i = 1 + \sum_{i=1}^{s} P_{a_i}$. Dividing now (36.2) by (36.21), we derive the conditional probability mass function of $\{N_i\}$ $(i \neq a_1, a_2, \ldots, a_s)$, given $N_{a_1}, N_{a_2}, \ldots, N_{a_s}$, as

$$P[\{n_j\}\ (j \neq a_1, a_2, \ldots, a_s) \mid n_{a_1}, n_{a_2}, \ldots, n_{a_s}]$$

$$= \frac{\Gamma\left(n + \sum_{i=1}^{s} n_{a_i} + \sum_{i \neq \{a_j\}} n_i\right)}{\Gamma\left(n + \sum_{i=1}^{s} n_{a_i}\right) \prod_{i \neq \{a_j\}} n_i!} \left(\frac{Q}{Q'}\right)^{-n-\sum_{i=1}^{s} n_{a_i}} \prod_{i \neq \{a_j\}} \left\{\frac{(P_i/Q')}{(Q/Q')}\right\}^{n_i}. \tag{36.22}$$

It is clear that this conditional distribution is negative multinomial with parameters

$$n + \sum_{i=1}^{s} n_{a_i},\ \frac{P_i}{Q'}\ (i = 1, 2, \ldots, k \text{ excluding } a_1, a_2, \ldots, a_s).$$

Combining (36.20) and (36.22), we see that the conditional joint distribution of any subset of the N_i's, given any other distinct subset, is also a negative multinomial distribution.

Tsui (1986) has listed the following structural properties of negative multinomial distributions, generalizing the results of Arbous and Sichel (1954) presented in the last section:

1. For the negative multinomial distribution (36.3), conditional on $\sum_{i=1}^k N_i = M$, $\boldsymbol{N} = (N_1, N_2, \ldots, N_k)'$ has a multinomial distribution and
$$E_{N|M}[N_i] = M\,\theta_i/\theta_\bullet \tag{36.23}$$
and
$$E_{N|M}[N_i^2] = M\,\frac{\theta_i}{\theta_\bullet}\left(1 - \frac{\theta_i}{\theta_\bullet}\right) + M^2\left(\frac{\theta_i}{\theta_\bullet}\right)^2, \tag{36.24}$$
where $E_{N|M}$ denotes expectation with respect to the conditional distribution of $\boldsymbol{N}|M$, $E[N_i] = \theta_i = np_i/p_0$, and $\theta_\bullet = \sum_{i=1}^k \theta_i = n(1-p_0)/p_0$;
2. The sum $M = \sum_{i=1}^k N_i$ has a negative binomial distribution with probability mass function
$$\Pr[M = m] = \frac{\Gamma(m+n)}{m!\Gamma(n)}\,p_0^k(1-p_0)^m, \quad m = 0, 1, 2, \ldots \tag{36.25}$$
and
$$E[M] = n(1-p_0)/p_0 = \theta_\bullet\,. \tag{36.26}$$

These properties were used by Tsui (1986) in his construction of admissible estimators of $\boldsymbol{\theta} = (\theta_1, \theta_2, \ldots, \theta_k)'$ for a large class of quadratic loss functions; see also Section 5.

Panaretos (1981) presented a characterization of negative multinomial distributions which is based on the assumption that the conditional distribution of two random vectors is multivariate inverse hypergeometric (see Chapter 39, Section 4.1). It makes use of a multivariate analog of the Rao-Rubin (1964) characterization condition [see also Shanbhag (1977)] derived earlier by Panaretos (1977). This characterization result is as follows. Let $\boldsymbol{X}$ and $\boldsymbol{Y}$ be two k-dimensional discrete random variables with

$$\Pr[\boldsymbol{Y} = \boldsymbol{r} \mid \boldsymbol{X} = \boldsymbol{n}]$$
$$= \frac{B(m + r_1 + \cdots + r_k, \rho + (n_1 - r_1) + \cdots + (n_k - r_k))}{B(m, \rho)} \prod_{i=1}^k \binom{n_i}{r_i},$$
$$r_i = 0, 1, \ldots, n_i \ (i = 1, 2, \ldots, k), \ m > 0, \ \rho > 0$$

(i.e., it is a multivariate inverse hypergeometric distribution with parameters m and ρ). Then, with $\boldsymbol{X}_{(i)} = (X_1, \ldots, X_i)'$ and $\boldsymbol{Y}_{(i)} = (Y_1, \ldots, Y_i)'$ for $i = 2, 3, \ldots, k$, the condition

$$\Pr[\boldsymbol{Y} = \boldsymbol{r}] = \Pr[\boldsymbol{Y} = \boldsymbol{r} \mid \boldsymbol{X} = \boldsymbol{Y}] = \Pr[\boldsymbol{Y} = \boldsymbol{r} \mid \boldsymbol{X}_{(i)} > \boldsymbol{Y}_{(i)}],$$
$$i = 2, 3, \ldots, k,$$

where $\boldsymbol{X}_{(i)} > \boldsymbol{Y}_{(i)}$ denotes ($X_j = Y_j$ for $j = 1, 2, \ldots, i-1$ and $X_i > Y_i$), holds iff

$$\Pr[\boldsymbol{X} = \boldsymbol{n}] = \frac{\Gamma(m + \rho + n_1 + \cdots + n_k)}{\Gamma(m + \rho)} p_0^{m+\rho} \prod_{i=1}^{k} \{p_i^{n_i}/n_i!\},$$
$$n_i = 0, 1, \ldots, \ 0 < p_i < 1, \ \sum_{i=1}^{k} p_i < 1, \ p_0 = 1 - \sum_{i=1}^{k} p_i$$

(i.e., it is a negative multinomial with parameters $m + \rho, p_1, \ldots, p_k$). Janardan (1974) has provided another characterization of the negative multinomial distribution. Panaretos (1981) has also presented characterizations for truncated forms of negative multinomial distributions.

4 MOMENTS AND CUMULANTS

Since N_i's are marginally negative binomial variables, the moments of individual N_i's are those of negative binomial (n, P_i) variables. The joint factorial moment is

$$\mu'_{(r_1, r_2, \ldots, r_k)} = E\left[\prod_{i=1}^{k} N_i^{(r_i)}\right]$$
$$= n^{\left[\sum_{i=1}^{k} r_i\right]} \prod_{i=1}^{k} P_i^{r_i}, \tag{36.27}$$

where $a^{[b]} = a(a+1)\cdots(a+b-1)$. From (36.27) it follows that the correlation coefficient between N_i and N_j $(i \neq j)$ is

$$\rho_{ij} = \sqrt{\frac{P_i P_j}{(1 + P_i)(1 + P_j)}}. \tag{36.28}$$

It is interesting to note that the correlation is positive, while for the multinomial it is negative [cf. Eq. (35.9)].

Further, the multiple regression of N_i on $N_{a_1}, N_{a_2}, \ldots, N_{a_s}$ $(i \neq a_1, a_2, \ldots, a_s)$ is

$$E\left[N_i \,\middle|\, \bigcap_{j=1}^{s} (N_{a_j} = n_{a_j})\right] = \left(n + \sum_{j=1}^{s} n_{a_j}\right) P_i/Q'. \tag{36.29}$$

From (36.29) we obtain, in particular, the regression of N_i on N_j $(i \neq j)$ as

$$E[N_i \mid N_j = n_j] = (n + n_j)P_i/(1 + P_j)\,. \tag{36.30}$$

The regressions are all linear. Observe that the common regression coefficient

$$P_i/Q' = P_i \Bigg/ \left(1 + \sum_{j=1}^{s} P_{a_j}\right)$$

decreases as the number(s) of given values $n_{a_1}, n_{a_2}, \ldots, n_{a_s}$ increases.

The moment generating function of the negative multinomial distribution is

$$\left(Q - \sum_{i=1}^{k} P_i e^{t_i}\right)^{-n}, \tag{36.31}$$

and the cumulant generating function is

$$-n \log \left(Q - \sum_{i=1}^{k} P_i e^{t_i}\right). \tag{36.32}$$

The factorial moment generating function is

$$\left(1 - \sum_{i=1}^{k} P_i t_i\right)^{-n}, \tag{36.33}$$

and the factorial cumulant generating function is

$$-n \log \left(1 - \sum_{i=1}^{k} P_i t_i\right). \tag{36.34}$$

Wishart (1949) provided formulas for several mixed cumulants (in an obvious notation) including

$$\begin{aligned}
\kappa_{...21.} &= nP_iP_j(1 + 2P_i), \quad \kappa_{.111.} = 2nP_iP_jP_g, \\
\kappa_{..31.} &= nP_iP_j\{1 + 6P_i(1 + P_i)\}, \\
\kappa_{..22.} &= nP_iP_j\{(1 + 2P_i)(1 + 2P_j) + 2P_iP_j\}, \\
\kappa_{.211.} &= 2nP_iP_jP_g(1 + 3P_i), \quad \kappa_{.1111.} = 6nP_iP_jP_gP_h\,.
\end{aligned} \tag{36.35}$$

These values can be obtained from Guldberg's (1935) formula [cf. (35.21)]

$$\kappa_{r_1,r_2,\ldots,r_{i-1},r_i+1,r_{i+1},\ldots,r_k} = P_i \frac{\partial}{\partial P_i} \kappa_{r_1,r_2,\ldots,r_i,\ldots,r_k}\,.$$

Sagae and Tanabe (1992) presented a symbolic Cholesky-type decomposition of the variance–covariance matrix of a negative multinomial distribution.

5 ESTIMATION

Given a set of observed values of $n_1, n_2, \ldots, n_k$, the likelihood equations for p_i $(i = 1, 2, \ldots, k)$ and n are

$$\frac{n_i}{\hat{p}_i} = \frac{\hat{n} + \sum_{j=1}^{k} n_j}{1 + \sum_{j=1}^{k} \hat{p}_j}, \qquad i = 1, 2, \ldots, k, \tag{36.36}$$

and

$$\sum_{j=0}^{m-1} \frac{1}{\hat{n} + j} = \log\left(1 + \sum_{j=1}^{k} \hat{p}_j\right), \tag{36.37}$$

where $m = \sum_{j=1}^{k} n_j$. From (36.36) it is clear that

$$\hat{p}_1 : \hat{p}_2 : \cdots : \hat{p}_k = n_1 : n_2 : \cdots : n_k\,. \tag{36.38}$$

Putting $c = n_i/\hat{p}_i$ we find from (36.36) that $c = \hat{n}$, using this we can rewrite (36.37) as

$$\sum_{j=0}^{m-1} \frac{1}{\hat{n} + j} = \log\left(1 + \frac{m}{\hat{n}}\right). \tag{36.39}$$

It is impossible to solve (36.39) with a positive value for $\hat{n}$ since the left-hand side is always greater than the right-hand side. The reason can be appreciated on studying the comments on the method of maximum likelihood applied to the univariate negative binomial distribution. It was, in fact, pointed out in Chapter 5 (pp. 218–219) that if $s^2 \le \bar{x}$ (where $\bar{x}$ and s^2 are the sample mean and variance, respectively) the equation for $\hat{n}$ might not be solvable. Here we are, essentially, attempting to estimate the value of n, for a negative binomial distribution with parameters n, $\sum_{i=1}^{k} p_i$, using only one observed value (m) so that necessarily $s^2 = 0$.

In order to obtain proper estimators, more than one set of observed values of $n_1, n_2, \ldots, n_k$ must be available. If t sets $\{n_{1\ell}, n_{2\ell}, \ldots, n_{k\ell}\}$, $\ell = 1, 2, \ldots, t$, are available, the likelihood equations are

$$\frac{n_i}{\hat{p}_i} = \frac{t\hat{n} + \sum_{\ell=1}^{t} m_\ell}{1 + \sum_{j=1}^{k} \hat{p}_j}, \quad i = 1, 2, \ldots, k\,, \tag{36.40}$$

and

$$\sum_{\ell=1}^{t} \sum_{j=0}^{m_\ell - 1} \frac{1}{\hat{n} + j} = t \log\left(1 + \sum_{j=1}^{k} \hat{p}_j\right), \tag{36.41}$$

where $m_\ell = \sum_{j=1}^{k} n_{j\ell}$ and $n_j = \sum_{\ell=1}^{t} n_{j\ell}$.

Arguments similar to those used in analyzing the consequences of (36.36) and (36.37) lead to the formulas

$$\hat{p}_j = \frac{n_j}{t\hat{n}} \tag{36.42}$$

and

$$\sum_{j=1}^{\infty} \frac{F_j}{\hat{n} + j - 1} = \log\left(1 + \frac{\sum_{\ell=1}^{m} m_\ell}{t\hat{n}}\right), \tag{36.43}$$

where F_j is the proportion of m_ℓ's which are greater than or equal to j. For a discussion of the existence for a solution of $\hat{n}$ in (36.43), see Chapter 5.

More details on the estimation of parameters in the bivariate case can be found in Subrahmaniam and Subrahmaniam (1973) and Kocherlakota and Kocherlakota (1992).

Tsui (1986) discussed the problem of simultaneous estimation of the means $\theta_1, \theta_2, \ldots, \theta_k$ of $N_1, N_2, \ldots, N_k$, respectively. Since N_i's are dependent, Tsui has shown that the maximum likelihood estimator of $\boldsymbol{\theta} = (\theta_1, \ldots, \theta_k)'$ which is

$$\hat{\boldsymbol{\theta}}^0(\boldsymbol{N}) = \boldsymbol{N} = (N_1, N_2, \ldots, N_k)'$$

is uniformly dominated by a class of simultaneous means estimators

$$\hat{\boldsymbol{\theta}}(\boldsymbol{N}) = (\hat{\theta}_1(\boldsymbol{N}), \hat{\theta}_2(\boldsymbol{N}), \ldots, \hat{\theta}_k(\boldsymbol{N}))'$$

under a large class of loss functions of the form

$$L_K(\boldsymbol{\theta}, \hat{\boldsymbol{\theta}}(\boldsymbol{N})) = \sum_{i=1}^{k} K(\theta_i)(\theta_i - \hat{\theta}_i(\boldsymbol{N}))^2/\theta_i\ ; \tag{36.44}$$

here, $K(\cdot)$ is a nonincreasing positive function.

Some of the improved estimators $\hat{\boldsymbol{\theta}}$ have the form

$$\hat{\boldsymbol{\theta}}(\boldsymbol{N}) = \left(1 - \frac{c}{M + b}\right)\boldsymbol{N}, \tag{36.45}$$

where $b \geq k - 1, 0 \leq c \leq 2(k - 1)$, and $M = \sum_{i=1}^{k} N_i$ as before. The estimators given by (36.45) are proposed in Clevenson and Zidek (1975) for the case of independent Poisson distributions; see Chapter 4, Section 7.2. To be more specific, suppose $\boldsymbol{N} = (N_1, N_2, \ldots, N_k)'$ has the negative multinomial distribution (36.3). Let $\hat{\boldsymbol{\theta}}^*(\boldsymbol{N})$ be an estimator of the mean vector $\boldsymbol{\theta} = (\theta_1, \theta_2, \ldots, \theta_k)'$ of $\boldsymbol{N}$ defined by

$$\hat{\boldsymbol{\theta}}^*(\boldsymbol{N}) = \left(1 - \frac{\phi(M)}{M + b}\right)\boldsymbol{N}, \tag{36.46}$$

where $M = \sum_{i=1}^{k} N_i$, $b > 0$, and the real-valued function $\phi(\cdot)$ satisfies

$$0 \leq \phi(z) \leq \min\{2b, 2(k - 1)\}, \tag{36.47}$$

where $\phi(\cdot)$ is nondecreasing and $\phi(\cdot) \not\equiv 0$. Then under the loss function (36.44), $\hat{\boldsymbol{\theta}}^*(\boldsymbol{N})$ dominates $\hat{\boldsymbol{\theta}}^0(\boldsymbol{N})$ uniformly. In the case when a simultaneous means estimation problem involves several independent negative multinomial distributions, it is convenient to resort to the so-called *multivariate negative multinomial* setting. In this case, the analogous result is valid except that the value of k in Eq. (36.47) must be replaced by $k^* = \min\{k_i\}$. Somewhat stronger results are obtained if $K(\cdot) \equiv 1$.

6 DIRICHLET (BETA-) COMPOUND NEGATIVE MULTINOMIAL DISTRIBUTION

If a negative multinomial distribution is compounded with a Dirichlet distribution, the resulting distribution is called a *Dirichlet (or beta-) compound negative multinomial distribution.* It is given by

$$\text{Negative multinomial}(n, p_1, \ldots, p_k) \bigwedge_{\left(\frac{P_i}{Q}, \frac{1}{Q}\right)} \text{Dirichlet}(\alpha_1, \alpha_2, \ldots, \alpha_{k+1}), \tag{36.48}$$

where $Q = 1 + \sum_{i=1}^k P_i$ [Mosimann (1963)]. Note that since there is no exact relationship among the P_i's, it is necessary to have $k + 1$ parameters in the Dirichlet distribution.

The probability mass function obtained from (36.48) is

$$\begin{aligned} P(n_1, n_2, \ldots, n_k) &= \frac{\Gamma(n + \sum_{i=1}^k n_i)\, \Gamma(\sum_{i=1}^{k+1} \alpha_i)\, \Gamma(n + \alpha_{k+1})}{\{\prod_{i=1}^k n_i!\}\Gamma(n)\, \Gamma(\alpha_{k+1})\, \Gamma(n + \sum_{i=1}^{k+1} \alpha_i + \sum_{i=1}^k n_i)} \prod_{i=1}^k \frac{\Gamma(n_i + \alpha_i)}{\Gamma(\alpha_i)} \\ &= \frac{n^{[\sum_{i=1}^k n_i]}}{(\sum_{i=1}^k \alpha_i)^{[n+\sum_{i=1}^k n_i]}}\, \alpha_{k+1}^{[n]} \prod_{i=1}^k \frac{\alpha_i^{[n_i]}}{n_i!}. \end{aligned} \tag{36.49}$$

If α_i's are all integers, the marginal distributions are generalized hypergeometric distributions. The properties of the Dirichlet-compound negative multinomial distribution (36.49) parallel very closely those of the Dirichlet-compound multinomial distribution discussed earlier in Chapter 35, Section 13. For example, the distribution of a subset $N_{a_1}, N_{a_2}, \ldots, N_{a_s}$ is also of the form (36.49) with parameters $n, \alpha_{a_1}, \alpha_{a_2}, \ldots, \alpha_{a_s}, \alpha_{k+1}$; the conditional distribution of the remaining N's, given $N_{a_1}, N_{a_2}, \ldots, N_{a_s}$, is also of the form (36.49) with parameters $n + \sum_{i=1}^s n_{a_i}$, $\{\alpha_j : j \neq a_i \text{ for } i = 1, 2, \ldots, s\}$, $\alpha_{k+1} + \sum_{i=1}^s \alpha_{a_i}$, and so on.

Mosimann (1963) presented a table of means, variances, covariances, and correlation coefficients. He also gave formulas for these quantities which can be used with a general compounding distribution applied to either the multinomial or the negative multinomial distributions.

7 MULTIVARIATE BERNOULLI, BINOMIAL, AND MULTINOMIAL DISTRIBUTIONS

The joint distributions of k variables $X_1, X_2, \ldots, X_k$, each of which can take only values 0 or 1 [and so has a Bernoulli distribution; see Chapter 3], are known as *multivariate Bernoulli distributions*. A standard notation for it is

$$\Pr\left[\bigcap_{i=1}^{k}(X_i = x_i)\right] = p_{x_1 x_2 \ldots x_k},\ x_i = 0, 1;\ i = 1, 2, \ldots, k\,. \tag{36.50}$$

The marginal distribution of X_i is

$$\Pr[X_i = x_i] = \sum_{x_1}\cdots\sum_{x_{i-1}}\sum_{x_{i+1}}\cdots\sum_{x_k} p_{x_1 x_2 \ldots x_k} = p_i \text{ (say)}. \tag{36.51}$$

Teugels (1990) has noted that there is a one-to-one correspondence between the probabilities in (36.50) and the integers $\xi = 1, 2, 3, \ldots, 2^k$ with

$$\xi(\boldsymbol{x}) = \xi(x_1, x_2, \ldots, x_k) = 1 + \sum_{i=1}^{k} 2^{i-1} x_i\,. \tag{36.52}$$

Writing now $p_{x_1 x_2 \ldots x_k}$ as $p_{\xi(\boldsymbol{x})}$, the vector $\boldsymbol{p}' = (p_1, p_2, \ldots, p_{2^k})$ can be interpreted as

$$\begin{aligned}\boldsymbol{p} &= E\left[\begin{pmatrix}1 - X_k\\ X_k\end{pmatrix} \otimes \begin{pmatrix}1 - X_{k-1}\\ X_{k-1}\end{pmatrix} \otimes \cdots \otimes \begin{pmatrix}1 - X_1\\ X_1\end{pmatrix}\right]\\ &= E\left[\bigotimes_{i=k}^{1}\begin{pmatrix}1 - X_i\\ X_i\end{pmatrix}\right],\end{aligned} \tag{36.53}$$

where $\otimes$ is the Kronecker product operator defined by

$$\boldsymbol{A}_{r_1\times s_1} \otimes \boldsymbol{B}_{r_2\times s_2} = \begin{pmatrix} a_{11}\boldsymbol{B} & a_{12}\boldsymbol{B} & \cdots & a_{1s_1}\boldsymbol{B}\\ \vdots & \vdots & & \vdots\\ a_{r_1 1}\boldsymbol{B} & a_{r_1 2}\boldsymbol{B} & \cdots & a_{r_1 s_1}\boldsymbol{B}\end{pmatrix}_{r_1 r_2 \times s_1 s_2}.$$

Note that $\left[\binom{a_k}{b_k} \otimes \binom{a_{k-1}}{b_{k-1}} \otimes \cdots \otimes \binom{a_1}{b_1}\right] = \prod_{i=1}^{k} a_i^{1-x_i}\, b_i^{x_i}$ $(x_i = 0, 1)$. We thus have

$$\begin{aligned}\boldsymbol{\mu} &= \left(\mu'_{\langle 1\rangle}, \mu'_{\langle 2\rangle}, \ldots, \mu'_{\langle 2^k\rangle}\right)\\ &= E\left[\begin{pmatrix}1\\ X_k\end{pmatrix} \otimes \begin{pmatrix}1\\ X_{k-1}\end{pmatrix} \otimes \cdots \otimes \begin{pmatrix}1\\ X_1\end{pmatrix}\right] = E\left[\bigotimes_{i=k}^{1}\begin{pmatrix}1\\ X_i\end{pmatrix}\right],\end{aligned} \tag{36.54}$$

where

$$\mu'_{\langle \xi(\boldsymbol{x})\rangle} = E\left[\prod_{i=1}^{k} X_i^{x_i}\right].$$

Note also that $\mu'_{\langle 1 \rangle} = 1$ because $\xi(\boldsymbol{x}) = 1$ corresponds to $x_j = 0$ for all j. Also,

$$E[X_j] = p_j \,. \tag{36.55}$$

The central moments

$$\sigma_{\langle \xi(\boldsymbol{x}) \rangle} = E\left[\prod_{i=1}^{k} (X_i - p_i)^{x_i}\right] \tag{36.56}$$

generate the so-called "dependency vector"

$$\boldsymbol{\sigma} = (\sigma_{\langle 1 \rangle}, \sigma_{\langle 2 \rangle}, \ldots, \sigma_{\langle 2^k \rangle}) \,.$$

We have

$$\begin{aligned} \boldsymbol{\sigma} &= E\left[\begin{pmatrix} 1 \\ X_k - p_k \end{pmatrix} \otimes \begin{pmatrix} 1 \\ X_{k-1} - p_{k-1} \end{pmatrix} \otimes \cdots \otimes \begin{pmatrix} 1 \\ X_1 - p_1 \end{pmatrix}\right] \\ &= E\left[\bigotimes_{i=k}^{1} \begin{pmatrix} 1 \\ X_i - p_i \end{pmatrix}\right] \end{aligned} \tag{36.57}$$

and also the relations

$$\boldsymbol{p} = \begin{pmatrix} 1 & -1 \\ 0 & 1 \end{pmatrix} \otimes^k \boldsymbol{\mu} \,, \tag{36.58}$$

$$\boldsymbol{\mu} = \begin{pmatrix} 1 & 1 \\ 0 & 1 \end{pmatrix} \otimes^k \boldsymbol{p} \,, \tag{36.59}$$

and

$$\boldsymbol{p} = \bigotimes_{j=k}^{1} \begin{pmatrix} q_j & -1 \\ p_j & 1 \end{pmatrix} \boldsymbol{\sigma} \tag{36.60}$$

and

$$\boldsymbol{\sigma} = \bigotimes_{j=k}^{1} \begin{pmatrix} 1 & 1 \\ -p_j & q_j \end{pmatrix} \boldsymbol{p} \,. \tag{36.61}$$

The probability generating function of the multivariate Bernoulli distribution in (36.50), which is

$$\sum_{x_1=0}^{1} \sum_{x_2=0}^{1} \cdots \sum_{x_k=0}^{1} p_{x_1 x_2 \ldots x_k} \prod_{i=1}^{k} t_i^{x_i} \,, \tag{36.62}$$

can be expressed in several different forms including

$$G(t_1, t_2, \ldots, t_k) = \bigotimes_{j=k}^{1} (1, t_j) \boldsymbol{p} \,, \tag{36.63}$$

$$G(t_1, t_2, \ldots, t_k) = \bigotimes_{j=k}^{1} (1, t_j - 1) \boldsymbol{\mu} \,, \tag{36.64}$$

and

$$G(t_1, t_2, \ldots, t_k) = \sum_{h=1}^{2^k} \theta_h(t_1, t_2, \ldots, t_k)\sigma_{\langle \xi(\boldsymbol{x}) \rangle}, \tag{36.65}$$

where

$$\theta_j(t_1, t_2, \ldots, t_k) = \prod_{i=1}^{k} (q_i + p_i t_i)^{1-x_i}(t_i - 1)^{x_i}$$

and $\xi(\boldsymbol{x}) = 1 + \sum_{i=1}^{k} x_i 2^{i-1}$. [Note that $t_i^{x_i} = (1 - x_i) + t_i x_i = (1, t_i)\binom{1-x_i}{x_i}$ since $x_i = 0$ or 1.]

If $\boldsymbol{X}_j = (X_{1j}, X_{2j}, \ldots, X_{kj})'$, $j = 1, 2, \ldots, n$, are n mutually independent sets of variables each having a multivariate Bernoulli distribution with the same values of $\boldsymbol{p}$ for each i, the distribution of $\boldsymbol{X}_{\bullet} \equiv (X_{1\bullet}, X_{2\bullet}, \ldots, X_{k\bullet})'$, where $X_{i\bullet} = \sum_{j=1}^{n} X_{ij}$, is a *multivariate binomial* distribution [Steyn (1975)]. The marginal distribution of $X_{i\bullet}$ is binomial (n, p_i). In fact, Doss and Graham (1975a) have provided a characterization of multivariate binomial distribution through univariate marginal distributions. Teugels (1990) provides an analysis for these distributions similar to that for the multivariate Bernoulli by using the integer representation

$$\xi(\boldsymbol{x}_{\bullet}) = 1 + \sum_{i=1}^{k} x_{i\bullet}(n+1)^{i-1}; \tag{36.66}$$

see Kemp and Kemp (1987). Lancaster (1976) has nicely demonstrated the usefulness of Kravčuk polynomials in studying the structure of the multivariate binomial distribution as well as some of its properties. Multivariate binomial models are widely applicable. They are natural choices to describe the results of experiments in which each trial results in a number, k (say), of binary (0,1) responses. If the probability of the combination $j_1, j_2, \ldots, j_k$ of responses is the same at each trial (with, of course, $\sum_{j_1=0}^{1} \cdots \sum_{j_k=0}^{1} p_{j_1 \cdots j_k} = 1$), then the numbers of responses $N_{j_1 \cdots j_k}$ in n independent trials have a joint multinomial distribution with parameters $(n; \{p_{j_1 \cdots j_k}\})$. The marginal distribution of the number of positive (1) responses for the hth characteristic is binomial (n, p_h) with

$$p_h = \sum_{j_1=0}^{1} \cdots \sum_{j_{h-1}=0}^{1} \sum_{j_{h+1}=0}^{1} \cdots \sum_{j_k=0}^{1} p_{j_1 \cdots j_{h-1} 1 j_{h+1} \cdots j_k}, \tag{36.67}$$

and their joint distribution is multivariate binomial. The 2^k-order arrays define the dependence structure. Westfall and Young (1989) have described analysis of data from a clinical trial comparing t treatments with k responses for each individual given a treatment, using multivariate binomial models of the type just described, with special reference to tests of hypotheses on the p-arrays (of which there are t) appropriate to the treatment. Particular attention is paid to the consequences of applying multiple significance tests in this context.

Matveychuk and Petunin (1990, 1991) and Johnson and Kotz (1991, 1994) have discussed a generalized Bernoulli model defined in terms of placement statistics from two random samples. These authors have also discussed applications of their results to developing tests of homogeneity for two populations.

A bivariate distribution $P(x, y)$ in a convex set C is extreme and called an *extremal distribution* if every convex decomposition

$$P(x, y) = \alpha\, P_1(x, y) + (1-\alpha) P_2(x, y), \quad 0 < \alpha < 1,$$

with $P_1(x, y)$ and $P_2(x, y)$ in C, implies

$$P_1(x, y) = P_2(x, y) = P(x, y) \text{ for every } x \text{ and } y.$$

Then, by using the extreme bivariate Bernoulli distributions given by (in an obvious notation)

$$E_1 = \begin{pmatrix} p_{1+} & 0 \\ p_{+1} - p_{1+} & 1 - p_{+1} \end{pmatrix} \qquad \text{for } p_{+1} \geq p_{1+},$$

$$E_2 = \begin{pmatrix} 0 & p_{1+} \\ p_{+1} & 1 - p_{1+} - p_{+1} \end{pmatrix} \qquad \text{for } \tfrac{1}{2} \geq p_{+1} \geq p_{1+},$$

$$E_3 = \begin{pmatrix} p_{+1} & p_{1+} - p_{+1} \\ 0 & 1 - p_{+1} \end{pmatrix} \qquad \text{for } p_{1+} \geq p_{+1},$$

and

$$E_4 = \begin{pmatrix} p_{1+} + p_{+1} - 1 & 1 - p_{+1} \\ 1 - p_{1+} & 0 \end{pmatrix} \quad \text{for } p_{1+} \geq p_{+1} \geq \frac{1}{2},$$

Oluyede (1994) generates families of bivariate binomial distributions. For example, the bivariate binomial distribution generated by E_1 has joint probability mass function

$$\begin{aligned} &\Pr_{E_1}[X = x, Y = y] \\ &\quad = \frac{n!}{x!(y-x)!(n-y)!}\, p_{1+}^{x} (p_{+1} - p_{1+})^{y-x} (1 - p_{+1})^{n-y}, \quad 0 \leq x \leq y \leq n. \end{aligned} \tag{36.68}$$

Evidently, the regression is $E[X \mid Y = y] = y\, p_{1+}/p_{+1}$ which is linear.

Similarly, the bivariate binomial distribution generated by E_1 or E_2 (with unequal marginal indices m and n) has joint probability mass function

$$\begin{aligned} &\Pr_{E_1}[X = x, Y = y] \\ &\quad = \sum_i \sum_j \frac{k!}{(x-i)!(y-x+i-j)!(k-y+j)!}\, p_{1+}^{x-i} (p_{+1} - p_{1+})^{y-x+i-j} \\ &\qquad \cdot (1 - p_{1+})^{k-y+j} \binom{n-k}{i} p_{1+}^{i} (1 - p_{1+})^{n-k-i} \\ &\qquad \cdot \binom{m-k}{j} p_{+1}^{j} (1 - p_{+1})^{m-k-j}, \quad 0 \leq x \leq y \leq k \leq \min(m, n). \end{aligned} \tag{36.69}$$

The descending factorial moment generating function of the distribution (36.69) is

$$\{1 - p_{+1} + (p_{+1} - p_{1+})t_2 + p_{1+}t_1t_2\}^k(1 - p_{1+} + p_{1+}t_1)^{n-k} \cdot (1 - p_{+1} + p_{+1}t_2)^{m-k} . \qquad (36.70)$$

The regression function is

$$E[X \mid Y = y] = (n-k)p_{1+} + y\,\frac{p_{1+}}{p_{+1}} - \left(\frac{p_{1+}}{p_{+1}}\right)\frac{\sum_{r=y-k}^{y} r\binom{m-k}{r}p_{+1}^r(1-p_{+1})^{m-k-r}}{\sum_{r=y-k}^{y}\binom{m-k}{r}p_{+1}^r(1-p_{+1})^{m-k-r}}, \qquad (36.71)$$

which reduces to $y\,p_{1+}/p_{+1}$ (as given above) when $k = m = n$.

A family of bivariate distributions introduced by Sarmanov (1966) and discussed further by Lee (1996) also needs to be mentioned here. For example, it gives rise to a family of bivariate distributions with binomial marginals as one with the joint probability mass function as

$$P(x_1, x_2) = \binom{n_1}{x_1}\binom{n_2}{x_2}\theta_1^{x_1}(1-\theta_1)^{n_1-x_1}\theta_2^{x_2}(1-\theta_2)^{n_2-x_2} \times \{1 + \omega\phi_1(x_1)\phi_2(x_2)\}, \quad x_1 = 0, 1, \ldots, n_1;\ x_2 = 0, 1, \ldots, n_2$$

[see Lee (1996, p. 1214)], where

$$\phi_i(x_i) = e^{-x_i} - \{\theta_i e^{-1} + (1-\theta_i)\}^{n_i} \qquad \text{for } i = 1, 2 .$$

A *multivariate multinomial distribution* can be constructed by replacing each X_{ij} by a multinomial variable. It is particularly suitable for representing data which can be classified in multidimensional contingency tables. Consider an m-way $h_1 \times h_2 \times \cdots \times h_m$ cross-classified contingency table, the number of factor levels for the ith classification factor being h_i. For example, in taking measurements on individuals sampled from a human population, factors of classification may be height, eye color, hair color, age, and sex. In this case, we have $m = 5$. If eye color is assessed as belonging to one of eight groups (levels), then $h_2 = 8$, while $h_5 = 2$ corresponds to the "levels" male and female. Let us denote the probability of obtaining the combination of levels $a_1, a_2, \ldots, a_m$ by $p_{a_1 a_2 \ldots a_m}$, and *assume that this is constant for all individuals selected.* Denoting the number of individuals among a sample of size n with this combination of levels by $N_{a_1 a_2 \ldots a_m}$, the marginal distribution of $N_1^{(1)}, N_2^{(1)}, \ldots, N_{h_1}^{(1)}$, where

$$N_i^{(1)} = \sum_{a_2} \cdots \sum_{a_m} N_{ia_2 \ldots a_m}$$

will be multinomial $(n; p_{i(1)}, \ldots, p_{i(h_i)})$, where

$$p_{i(1)} = \sum_{a_2} \cdots \sum_{a_m} p_{ia_2 \ldots a_m} .$$

We then naturally wish to consider the joint distribution of the m sets of variables

$$n_1^{(j)}, n_2^{(j)}, \ldots, n_{h_j}^{(j)}, \qquad j = 1, 2, \ldots, m .$$

Each set has a multinomial distribution; the joint distribution is called a *multivariate multinomial distribution*.

The joint probability generating function of the $N_{a_1 a_2 \ldots a_m}$ is

$$\left(\sum_{a_1} \sum_{a_2} \cdots \sum_{a_m} p_{a_1 a_2 \ldots a_m} t_{a_1 a_2 \ldots a_m} \right)^n \tag{36.72}$$

and the joint probability generating function of the $n_i^{(j)}$'s is

$$\left(\sum_{a_1} \sum_{a_2} \cdots \sum_{a_m} p_{a_1 a_2 \ldots a_m} t_{a_1}^{(1)} t_{a_2}^{(2)} \ldots t_{a_m}^{(m)} \right)^n \tag{36.73}$$

in an obvious notation.

Wishart (1949) presented general formulas for calculating cumulants and cross-cumulants (and hence moments and product moments) of the distribution. He devoted particular attention to the case $m = 2$, $h_1 = h_2 = 2$, which corresponds to 2×2 contingency tables.

Steyn (1963) constructed another distribution which he also referred to as a "multivariate multinomial distribution." He supposed that members of an infinite population fall into one of $k + 1$ cells. If an individual is chosen from the ith cell, it causes an event E to occur i times ($i = 0, 1, \ldots, k$). The probability generating function for the number of times event E occurs when n individuals are chosen is

$$(p_0 + p_1 t + p_2 t^2 + \cdots + p_k t^k)^n . \tag{36.74}$$

The corresponding distribution was called by Steyn (1951) a "univariate multinomial distribution." It is clearly different from the multinomial distribution discussed in the last chapter, which is a multivariate distribution. As a natural generalization of (36.74), Steyn (1963) then defined a "multivariate multinomial distribution" through the probability generating function

$$\left(p_0 + \sum_i \sum_j p_{ij} t_i^j \right)^n , \tag{36.75}$$

where

$$p_0 \geq 0, \quad p_{ij} \geq 0, \qquad p_0 + \sum_i \sum_j p_{ij} = 1$$
$$(i = 1, 2, \ldots, k;\ j = 1, 2, \ldots, n_i)$$

and n is a positive integer.

Tallis (1962, 1964) constructed a "generalized multinomial distribution"[1] as the joint distribution of n discrete variables $X_1, X_2, \ldots, X_n$, each having the same marginal distribution

$$\Pr[X_j = i] = p_i, \ i = 0, 1, \ldots, k, \ j = 1, 2, \ldots, n$$

and such that the correlation between any two different X's has a specified value ρ. This joint distribution has joint probability generating function

$$\rho \left\{ \sum_{i=0}^{k} p_i \left(\prod_{j=1}^{n} t_j \right)^i \right\} + (1 - \rho) \prod_{j=1}^{n} \left\{ \sum_{i=1}^{k} p_i t_j^i \right\} . \tag{36.76}$$

Tallis (1962) described how the parameters $p_0, p_1, \ldots, p_k$ and ρ may be estimated, based on N sets of observed values of $X_1, X_2, \ldots, X_n$. He also discussed compound distributions in which n is allowed to be a random variable with the ratios $p_0 : p_1 : p_2 \ldots$ fixed, and also with n fixed and $(p_0, p_1, \ldots, p_k)$ having a Dirichlet distribution. Some general results, with both n and $(p_0, p_1, \ldots, p_k)$ varying, were also obtained by Tallis.

A *generalized multinomial distribution* for the numbers $(N_1, N_2, \ldots, N_k)$ of observations in a set of n $(= \sum_{i=1}^k N_i)$ observations falling into interval (a_{i-1}, a_i) $(i = 1, 2, \ldots, k)$, of form

$$\Pr\left[\bigcap_{i=1}^{k} (N_i = n_i) \right] = \binom{n}{n_1, n_2, \ldots, n_k} \prod_{i=1}^{k} \prod_{j=1}^{n_i} p_{ij} , \tag{36.77}$$

was suggested by Al-Husseini (1989). Beaulieu (1991) pointed out that the factor $\binom{n}{n_1, n_2, \ldots, n_k}$ is inappropriate since the values of the p's varied. In a note on Beaulieu's paper, Al-Husseini suggested the form

$$\Pr\left[\bigcap_{i=1}^{k} (N_i = n_i) \right] = \sum_{h=1}^{J(\boldsymbol{n})} \prod_{i=1}^{k} \prod_{j=1}^{n_i} p_{ijh} , \tag{36.78}$$

where $J(\boldsymbol{n}) = \binom{n}{n_1, n_2, \ldots, n_k}$ and $\prod_{j=1}^{0} p_{ijh} = 1$.

Tracy and Doss (1980) investigated *multivector multinomial distributions*. (A "multivector distribution" is the joint distribution of a number of vectors; the distribution of each vector is "vector marginal.") In fact, these are joint multivariate multinomial distributions as defined in (36.73). They also obtained the following characterizations:

[1] Note that Tallis used the word "multinomial" to describe the *univariate* distribution of each X_i.

(i) A k-vector linear exponential distribution of a multivector $(\boldsymbol{X}_1, \ldots, \boldsymbol{X}_k)'$ defined by

$$dP_{\boldsymbol{\theta}_1,\ldots,\boldsymbol{\theta}_k}(\boldsymbol{x}_1, \ldots, \boldsymbol{x}_k) = \exp\left(\sum_{i=1}^{k} \boldsymbol{\theta}_i' \boldsymbol{x}_i\right) d\Pi(\boldsymbol{x}_1, \ldots, \boldsymbol{x}_k)/f(\boldsymbol{\theta}_1, \ldots, \boldsymbol{\theta}_k)$$

[see Doss and Graham (1975b)] is multivector multinomial distribution if and only if its vector marginals are multinomial.

(ii) Defining a k-vector negative multinomial distribution as one with probability generating function

$$\left(p_0 + \sum_{j,\alpha} a_{j\alpha} t_{j\alpha} + \sum_{i,j,\alpha,\beta} a_{ij\alpha\beta} t_{j\alpha} t_{i\beta} + \cdots + \sum_{\alpha_1,\ldots,\alpha_k} t_{1\alpha_1} t_{2\alpha_2} \cdots t_{k\alpha_k}\right)^{-n}, \tag{36.79}$$

a k-vector linear exponential distribution is multivector negative multinomial distribution if and only if its vector marginals are negative multinomial.

Panaretos and Xekalaki (1986) studied *cluster multinomial distributions* which can be generated by the following urn model: "From an urn containing balls of m different colors with balls of color i are numbered 1 to k_i $(i = 1, 2, \ldots, m)$, n balls are drawn at random with replacement." Denote the number of balls of color i in the sample that bear the number j by X_{ij}, so that

$$X_i = \sum_{j=1}^{k_i} jX_{ij}$$

represents the sum of numbers on balls of color i $(i = 1, 2, \ldots, m)$. Denoting by p_{ij} the probability that a ball of color i will bear the number j, with of course $\sum_{j=1}^{k_i} p_{ij} = 1$, the joint probability mass function of $X_1, X_2, \ldots, X_m$ is

$$\Pr\left[\bigcap_{i=1}^{m}(X_i = x_i)\right] = \sum \binom{n}{\xi_{11}, \ldots, \xi_{1k_1}, \xi_{21}, \ldots, \xi_{2k_2}, \ldots, \xi_{m1}, \ldots, \xi_{mk_m}}. \tag{36.80}$$

The distribution of the number (say Y_h) of N's equal to h when $\boldsymbol{N}$ has a multinomial distribution—that is, the numbers of cells represented h times in a sample of size n—was studied initially by Price (1946). He, however, did not study the joint distribution of $Y_1, Y_2, \ldots, Y_n$. See Charalambides (1981) (described in Chapter 39) for a discussion of the joint distribution of $Y_1, Y_2, \ldots, Y_n$ when sampling without replacement.

For $j = 1, 2, \ldots, p$, let $\boldsymbol{N}_j = (N_{j1}, N_{j2}, \ldots, N_{jk_j})'$ have a negative multinomial distribution with parameters $(k_j; p_{j0}, p_{j1}, \ldots, p_{jk_j})$, where $k_j > 0$, $p_{ji} > 0$ and $p_{j0} = 1 - \sum_{i=1}^{k_j} p_{ji} > 0$. Further, let the variables $\boldsymbol{N}_1, \boldsymbol{N}_2, \ldots, \boldsymbol{N}_p$ be mutually independent. Then the random vector $\boldsymbol{N} = (\boldsymbol{N}_1, \boldsymbol{N}_2, \ldots, \boldsymbol{N}_p)'$ is said to have a *multivariate negative multinomial distribution*. Let $\boldsymbol{\theta}_j = (\theta_{j1}, \theta_{j2}, \ldots, \theta_{jk_j})'$ be the expected value of $\boldsymbol{N}_j$, and let $\boldsymbol{\theta} = (\boldsymbol{\theta}_1, \boldsymbol{\theta}_2, \ldots, \boldsymbol{\theta}_p)'$. The problem of estimating

$\boldsymbol{\theta}$ based on $\boldsymbol{N}$ has been studied by Tsui (1986) along the lines of his estimation method for the negative multinomial distribution discussed earlier in Section 5. The corresponding loss function used in this case is [see (36.44)]

$$L_K(\boldsymbol{\theta}, \hat{\boldsymbol{\theta}}) = \sum_{j=1}^{p} \sum_{i=1}^{k_j} K(\theta_{ji})(\theta_{ji} - \hat{\theta}_{ji})^2/\theta_{ji} \,, \tag{36.81}$$

where $\hat{\boldsymbol{\theta}}(\boldsymbol{N})$ is an estimator of $\boldsymbol{\theta}$.

Cox and Wermuth (1994) introduced a *quadratic exponential binary distribution* in which

$$\Pr\left[\bigcap_{j=1}^{k}(N_j = n_j)\right] \propto \exp\left\{\sum_{j=1}^{k} \alpha_j i_j + \sum\sum_{j<j'=1}^{k} \alpha_{jj'} i_j i_{j'}\right\}, \qquad n_j = 0, 1\,, \tag{36.82}$$

where $i_j = 2n_j - 1$. They pointed out that in a sense this is the binary analogue of the multivariate normal distribution, although the analogy is not precisely correct in terms of conditional and marginal distributions. They suggested ways of approximately alleviating this drawback.

8 COUNTABLE MIXTURES OF BIVARIATE BINOMIAL DISTRIBUTIONS

Hamdan and Tsokos (1971) introduced a *bivariate binomial distribution* (actually, a *bivariate compound Poisson distribution*) with joint probability generating function

$$G(t_1, t_2 \mid n) = (p_{00} + p_{10}t_1 + p_{01}t_2 + p_{11}t_1t_2)^n, \tag{36.83}$$

where $0 < p_{00}, p_{10}, p_{01}, p_{11} < 1$ and $p_{00} = 1 - (p_{10} + p_{01} + p_{11})$, $n = 0, 1, 2, \ldots$, assuming that the exponent is a Poisson random variable. Special cases of (36.83) include *trinomial distribution* with probability generating function

$$(p_{00} + p_{10}t_1 + p_{01}t_2)^n \tag{36.84}$$

and bivariate binomial distribution with probability generating function

$$(pt_1 + qt_2)^n\,, \qquad q = 1 - p\,. \tag{36.85}$$

Application of (36.85) for random n in earlier literature are the following:

1. If n is a random variable with power series distribution, then Patil (1968) derived the *bivariate sum-symmetric power series distribution*.
2. Rao *et al.* (1973) assumed that n is a random variable representing the family (sibship) size and obtained a model for the distribution of the two genders of children in a family (see also Section 2).

Papageorgiou and David (1994) made the assumption that n is a *general* discrete random variable with probability mass function $\Pr[N = n] = P_N(n)$ and probability generating function $G_N(t)$. Then, from (36.83) we obtain

$$G(t_1, t_2) = \sum_{n=0}^{\infty} P_N(n)P(t_1, t_2 \mid n) = G_N(p_{00} + p_{10}t_1 + p_{01}t_2 + p_{11}t_1t_2)\,; \qquad (36.86)$$

see Chapter 1 for details. Then, by using the relationship between a bivariate probability generating function and the corresponding bivariate probability mass function described in Chapter 34 [cf. (34.37)], we readily have

$$P(r, s) = \frac{1}{r!s!} \left. \frac{\partial^{r+s} G(t_1, t_2)}{\partial t_1^r \partial t_2^s} \right|_{t_1=t_2=0} . \qquad (36.87)$$

Similarly, the descending factorial moment of order (r, s) is given by

$$\mu'_{(r,s)} = \left. \frac{\partial^{r+s} G(t_1, t_2)}{\partial t_1^r \partial t_2^s} \right|_{r=s=1} . \qquad (36.88)$$

Now, from (36.86) we have

$$\frac{\partial^r G(t_1, t_2)}{\partial t_1^r} = G_N^{(r)}(p_{00} + p_{10}t_1 + p_{01}t_2 + p_{11}t_1t_2)(p_{10} + p_{11}t_2)^r$$

which, when used with Leibnitz's formula

$$\frac{d^n}{dx^n}\{\phi(x)\psi(x)\} = \sum_{i=0}^{n} \binom{n}{i} \phi^{(i)}(x)\psi^{(n-i)}(x)\,,$$

yields

$$\begin{aligned} \frac{\partial^{r+s} G(t_1, t_2)}{\partial t_1^r \partial t_2^s} &= \sum_{i=0}^{\min(r,s)} i! \binom{s}{i} G_N^{(r+s-i)}(p_{00} + p_{10}t_1 + p_{01}t_2 + p_{11}t_1t_2) \\ &\quad \times (p_{01} + p_{11}t_1)^{s-i} \binom{r}{i} (p_{10} + p_{11}t_2)^{r-i} p_{11}^i\,. \end{aligned} \qquad (36.89)$$

In the above equations, $G_N^{(k)}$ is the kth derivative of G_N. Equation (36.89), when used in (36.87), readily gives the joint probability mass function as

$$P(r, s) = \sum_{i=0}^{\min(r,s)} \frac{1}{(r-i)!(s-i)!i!} G_N^{(r+s-i)}(p_{00}) p_{10}^{r-i} p_{01}^{s-i} p_{11}^i\,. \qquad (36.90)$$

Similarly, when Eq. (36.89) is used in (36.88), it gives the descending factorial moment of order (r, s) as

$$\mu'_{(r,s)} = \sum_{i=0}^{\min(r,s)} i! \binom{r}{i} \binom{s}{i} G_N^{(r+s-i)}(1) p_{1\bullet}^{r-i} p_{\bullet 1}^{s-i} p_{11}^i\,, \qquad (36.91)$$

where $p_{1\bullet} = p_{10} + p_{11}$ and $p_{\bullet 1} = p_{01} + p_{11}$. Since $G_N^{(r+s-i)}(1)$ is the descending factorial moment of order $r + s - i$ of the random variable N, (36.91) can be rewritten as

$$\mu'_{(r,s)} = \sum_{i=0}^{\min(r,s)} i!\binom{r}{i}\binom{s}{i}\mu'_{(r+s-i)}(N)p_{1\bullet}^{r-i}p_{\bullet 1}^{s-i}p_{11}^{i}\,. \tag{36.92}$$

We further observe that the bivariate factorial cumulant generating function is

$$\begin{aligned}
K^*(t_1, t_2) &= \ln G(t_1 + 1, t_2 + 1) \\
&= \ln G_N\left(p_{00} + p_{10}(t_1 + 1) + p_{01}(t_2 + 1) + p_{11}(t_1 + 1)(t_2 + 1)\right) \\
&\qquad\qquad\qquad\qquad\qquad\qquad\qquad\qquad\qquad\qquad \text{[from (36.86)]} \\
&= \ln G_N\left((p_{10} + p_{11})t_1 + (p_{01} + p_{11})t_2 + p_{11}t_1t_2\right. \\
&\qquad \left. + (p_{00} + p_{10} + p_{01} + p_{11})\right) \\
&= \ln G_N\left(p_{1\bullet}t_1 + p_{\bullet 1}t_2 + p_{11}t_1t_2 + 1\right) \\
&= K_N^*\left(p_{1\bullet}t_1 + p_{\bullet 1}t_2 + p_{11}t_1t_2\right),
\end{aligned} \tag{36.93}$$

where $K_N^*(t)$ is the factorial cumulant generating function of the random variable N. From (36.93), we obtain the relation

$$\kappa'_{(r,s)} = \sum_{i=0}^{\min(r,s)} i!\binom{r}{i}\binom{s}{i}p_{1\bullet}^{r-i}p_{\bullet 1}^{s-i}p_{11}^{i}\kappa'_{(r+s-i)}(N)\,, \tag{36.94}$$

where $\kappa'_{(r+s-i)}(N)$ is the $(r + s - i)$th factorial cumulant of the random variable N.

As remarked earlier in Chapter 34 (Section 2), Kocherlakota and Kocherlakota (1992, p. 5) have defined the bivariate factorial cumulant generating function as $\ln G(t_1, t_2)$ instead of $\ln G(t_1 + 1, t_2 + 1)$, which can cause some confusion.

We shall now consider some special cases of (36.90) that are of interest:

1. If N has a Poisson distribution with probability generating function

$$G_N(t) = e^{\lambda(t-1)}, \qquad \lambda > 0\,,$$

we obtain the distribution due to Hamdan and Tsokos (1971) with joint probability mass function

$$P(r, s) = e^{-\lambda(p_{10}+p_{01}+p_{11})} \sum_{i=0}^{\min(r,s)} \frac{1}{(r - i)!(s - i)!i!}(\lambda p_{10})^{r-i}(\lambda p_{01})^{s-i}(\lambda p_{11})^{i} \tag{36.95}$$

and with the corresponding probability generating function

$$G(t_1, t_2) = e^{\lambda p_{10}(t_1-1)+\lambda p_{01}(t_2-1)+\lambda p_{11}(t_1t_2-1)}. \tag{36.96}$$

In this case, we note that

$$K_N^*(t) = \ln G_N(t+1) = \lambda t$$

and hence $\kappa'_{(1)}(N) = \lambda$ and $\kappa'_{(r)}(N) = 0$ for $r \geq 2$. Therefore, the only nonzero (r, s)th factorial cumulants are

$$\kappa'_{(1,0)} = \lambda p_{1\bullet}, \; \kappa'_{(0,1)} = \lambda p_{\bullet 1} \text{ and } \kappa'_{(1,1)} = \lambda p_{11}\,.$$

2. If N has a binomial(m, p) distribution with probability generating function

$$G_N(t) = (q + pt)^m, \qquad q = 1 - p\,,$$

we obtain the bivariate distribution with joint probability mass function

$$P(r, s) = \sum_{i=0}^{\min(r,s)} \frac{m!}{(r-i)!(s-i)!i!(m-r-s+i)!} (pp_{10})^{r-i} (pp_{01})^{s-i} (pp_{11})^i \times (q + pp_{00})^{m-r-s+i} \qquad (36.97)$$

and with the corresponding probability generating function

$$G(t_1, t_2) = (q + pp_{00} + pp_{10}t_1 + pp_{01}t_2 + pp_{11}t_1t_2)^m. \qquad (36.98)$$

The expressions for $\kappa'_{(r,s)}$ in this case are quite complicated.

3. Papageorgiou and David (1994) have similarly provided explicit expressions for the bivariate distributions resulting from (36.90) when N has negative binomial, Hermite, Neyman A Type I, and modified logarithmic series distributions; see Chapters 9 and 8.

For the bivariate distribution with probability generating function as in (36.86), Papageorgiou and David (1994) have shown that the conditional distribution of X_2 given $X_1 = x_1$ is the convolution of (a) a binomial random variable Y_1 with parameters x_1 and $p_{11}/(p_{10} + p_{11})$ and (b) a random variable Y_2 with probability generating function given by

$$\frac{G_N^{(x_1)}(p_{00} + p_{01}t)}{G_N^{(x_1)}(p_{00} + p_{01})}\,. \qquad (36.99)$$

This follows from the result of Subrahmaniam (1966): For a bivariate random variable $(X_1, X_2)'$ with probability generating function $G(t_1, t_2)$, the conditional probability generating function $G_{X_2|X_1=x_1}(t)$ is given by

$$G_{X_2|X_1=x_1}(t) = \frac{G^{(x_1,0)}(0, t)}{G^{(x_1,0)}(0, 1)}\,, \qquad (36.100)$$

where $G^{(r,s)}(t_1, t_2) = \frac{\partial^{r+s}}{\partial t_1^r \partial t_2^s} G(t_1, t_2)$ [cf. (34.41)]. In particular, the random variable Y_2 is

(i) Poisson(λp_{01}) if the random variable N is Poisson(λ),

(ii) Binomial$\left(m - x_1, \frac{pp_{01}}{q+pp_{00}+pp_{01}}\right)$ if the random variable N is Binomial(m, p), and

(iii) Negative binomial$\left(\alpha + x_1, \frac{\beta p_{01}}{\gamma-\beta p_{00}-\beta p_{01}}\right)$ if the random variable N is Negative binomial(α, β).

Papageorgiou and David (1994) have applied these distributions to model the joint distribution of the numbers of children of the two genders in a family, a problem considered in some detail earlier by Rao *et al.* (1973) and Sinha (1985); see Section 2.

9 MULTIVARIATE GEOMETRIC DISTRIBUTIONS OF ORDER s [2]

Univariate geometric distributions of order s are described in Chapter 10, Section 6. They arise as distributions of the waiting time (T), in terms of numbers of trials, each trial resulting in observed values of independent random variables $X_1, X_2, \ldots$ having the Bernoulli distribution

$$\Pr[X_i = 1] = p, \quad \Pr[X_i = 0] = 1 - p = q, \tag{36.101}$$

until s consecutive 1's are observed. The probability generating function of this distribution is

$$G_T(t) = \frac{p^s t^s (1 - pt)}{1 - t + qp^s t^{s+1}}; \tag{36.102}$$

see, for example, Bizley (1962), Feller (1957), and Chapter 10 (p. 426).

Aki and Hirano (1994, 1995) have constructed *multivariate geometric distributions of order s* by introducing the number, M_0, of 0's appearing in the sequence $X_1, X_2, \ldots$ before s successive 1's are obtained. The joint probability generating function of T and M_0 is

$$\begin{aligned} G_{T,M_0}(t, t_0) &= \frac{p^s t^s}{1 - \sum_{i=0}^{s-1} p^i q t_0 t^{i+1}} \\ &= \frac{p^s t^s}{1 - qt_0 t \sum_{i=0}^{s-1} (pt)^i} \\ &= \frac{p^s t^s (1 - pt)}{1 - pt - qt_0 t\{1 - (pt)^s\}}. \end{aligned} \tag{36.103}$$

[2]These distributions are widely known as *order k* distributions. However, we have deliberately used the name *order s* since k has been used throughout this book to denote the dimension of a multivariate distribution. This problem arises partly due to the fact that the letter k, which was used in the original derivation of the distribution, has been (unfortunately) used in naming the distribution. We have also come to learn that the word *order* was originally used when it literally meant *run* or *consecutive*; hence, either of these two words would have been a better choice.

Of course, $G_{T,M_0}(t, 1) = G_T(t)$ as given by (36.102), and

$$G_{M_0}(t_0) = G_{T,M_0}(1, t_0) = \frac{p^s}{1 - t_0(1 - p^s)} . \tag{36.104}$$

Aki and Hirano (1995) extended this distribution by supposing that each X_i has the distribution

$$\Pr[X_i = j] = p_j, \qquad j = 1, 2, \ldots, m . \tag{36.105}$$

We again let T denote the waiting time until s consecutive 1's are observed, but now consider the numbers $M_2, M_3, \ldots, M_m$ of 2's, 3's, ..., m's, respectively, observed before s consecutive 1's are obtained. The joint probability generating function of $T, M_2, \ldots, M_m$ is

$$G_{T,M_2,\ldots,M_m}(t, t_2, \ldots, t_m)$$
$$= \frac{(p_1 t)^s (1 - p_1 t)}{1 - p_1 t - t(\sum_{j=2}^m p_j t_j)\{1 - (p_1 t)^s\}} . \tag{36.106}$$

Moments of this distribution are obtained as

$$E[T] = \frac{1 - p_1^s}{p_1^s(1 - p_1)} , \tag{36.107}$$

$$E[M_j] = p_j\, E[T], \qquad j = 2, \ldots, m , \tag{36.108}$$

$$\text{var}(T) = \frac{1 - (2k + 1)(1 - p_1)p_1^s - p_1^{2s+1}}{p_1^{2s}(1 - p_1)^2} , \tag{36.109}$$

$$\text{var}(M_j) = \frac{p_j(1 - p_1)^s\{p_j(1 - p_1)^s + (1 - p_1)p_1^s\}}{p_1^{2s}(1 - p_1)^2} , \quad j = 2, \ldots, m , \tag{36.110}$$

$$\text{cov}(T, M_j) = \frac{p_j\{1 - p_1^s - k(1 - p_1)p_1^s\}}{p_1^{2s}(1 - p_1)^2} \geq 0, \quad j = 2, \ldots, m , \tag{36.111}$$

and

$$\text{cov}(M_i, M_j) = \frac{p_i p_j (1 - p_1^s)^2}{p_1^{2s}(1 - p_1)^2} \geq 0 \quad \text{for } i \neq j = 2, \ldots, m . \tag{36.112}$$

Note that

$$\text{var}(M_j) = E[M_j]\{E[M_j] + 1\} \tag{36.113}$$

and

$$\text{cov}(M_i, M_j) = E[M_i]E[M_j] . \tag{36.114}$$

The joint probability generating function of $T, M_1, M_2, \ldots, M_m$ is

$$G_{T,M_1,M_2,\ldots,M_m}(t, t_1, t_2, \ldots, t_m) = \frac{(1 - p_1t_1t)(p_1t_1t)^s}{1 - p_1t_1t - t(\sum_{j=2}^{m} p_jt_j)\{1 - (p_1t_1t)^s\}}. \tag{36.115}$$

Of course, we have $G_{T,M_1,M_2,\ldots,M_m}(1, t_1, 1, \ldots, 1) = G_{M_1}(t_1)$, and so on. For some further results on this multivariate distribution and also some related distributions, one may refer to Balakrishnan (1996) and Johnson and Balakrishnan (1997); also see Chapter 42.

Further generalizations studied by Aki and Hirano (1995) include the joint distribution of $T, M_1, M_2, \ldots, M_m, T_{1(2)}, T_{1(3)}, \ldots, T_{1(s-1)}$, where $T_{1(\ell)}$ is the number of (possibly overlapping) runs of 1's of length ℓ in the first T trials. Philippou and Antzoulakos (1990) and Philippou, Antzoulakos, and Tripsiannis (1989, 1990) have provided comprehensive reviews on multivariate distributions of order s. See also Antzoulakos and Philippou (1991, 1994) for some work on multivariate negative binomial distributions of order s.

These authors have discussed specifically the following distributions:

(i) A random vector $\boldsymbol{X} = (X_1, \ldots, X_k)'$ is said to have a *Type I multivariate negative binomial distribution of order s* with parameters $r, q_1, \ldots, q_k$, denoted by $\mathrm{MNB}_{s,\mathrm{I}}(r; q_1, \ldots, q_k)$, if for $x_i = 0, 1, 2, \ldots$ $(1 \le i \le m)$,

$$P_{\boldsymbol{X}}(\boldsymbol{x}) = p^{sr} \sum_{\substack{\sum_j jx_{ij}=x_i,\\ i=1,\ldots,k}} \frac{\Gamma(r + \sum_i \sum_j x_{ij})}{\Gamma(r)\prod_i \prod_j x_{ij}!} \prod_i p^{x_i}\left(\frac{q_i}{p}\right)^{\sum_j x_{ij}}, \tag{36.116}$$

where $r > 0$, $0 < q_i < 1$ for $1 \le i \le k$, $q_1 + \cdots + q_k < 1$, and $p = 1 - q_1 - \cdots - q_k$.

(ii) A random vector $\boldsymbol{X} = (X_1, \ldots, X_k)'$ is said to have a *Type II multivariate negative binomial distribution of order s* with parameters $r, q_1, \ldots, q_k$, denoted by $\mathrm{MNB}_{s,\mathrm{II}}(r; q_1, \ldots, q_k)$, if for $x_i = 0, 1, 2, \ldots$ $(1 \le i \le k)$,

$$P_{\boldsymbol{X}}(\boldsymbol{x}) = p^{r} \sum_{\substack{\sum_j jx_{ij}=x_i,\\ i=1,\ldots,k}} \frac{\Gamma(r + \sum_i \sum_j x_{ij})}{\Gamma(r)\prod_i \prod_j x_{ij}!} \prod_i \left(\frac{q_i}{s}\right)^{\sum_j x_{ij}}, \tag{36.117}$$

where $r > 0$, $0 < q_i < 1$ for $1 \le i \le k$, $q_1 + \cdots + q_k < 1$, and $p = 1 - q_1 - \cdots - q_k$.

(iii) A random vector $\boldsymbol{X} = (X_1, \ldots, X_k)'$ is said to have a *multivariate negative binomial distribution of order s* with parameters $r, q_{11}, \ldots, q_{ks}$ $(r > 0, 0 < q_{ij} < 1$ for $1 \le i \le k$ and $1 \le j \le s$, and $q_{11} + \cdots q_{ks} < 1)$, denoted by $\mathrm{MNB}_k(r; q_{11}, \ldots, q_{ks})$, if for $x_i = 0, 1, 2, \ldots$ $(1 \le i \le k)$,

$$P_{\boldsymbol{X}}(\boldsymbol{x}) = p^{r} \sum_{\substack{\sum_j jx_{ij}=x_i,\\ i=1,\ldots,k}} \frac{\Gamma(r + \sum_i \sum_j x_{ij})}{\Gamma(r)\prod_i \prod_j x_{ij}!} \prod_i \prod_j q_{ij}^{x_{ij}}, \tag{36.118}$$

where $p = 1 - q_{11} - \cdots - q_{ks}$.

For $s = 1$ the distribution (36.118) reduces to the usual negative multinomial distribution in (36.2), and for $m = 1$ it reduces to the multiparameter negative binomial distribution of order s of Philippou (1988). Further, for $r = 1$ it reduces to a multivariate geometric distribution of order s with parameters $q_{11}, \ldots, q_{ks}$.

For more details on multivariate distributions of order s, one may refer to Chapter 42.

BIBLIOGRAPHY

Aki, S., and Hirano, K. (1994). Distributions of numbers of failures and successes until the first consecutive k successes, *Annals of the Institute of Statistical Mathematics*, **46**, 193–202.

Aki, S., and Hirano, K. (1995). Joint distributions of numbers of success-runs and failures until the first k consecutive successes, *Annals of the Institute of Statistical Mathematics*, **47**, 225–235.

Al-Husseini, E. K. (1989). Generalized multinomial detectors for communication signals, *IEEE Transactions on Communications*, **37**, 1099–1102.

Antzoulakos, D. L., and Philippou, A. N. (1991). A note on the multivariate negative binomial distributions of order k, *Communications in Statistics—Theory and Methods*, **20**, 1389–1399.

Antzoulakos, D. L., and Philippou, A. N. (1994). Expressions in terms of binomial coefficients for some multivariate distributions of order k, in *Runs and Patterns in Probability* (Eds. A. P. Godbole and S. G. Papastavridis), pp. 1–14, Amsterdam: Kluwer.

Arbous, A. G., and Sichel, H. S. (1954). New techniques for the analysis of absenteeism data, *Biometrika*, **41**, 77–90.

Basu, A. P., and Dhar, S. K. (1995). Bivariate geometric distribution, *Journal of Applied Statistical Science*, **2**, 33–44.

Balakrishnan, N. (1996). On the joint distributions of numbers of success-runs and failures until the first consecutive k successes, *submitted for publication*.

Balakrishnan, N., and Johnson, N. L. (1997). A recurrence relation for discrete multivariate distributions and some applications to multivariate waiting time problems, in *Advances in the Theory and Practice of Statistics—A Volume in Honor of Samuel Kotz* (Eds. N. L. Johnson and N. Balakrishnan), New York: John Wiley & Sons (to appear).

Bates, G. E., and Neyman, J. (1952). Contributions to the theory of accident proneness, *University of California, Publications in Statistics,* **1,** 215–253.

Beaulieu, N. C. (1991). On the generalized multinomial distribution, optimal multinomial detectors and generalized weighted partial decision detectors, *IEEE Transactions on Communications*, **39**, 193–194 (comment by E. K. Al-Husseini, p. 195).

Bizley, M. T. L. (1962). Patterns in repeated trials, *Journal of the Institute of Actuaries*, **88**, 360–366.

Charalambides, C. A. (1981). On a restricted occupancy model and its applications, *Biometrical Journal*, **23**, 601–610.

Clevenson, M. L., and Zidek, J. V. (1975). Simultaneous estimation of the means of independent Poisson laws, *Journal of the American Statistical Association,* **70,** 698–705.

Cox, D. R., and Wermuth, N. (1994). A note on the quadratic exponential binary distribution, *Biometrika,* **81,** 403–408.

Doss, D. C., and Graham, R. C. (1975a). A characterization of multivariate binomial distribution by univariate marginals, *Calcutta Statistical Association Bulletin*, **24**, 93–99.

Doss, D. C., and Graham, R. C. (1975b). Construction of multivariate linear exponential distributions from univariate marginals, *Sankhyā, Series A*, **37**, 257–268.

Feller, W. (1957). *An Introduction to Probability Theory and Its Applications*, second edition, New York: John Wiley & Sons.

Froggatt, P. (1970). Application of discrete distribution theory to the study of noncommunicable events in medical epidemiology, in *Random Counts in Scientific Work, Vol. 2* (Ed. G. P. Patil), pp. 15–40, Dordrecht: Reidel.

Guldberg, S. (1935). Recurrence formulae for the semi-invariants of some discontinuous frequency functions of *n* variables, *Skandinavisk Aktuarietidskrift*, **18,** 270–278.

Guo, G. (1995). Negative multinomial regression models for clustered event counts, *Technical Report*, Department of Sociology, University of North Carolina, Chapel Hill, NC.

Hamdan, M. A., and Tsokos, C. P. (1971). A model for physical and biological problems: The bivariate-compounded Poisson distribution, *International Statistical Review*, **39**, 59–63.

Hawkes, A. G. (1972). A bivariate exponential distribution with applications to reliability, *Journal of the Royal Statistical Society, Series B*, **34**, 129–131.

Janardan, K. G. (1974). A characterization of multinomial and negative multinomial distributions, *Skandinavisk Aktuarietidskrift*, **57**, 58–62.

Johnson, N. L., and Kotz, S. (1991). Some generalizations of Bernoulli and Pólya–Eggenberger contagion models, *Statistical Papers*, **32**, 1–17.

Johnson, N. L., and Kotz, S. (1994). Further comments on Matveychuk and Petunin's generalized Bernoulli model, and nonparametric tests of homogeneity, *Journal of Statistical Planning and Inference*, **41**, 61–72.

Joshi, S. W. (1975). Integral expressions for tail probabilities of the negative multinomial distribution, *Annals of the Institute of Statistical Mathematics,* **27,** 95–97.

Kemp, C. D., and Kemp, A. W. (1987). Rapid generation of frequency tables, *Applied Statistics*, **36,** 277–282.

Khatri, C. G., and Mitra, S. K. (1968). Some identities and approximations concerning positive and negative multinomial distributions, *Technical Report 1/68,* Indian Statistical Institute, Calcutta, India.

Kocherlakota, S., and Kocherlakota, K. (1992). *Bivariate Discrete Distributions*, New York: Marcel Dekker.

Lancaster, H. O. (1976). Multivariate binomial distributions, in *Studies in Probability and Statistics: Papers in Honour of Edwin J. G. Pitman* (Ed. E. J. Williams), pp. 13–19, Amsterdam: North-Holland.

Lee, M.-L. T. (1996). Properties and applications of the Sarmanov family of bivariate distributions, *Communications in Statistics—Theory and Methods*, **25**, 1207–1222.

Lundberg, O. (1940). *On Random Processes and Their Application to Sickness and Accident Statistics*, Uppsala, Sweden: Almqvist and Wiksell.

Marshall, A. W., and Olkin, I. (1967). A multivariate exponential distribution, *Journal of the American Statistical Association*, **62**, 30–44.

Marshall, A. W., and Olkin, I. (1995). Multivariate exponential and geometric distributions with limited memory, *Journal of Multivariate Analysis*, **53**, 110–125.

Matveychuk, S. A., and Petunin, Y. T. (1990). A generalization of the Bernoulli model arising in order statistics. I, *Ukrainian Mathematical Journal*, **42**, 518–528.

Matveychuk, S. A., and Petunin, Y. T. (1991). A generalization of the Bernoulli model arising in order statistics. II, *Ukrainian Mathematical Journal*, **43**, 779–785.

Mosimann, J. E. (1963). On the compound negative multinomial distribution and correlations among inversely sampled pollen counts, *Biometrika*, **50**, 47–54.

Nagaraja, H. N. (1983). Two characterizations of a bivariate geometric distribution, *Journal of the Indian Statistical Association*, **21**, 27–30.

Neyman, J. (1963). Certain chance mechanisms involving discrete distributions (Inaugural Address), *Proceedings of the International Symposium on Discrete Distributions, Montreal,* pp. 4–14.

Olkin, I., and Sobel, M. (1965). Integral expressions for tail probabilities of the multinomial and the negative multinomial distributions, *Biometrika,* **52,** 167–179.

Oluyede, B. (1994). A family of bivariate binomial distributions generated by extreme Bernoulli distributions, *Communications in Statistics—Theory and Methods*, **23**, 1531–1547.

Panaretos, J. (1977). A characterization of a general class of multivariate discrete distributions, in *Analytic Function Methods in Probability Theory, Colloquia Mathematica Societatis Janos Bolyai* (Ed. B. Gyires), pp. 243–252, Amsterdam: North-Holland.

Panaretos, J. (1981). A characterization of the negative multinomial distribution, in *Statistical Distributions in Scientific Work—4* (Eds. G. P. Patil, C. Taillie and B. A. Baldessari), pp. 331–341, Dordrecht: Reidel.

Panaretos, J., and Xekalaki, E. (1986). On generalized binomial and multinomial distributions and their relation to generalized Poisson distributions, *Annals of the Institute of Statistical Mathematics,* **38,** 223–231.

Papageorgiou, H., and David, K. M. (1994). On countable mixture of bivariate binomial distributions, *Biometrical Journal*, **36**, 581–601.

Papageorgiou, H., and Loukas, S. (1987). Estimation based on conditional distributions for the bivaiate negative binomial–Poisson distribution, *Proceedings of ASA Statistical Computing Section.*

Papageorgiou, H., and Loukas, S. (1988). Conditional even point estimation for bivariate discrete distributions, *Communications in Statistics—Theory and Methods*, **17**, 3403–3412.

Patil, G. P. (1968). On sampling with replacement from populations with multiple characters, *Sankhyā, Series B*, **30**, 355–364.

Phatak, A. G., and Sreehari, M. (1981). Some characterizations of a bivariate geometric distribution, *Journal of the Indian Statistical Association*, **19**, 141–146.

Philippou, A. N. (1988). On multiparameter distributions of order k, *Annals of the Institute of Statistical Mathematics*, **40**, 467–475.

Philippou, A. N., and Antzoulakos, D. L. (1990). Multivariate distributions of order k on a general sequence, *Statistics & Probability Letters*, **9**, 453–463.

Philippou, A. N., Antzoulakos, D. L., and Tripsiannis, G. A. (1989). Multivariate distributions of order k, *Statistics & Probability Letters*, **7**, 207–216.

Philippou, A. N., Antzoulakos, D. L., and Tripsiannis, G. A. (1990). Multivariate distributions of order k, Part II, *Statistics & Probability Letters*, **10**, 29–35.

Price, G. B. (1946). Distributions derived from the multinomial expansion, *American Mathematical Monthly*, **53**, 59–74.

Rao, B. R., Mazumdar, S., Waller, J. M., and Li, C. C. (1973). Correlation between the numbers of two types of children in a family, *Biometrics*, **29**, 271–279.

Rao, C. R., and Rubin, H. (1964). On a characterization of the Poisson distribution, *Sankhyā, Series A*, **26**, 295–298.

Sagae, M., and Tanabe, K. (1992). Symbolic Cholesky decomposition of the variance–covariance matrix of the negative multinomial distribution, *Statistics & Probability Letters*, **15**, 103–108.

Sarmanov, O. V. (1966). Generalized normal correlation and two-dimensional Frechet classes, *Soviet Doklady*, **168**, 596–599.

Shanbhag, D. N. (1977). An extension of the Rao-Rubin characterization of the Poisson distribution, *Journal of Applied Probability*, **14**, 640–646.

Sibuya, M., Yoshimura, I., and Shimizu, R. (1964). Negative multinomial distribution, *Annals of the Institute of Statistical Mathematics,* **16,** 409–426.

Sinha, A. K. (1985). On the symmetrical bivariate negative binomial distribution, in *Statistical Theory and Data Analysis* (Ed. K. Matusita), pp. 649–657, Amsterdam: North-Holland.

Sinoquet, H., and Bonhomme, R. (1991). A theoretical analysis of radiation interception in a two-species plant canopy, *Mathematical Biosciences*, **105**, 23–45.

Steyn, H. S. (1951). On discrete multivariate probability functions, *Proceedings, Koninklijke Nederlandse Akademie van Wetenschappen, Series A*, **54**, 23–30.

Steyn, H. S. (1963). On approximations for the discrete distributions obtained from multiple events, *Proceedings, Koninklijke Nederlandse Akademie van Wetenschappen, Series A,* **66,** 85–96.

Steyn, H. S. (1975). On extensions of the binomial distributions of Bernoulli and Poisson, *South African Statistical Journal*, **9**, 163–172.

Subrahmaniam, K. (1966). A test for "intrinsic" correlation in the theory of accident proneness, *Journal of the Royal Statistical Society, Series B*, **28**, 180–189.

Subrahmaniam, K., and Subrahmaniam, K. (1973). On the estimation of the parameters in the bivariate negative binomial distribution, *Journal of the Royal Statistical Society, Series B*, **35**, 131–146.

Tallis, G. M. (1962). The use of a generalized multinomial distribution in the estimation of correlation in discrete data, *Journal of the Royal Statistical Society, Series B,* **24,** 530–534.

Tallis, G. M. (1964). Further models for estimating correlation in discrete data, *Journal of the Royal Statistical Society, Series B*, **26**, 82–85.

Teugels, J. L. (1990). Some representations of the multivariate Bernoulli and binomial distributions, *Journal of Multivariate Analysis,* **32,** 256–268.

Tracy, D. S., and Doss, D. C. (1980). A characterization of multivector multinomial and negative multinomial distributions, in *Multivariate Statistical Analysis* (Ed. R. P. Gupta), pp. 281–289, Amsterdam: North-Holland.

Tsui, K.-W. (1986). Multiparameter estimation for some multivariate discrete distributions with possibly dependent components, *Annals of the Institute of Statistical Mathematics,* **38,** 45–56.

Westfall, P. M., and Young, S. S. (1989). *p* value adjustments for multiple tests in multivariate binomial models, *Journal of the American Statistical Association*, **84**, 780–786.

Wishart, J. (1949). Cumulants of multivariate multinomial distributions, *Biometrika*, **36**, 47–58.

CHAPTER 37

Multivariate Poisson Distributions

1 INTRODUCTION

If, in the multinomial $(n; p_1, p_2, \ldots, p_k)$ distribution [defined in (35.1)], the parameters $p_1, p_2, \ldots, p_{k-1}$ are allowed to tend to 0 (and so p_k tends to 1) as $n \to \infty$, in such a way that $np_i \to \theta_i$ $(i = 1, 2, \ldots, k-1)$, then the probability of obtaining the event $\bigcap_{i=1}^{k-1}(N_i = n_i)$ tends to

$$P(n_1, n_2, \ldots, n_{k-1}) = \prod_{i=1}^{k-1} \left\{e^{-\theta_i}\theta_i^{n_i}/n_i!\right\} . \tag{37.1}$$

This, however, is simply the joint distribution of $k-1$ mutually independent Poisson random variables with means $\theta_1, \theta_2, \ldots, \theta_{k-1}$. Consequently, it does not require any special analysis. Nevertheless, the name *Multiple Poisson Distribution* has been given to such distributions [see Patil and Bildikar (1967)], and formulas for the moments have been given by Banerjee (1959) and by Sibuya, Yoshimura, and Shimizu (1964).

The distribution in (37.1) can also be obtained as a limiting form of negative multinomial distributions. We shall not consider (37.1) any further, but we will discuss some other multivariate distributions with Poisson marginals, and distributions related thereto.

2 BIVARIATE POISSON DISTRIBUTIONS

A nontrivial class of *bivariate Poisson distributions* was constructed by Holgate (1964) [see also the earlier papers by Campbell (1938), Aitken and Gonin (1935), Aitken (1936), and Consael (1952), and comments at the end of this Section]. It is based on the joint distribution of the variables

$$X_1 = Y_1 + Y_{12} \qquad \text{and} \qquad X_2 = Y_2 + Y_{12} , \tag{37.2}$$

where Y_1, Y_2, and Y_{12} are mutually independent Poisson random variables with means θ_1, θ_2, and θ_{12}, respectively. The joint probability mass function of X_1 and X_2 is

$$\begin{aligned}\Pr[X_1 = x_1, X_2 = x_2] &= P(x_1, x_2) \\ &= e^{-(\theta_1+\theta_2+\theta_{12})} \sum_{i=0}^{\min(x_1,x_2)} \frac{\theta_1^{x_1-i}\theta_2^{x_2-i}\theta_{12}^{i}}{(x_1-i)!(x_2-i)!i!}\,. \end{aligned} \tag{37.3}$$

This form of the bivariate Poisson distribution was derived by Campbell (1938) by considering the factorial moment generating function corresponding to fourfold sampling with replacement; see also McKendrick (1926). More recently, Hamdan and Al-Bayyati (1969) presented the following simple algebraic derivation of (37.3):

Let each individual of a population be classified as either A or A^c and simultaneously as either B or B^c with the probabilities given by

	B	B^c	
A	p_{11}	p_{10}	p_1
A^c	p_{01}	p_{00}	q_1
	p_2	q_2	1

Under random sampling with replacement n times, the random variables X_1 and X_2 denoting the numbers of occurrences of A and B, respectively, have jointly a bivariate binomial distribution with probability mass function

$$\begin{aligned}&\Pr[X_1 = x_1, X_2 = x_2] \\ &= \sum_{i=0}^{\min(x_1,x_2)} \frac{n!}{i!(x_1-i)!(x_2-i)!(n+i-x_1-x_2)!}\, p_{11}^{i} p_{10}^{x_1-i} p_{01}^{x_2-i} p_{00}^{n-x_1-x_2+i}\,. \end{aligned} \tag{37.4}$$

Hamdan and Al-Bayyati (1969) then derived the bivariate Poisson distribution in (37.3) simply as the limit of (37.4) by letting $n \to \infty$ and $p_{11}, p_{01}, p_{10} \to 0$ in such a way that $np_1 \to \theta_1$, $np_2 \to \theta_2$, and $np_{11} \to \theta_{12}$.

From (37.2), it is clear that the marginal distributions of X_1 and X_2 are Poisson with means $\theta_1 + \theta_{12}$ and $\theta_2 + \theta_{12}$, respectively. In the numerical computation of values of $P(x_1, x_2)$, the following recurrence relations [Teicher (1954)] will be useful:

$$\begin{aligned} x_1\, P(x_1, x_2) &= \theta_1\, P(x_1-1, x_2) + \theta_{12}\, P(x_1-1, x_2-1) \\ x_2\, P(x_1, x_2) &= \theta_2\, P(x_1, x_2-1) + \theta_{12}\, P(x_1-1, x_2-1)\,. \end{aligned} \tag{37.5}$$

[If either x_1 or x_2 is negative, $P(x_1, x_2)$ is to be taken to be 0.] Hesselager (1996b) recently established some recurrence relations for certain bivariate counting distributions and their compound distributions from which (37.5) follows as a special case. In addition, his results also extend Panjer's (1981) recursion for a family of compound distributions to the bivariate case and those of Willmot (1993) and Hesselager (1996a) on a class of mixed Poisson distributions to the bivariate case.

The joint moment generating function of X_1 and X_2 is

$$E\left[e^{t_1X_1+t_2X_2}\right] = E\left[e^{t_1Y_1+t_2Y_2+(t_1+t_2)Y_{12}}\right]$$
$$= \exp\left\{-\theta_1(1-e^{t_1}) - \theta_2(1-e^{t_2}) - \theta_{12}(1-e^{t_1+t_2})\right\}, \quad (37.6)$$

and hence the joint cumulant generating function is

$$\theta_1(e^{t_1}-1) + \theta_2(e^{t_2}-1) + \theta_{12}(e^{t_1+t_2}-1). \quad (37.7)$$

The joint probability generating function is

$$\exp\{\theta_1(t_1-1) + \theta_2(t_2-1) + \theta_{12}(t_1t_2-1)\} \quad (37.8)$$
$$= \exp\{\phi_1(t_1-1) + \phi_2(t_2-1) + \phi_{12}(t_1-1)(t_2-1)\}, \quad (37.9)$$

where $\phi_1 = \theta_1 + \theta_{12}$, $\phi_2 = \theta_2 + \theta_{12}$ and $\phi_{12} = \theta_{12}$. A mixture of (37.9) with respect to ϕ_{12} still has Poisson marginal distributions. This, however, is not the same as mixing (37.8) with respect to θ_{12} even though $\phi_{12} = \theta_{12}$: when mixing (37.9), ϕ_1 and ϕ_2 are kept constant and so θ_1 and θ_2 vary with ϕ_{12} ($= \theta_{12}$).

The covariance between X_1 and X_2 is

$$\operatorname{cov}(X_1, X_2) = \operatorname{cov}(Y_1 + Y_{12},\ Y_2 + Y_{12}) = \operatorname{var}(Y_{12}) = \theta_{12}. \quad (37.10)$$

Hence, the correlation coefficient between X_1 and X_2 is

$$\operatorname{corr}(X_1, X_2) = \frac{\theta_{12}}{\sqrt{(\theta_1+\theta_{12})(\theta_2+\theta_{12})}}. \quad (37.11)$$

This cannot exceed $\theta_{12}\{\theta_{12} + \min(\theta_1, \theta_2)\}^{-1/2}$, or as Holgate (1964) pointed out,

$$\min\left[\sqrt{\frac{\theta_1+\theta_{12}}{\theta_2+\theta_{12}}},\ \sqrt{\frac{\theta_2+\theta_{12}}{\theta_1+\theta_{12}}}\right]; \quad (37.12)$$

that is, the correlation coefficient cannot exceed the square root of the ratio of the smaller to the larger of the means of the two marginal distributions. This is a limitation on the applicability of these bivariate Poisson distributions. Yet, it also provides a check on the appropriateness of the model in (37.3). Griffiths, Milne, and Wood (1980) have discussed some results and properties about correlation in bivariate Poisson distributions. These authors have also presented some results and examples relating to negative correlation. Samaniego (1976) has discussed a characterization of convoluted Poisson distributions and its application to estimation problems.

The conditional distribution of X_1, given X_2, is

$$\Pr[X_1 = x_1 \mid X_2 = x_2]$$
$$= P(x_1 \mid x_2)$$
$$= \frac{P(x_1, x_2)}{e^{-(\theta_2+\theta_{12})}(\theta_2+\theta_{12})^{x_2}/x_2!}$$

$$= e^{-\theta_1} \sum_{j=0}^{\min(x_1,x_2)} \binom{x_2}{j} \left(\frac{\theta_{12}}{\theta_2 + \theta_{12}}\right)^j \left(\frac{\theta_2}{\theta_2 + \theta_{12}}\right)^{x_2 - j} \frac{\theta_1^{x_1 - j}}{(x_1 - j)!} . \qquad (37.13)$$

This is the distribution of the sum of two mutually independent random variables, one distributed as Y_1 (Poisson with mean θ_1) and the other distributed as a binomial $(x_2, \theta_{12}/(\theta_2 + \theta_{12}))$ (the distribution of Y_{12}, given $X_2 = x_2$). Hence,

$$E[X_1 \mid X_2 = x_2] = \theta_1 + \frac{\theta_{12}}{\theta_2 + \theta_{12}} x_2 , \qquad (37.14)$$

that is, the regression of X_1 on X_2 is linear (and conversely). Furthermore,

$$\operatorname{var}(X_1 \mid X_2 = x_2) = \theta_1 + \frac{\theta_2 \theta_{12}}{(\theta_2 + \theta_{12})^2} x_2 , \qquad (37.15)$$

which reveals that the variation about the regression line is heteroscedastic.

Loukas and Kemp (1986) have noted that in (37.9) if $\phi_1 \to \infty$, $\phi_2 \to \infty$ and $\phi_{12}/\sqrt{\phi_1 \phi_2} \to \rho$, then the limiting distribution of

$$(Z_1, Z_2) = \left(\frac{X_1 - \phi_1}{\sqrt{\phi_1}} , \frac{X_2 - \phi_2}{\sqrt{\phi_2}} \right)$$

is a standardized bivariate normal distribution with correlation coefficient ρ; consequently, it follows that the limiting distribution of

$$(Z_1^2 - 2\rho Z_1 Z_2 + Z_2^2)/(1 - \rho^2)$$

is χ_2^2 (see Chapter 18). Rayner and Best (1995) have, however, pointed out that this χ^2 approximation may not be good for large correlations and hence suggested a revised goodness-of-fit statistic.

Given n independent pairs of observations (X_{1i}, X_{2i}), $i = 1, 2, \ldots, n$, from the bivariate Poisson distribution in (37.3), the maximum likelihood estimators $\hat{\theta}_{12}$, $\hat{\theta}_1$, and $\hat{\theta}_2$ of θ_{12}, θ_1, and θ_2, respectively, satisfy the equations

$$\sum_{i=1}^{n} \frac{1}{P(X_{1i}, X_{2i})} \frac{\partial P(X_{1i}, X_{2i})}{\partial \tau} = 0 \qquad (\tau = \theta_{12}, \theta_1, \theta_2) . \qquad (37.16)$$

From Teicher's recurrence relations in (37.5), we have

$$\frac{\partial P(x_1, x_2)}{\partial \theta_{12}} = P(x_1 - 1, x_2 - 1) - P(x_1, x_2) , \qquad (37.17)$$

$$\frac{\partial P(x_1, x_2)}{\partial \theta_1} = P(x_1 - 1, x_2) - P(x_1, x_2) , \qquad (37.18)$$

and

$$\frac{\partial P(x_1, x_2)}{\partial \theta_2} = P(x_1, x_2 - 1) - P(x_1, x_2) . \qquad (37.19)$$

Combining these three equations with (37.5), Holgate (1964) showed that (37.16) can be written in the form

$$\overline{X}_{t.} - \hat{\theta}_{12}\overline{R} = \hat{\theta}_t \qquad (t = 1, 2)\,, \tag{37.20}$$

$$\frac{1}{\hat{\theta}_1}(\overline{X}_{1.} - \hat{\theta}_{12}\overline{R}) + \frac{1}{\hat{\theta}_2}(\overline{X}_{2.} - \hat{\theta}_{12}\overline{R}) = 1 + \overline{R}\,, \tag{37.21}$$

where

$$\overline{X}_{t.} = \frac{1}{n}\sum_{i=1}^{n} X_{ti} \qquad (t = 1, 2)$$

and

$$\overline{R} = \frac{1}{n}\sum_{i=1}^{n}\{P(X_{1i} - 1, X_{2i} - 1)/P(X_{1i}, X_{2i})\}\,.$$

Hence,

$$\hat{\theta}_{12} + \hat{\theta}_t = \overline{X}_{t.} \qquad (t = 1, 2) \tag{37.22}$$

and

$$\overline{R} = 1\,. \tag{37.23}$$

Equation (37.23) is a polynomial in $\hat{\theta}_{12}$ which has to be solved numerically.

Evidently,

$$\mathrm{var}(\hat{\theta}_{12} + \hat{\theta}_t) = (\theta_{12} + \theta_t)/n \qquad (t = 1, 2) \tag{37.24}$$

and

$$\mathrm{corr}(\hat{\theta}_{12} + \hat{\theta}_1, \hat{\theta}_{12} + \hat{\theta}_2) = \mathrm{corr}(X_{1i}, X_{2i}) = \frac{\theta_{12}}{\sqrt{(\theta_1 + \theta_{12})(\theta_2 + \theta_{12})}}\,. \tag{37.25}$$

Also [Holgate (1964); Paul and Ho (1989)]

$$\mathrm{var}(\hat{\theta}_{12}) \doteq \frac{\theta_{12}^2\{(\theta_1 + \theta_2)(Q - 1) - 1\} + (\theta_1 - \theta_{12})(\theta_2 - \theta_{12})}{\{(\theta_1 + \theta_2)\theta_{12} + \theta_1\theta_2\}(Q - 1) - (\theta_1 + \theta_2)}\,, \tag{37.26}$$

where

$$Q = \sum_{x_1=0}^{\infty}\sum_{x_2=0}^{\infty}\left[\{P(x_1 - 1, x_2 - 1)\}^2/P(x_1, x_2)\right]\,.$$

[We thank Dr. Paul for drawing our attention to a misprint in Johnson and Kotz (1969, p. 300).]

Alternatively, θ_{12} may be estimated by the sample covariance

$$\frac{1}{n-1}\sum_{i=1}^{n}(X_{1i}-\overline{X}_{1.})(X_{2i}-\overline{X}_{2.})$$

[see Eq. (37.10)]. The variance of this estimator approximately equals

$$\frac{1}{n}\{\theta_{12}(1+\theta_{12})+(\theta_{12}+\theta_1)(\theta_{12}+\theta_2)\}\ . \tag{37.27}$$

This estimator is uncorrelated with $\hat{\theta}_1$ and $\hat{\theta}_2$; its efficiency tends to 1 as θ_{12} approaches 0, but decreases as θ_{12}/θ_i $(i = 1, 2)$ decreases. Of course, once an estimator of θ_{12} is obtained, (37.20) may be used readily to obtain estimators of θ_1 and θ_2.

Several other methods of estimation have been proposed, some of which we shall now describe. Papageorgiou and Kemp (1977) have proposed the *even-points* estimation, utilizing the sum A_1 (say) of frequency of occurrence of pairs (X_{1i}, X_{2i}) for which both variables are even or both variables are odd (i.e., are of the same "parity"). The even-point estimators so obtained are

$$\begin{aligned}\theta_t^+ &= \overline{X}_{t.} - \theta_{12}^+ \qquad (t = 1, 2),\\ \theta_{12}^+ &= \frac{1}{2}\left(\overline{X}_{1.}+\overline{X}_{2.}\right)+\frac{1}{4}\log\left(\frac{2}{n}A_1-1\right),\end{aligned} \tag{37.28}$$

where A_1 must exceed $n/2$. If $A_1 \le n/2$, it should be regarded as evidence that this bivariate Poisson distribution is inappropriate for the data at hand; see also Loukas, Kemp, and Papageorgiou (1986).

The method of *double zero* estimation utilizes instead the frequency E (say) of the combination $X_{1i} = X_{2i} = 0$, in place of A_1, to estimate θ_{12} as

$$\theta_{12}^{++} = \overline{X}_{1.}+\overline{X}_{2.}+\log(E/n)\,. \tag{37.29}$$

Papageorgiou and Loukas (1988) have proposed the method of *conditional even-points* estimation. This uses the sample means $\overline{X}_{1.}$ and $\overline{X}_{2.}$, together with the sums

$$A=\sum_{X_{2i}\ \text{even}} X_{2i}, \qquad B=\sum_{X_{2i}\ \text{odd}} X_{2i}$$

restricted to those pairs (X_{1i}, X_{2i}) with $X_{1i} = 0$. It is clear that

$$A+B=\sum_{i:X_{1i}=0} X_{2i}=\sum_{i=1}^{n}(1-I_{X_{1i}})X_{2i}\,,$$

where I_X is the indicator function

$$I_X=\begin{cases}1 & \text{if } X>0\\ 0 & \text{if } X=0\,.\end{cases}$$

Equation of observed and expected values leads to the estimators

$$\theta_t^* = \overline{X}_{t.} - \theta_{12}^* \qquad (t = 1, 2),$$

$$\theta_{12}^* = \overline{X}_{2.} + \frac{1}{2}\log\left(\frac{A - B}{A + B}\right) \qquad \text{if } A > B\,. \tag{37.30}$$

If $A < B$, this may be taken as evidence that this bivariate Poisson distribution is inappropriate.

One may alternatively use the sums A' and B' of X_{1i}'s over even and odd values, respectively, restricted to those pairs (X_{1i}, X_{2i}) with $X_{2i} = 0$, leading to

$$\theta_{12}^* = \overline{X}_{1.} + \frac{1}{2}\log\left(\frac{A' - B'}{A' + B'}\right) \qquad \text{if } A' > B'\,. \tag{37.31}$$

Once again, if $A' < B'$, we may regard it as an evidence that this bivariate Poisson distribution is inappropriate.

Paul and Ho (1989) have compared the following methods of estimation:

- **(i)** moments
- **(ii)** even-points (EP)
- **(iii)** conditional even-points (CEP)
- **(iv)** double zero

Through their study of asymptotic relative efficiencies, Paul and Ho have determined that the CEP method is more efficient than the EP method if the values of θ_1 and θ_2 differ greatly, but the EP method is superior when this is not the case and especially when θ_1 and θ_2 are small compared to θ_{12} (so that there is a high correlation between X_1 and X_2).

Maximum likelihood estimation of the parameters θ_1, θ_2 and θ_{12} has been investigated by Belov (1993).

Johnson and Brooks (1985) provided the following construction of a bivariate Poisson distribution with a negative correlation (in response to a problem posed in *American Mathematical Monthly*, November 1984). Let U be distributed uniformly on the unit interval, and let

$$X_1 = -\ln U \qquad \text{and} \qquad X_2 = -\ln(1 - U)\,.$$

It follows that X_1 and X_2 are distributed as standard exponential (see Chapter 19). Define $N_1(t)$ as a counting process with interarrival times according to X_1, and define $N_2(t)$ as a counting process with interarrival times according to X_2. Then $N_1(t)$ and $N_2(t)$ are both distributed as Poisson, and the vector $(N_1(t), N_2(t))'$ is distributed as bivariate Poisson. For intervals of length $t < -\ln 0.5$,

$$\Pr[N_1(t) > 0,\ N_2(t) > 0] = 0,$$

$$\Pr[N_1(t) = 0,\ N_2(t) = 0] = 2e^{-t} - 1 > 0\,,$$

and

$$\Pr[N_1(t) = k,\ N_2(t) = 0] = \Pr[N_1(t) = 0,\ N_2(t) = k] = e^{-t}\, t^k/k!\,, \qquad k = 1, 2, \ldots .$$

Further,

$$E[N_1(t)] = E[N_2(t)] = t\,,$$

$$\text{var}(N_1(t)) = \text{var}(N_2(t)) = t\,,$$

and

$$\text{corr}(N_1(t),\ N_2(t)) = -t\,.$$

Therefore, one may select a bivariate Poisson distribution with correlation ρ, where $\ln 0.5 < \rho < 0$, by simply setting the interval length $t = -\rho$. Note that for values of $t > -\ln 0.5$, $\Pr[N_1(t) = 0,\ N_2(t) = 0] = 0$, and the correlation ρ approaches 0 from $\ln 0.5$ as t increases; also, the joint probability mass function can be found by solving a set of linear equations using the symmetry of the bivariate distribution.

A test of "intrinsic correlation" in the theory of accident proneness has been discussed by Subrahmaniam (1966).

3 DISTRIBUTIONS RELATED TO BIVARIATE POISSON

Ahmad (1981) has noted that the probability generating function in (37.9) can be written in terms of confluent hypergeometric functions as

$$e^{\phi_{12}(t_1-1)(t_2-1)} \prod_{i=1}^{2} \left\{ {}_1F_1(1; 1; \phi_i t_i)/{}_1F_1(1; 1; \phi_i) \right\}\,, \tag{37.32}$$

where ${}_1F_1$ is as defined in Chapter 1, Eq. (1.121). Ahmad then suggested creating a generalized family of distributions by introducing extra parameters λ_1 and λ_2 in place of the second argument 1 in the functions ${}_1F_1$ on the right-hand side of (37.32). This leads to a *bivariate hyper-Poisson family* with probability generating function

$$e^{\phi_{12}(t_1-1)(t_2-1)} \prod_{i=1}^{2} \left\{ {}_1F_1(1; \lambda_i; \phi_i t_i)/{}_1F_1(1; \lambda_i; \phi_i) \right\}\,. \tag{37.33}$$

The marginal distributions in this case are hyper-Poisson (as defined in Chapter 4, Section 12.4).

Patil, Patel, and Kornes (1977) have discussed the bivariate truncated Poisson distribution. Shoukri and Consul (1982) have studied the properties and inferential issues concerning the bivariate modified power series distribution (see Chapter 38 for more details).

Compound bivariate Poisson distributions may be constructed by ascribing joint distributions to the expected values Θ_1 and Θ_2 of two mutually independent Poisson random variables X_1 and X_2. The probability generating function of X_1 and X_2 is

$$G(t_1, t_2) = e^{\Theta_1(t_1-1)+\Theta_2(t_2-1)} \qquad \text{[see (37.8)]}. \tag{37.34}$$

The probability generating function of the compound distribution is

$$E\left[\exp\{\Theta_1(t_1 - 1) + \Theta_2(t_2 - 1)\}\right] \tag{37.35}$$

(with respect to, of course, the joint distribution of Θ_1 and Θ_2). David and Papageorgiou (1994) have derived expressions for the moments of this distribution in terms of the moment generating function of the joint distribution of Θ_1 and Θ_2.

Zheng and Matis (1993) have extended the model (37.2) by defining, for $i = 1, 2$,

$$\begin{aligned} X_i &= Y_i + Y_{12} && \text{with probability } \omega_i \\ &= Y_i && \text{with probability } 1 - \omega_i\,. \end{aligned} \tag{37.36}$$

This model produces a mixture of four joint distributions with $(Y_1 + Y_{12}, Y_2 + Y_{12})$, $(Y_1 + Y_{12}, Y_2)$, $(Y_1, Y_2 + Y_{12})$, (Y_1, Y_2) in proportions $\omega_1\omega_2$, $\omega_1(1 - \omega_2)$, $(1 - \omega_1)\omega_2$, $(1 - \omega_1)(1 - \omega_2)$, respectively. The first- and second-order moments are

$$E[X_i] = E[Y_i] + \omega_i\, E[Y_{12}], \quad i = 1, 2, \tag{37.37}$$

$$\text{var}(X_i) = \text{var}(Y_i) + \omega_i \,\text{var}(Y_{12}) + \omega_i(1 - \omega_i)\{E[Y_{12}]\}^2, \; i = 1, 2, \tag{37.38}$$

and

$$\text{cov}(X_1, X_2) = \omega_1\omega_2 \,\text{var}(Y_{12}). \tag{37.39}$$

Zheng and Matis (1993) have noted that

$$0 < \text{cov}(X_1, X_2) < \min\{\text{var}(X_1), \text{var}(X_2)\}. \tag{37.40}$$

The marginal distributions are clearly mixtures of distributions of Y_i and $Y_i + Y_{12}$ $(i = 1, 2)$, respectively.

Of course, this type of extension can be introduced into more elaborate schemes. The above results apply for any distributions of Y_1, Y_2, and Y_{12} (not just Poisson) provided that they are independent (and also ω_1 and ω_2 act independently). In fact, Renshaw (1986) has used this model with Y_1, Y_2, and Y_{12} having independent negative binomial distributions (see Chapter 7) to represent spread of insect populations; see also Renshaw (1991) and Wehrly, Matis, and Otis (1993).

With the model (37.2), generalizations can be obtained by ascribing generalized Poisson distributions to some (or all) of the random variables Y_1, Y_2, and Y_{12}. If Consul's (1989) generalized Poisson distributions, also known as Lagrangian Poisson distributions (see Chapter 9, Section 11), with probability mass functions

$$\Pr[Y = y] = \theta(\theta + \lambda y)^{y-1} e^{-\theta-\lambda y}/y!\,,$$
$$y = 0, 1, 2, \dots, \quad 0 < \lambda < 1, \quad \theta > 0 \qquad (37.41)$$

are used, we obtain *bivariate generalized Poisson distributions* [see Stein and Juritz (1987)]. Stein and Juritz (1987) discussed four alternative methods of forming bivariate distributions with compound Poisson marginals. One of the methods is the trivariate reduction which was introduced by Mardia (1970). The bivariate Poisson distribution defined and studied by Holgate (1964) (see Section 2) is in fact based on this method. Consul (1989) has extended the definition in (37.41) to include negative values of λ, but has restricted values of y to ensure that $\theta + \lambda y$ is positive; see also Consul and Jain (1973), Consul and Shoukri (1985), and Consul (1994). However, if the values of $\Pr[Y = y]$ are added with respect to y, they no longer in general add to 1. We shall not consider here the case $\lambda < 0$. Denoting the values of (θ, λ) corresponding to Y_1, Y_2, and Y_{12} as (θ_1, λ_1), (θ_2, λ_2) and $(\theta_{12}, \lambda_{12})$, respectively, we have

$$\Pr\left[\bigcap_{i=1}^{2}(X_i = x_i)\right]$$
$$= \theta_{12}\theta_1\theta_2\, e^{-\theta_{12}-\theta_1-\theta_2-\lambda_1 x_1-\lambda_2 x_2}$$
$$\times \sum_{i=0}^{\min(x_1,x_2)} \frac{\{\theta_1 + (x_1 - i)\lambda_1\}^{x_1-i-1}}{(x_1 - i)!} \frac{\{\theta_2 + (x_2 - i)\lambda_2\}^{x_2-i-1}}{(x_2 - i)!}$$
$$\times \frac{(\theta_{12} + \lambda_{12} i)^{i-1}}{i!} e^{i(\lambda_1+\lambda_2-\lambda_{12})}. \qquad (37.42)$$

The moments and product moments of X_1 and X_2 can be derived easily from (37.2). In particular,

$$\operatorname{cov}(X_1, X_2) = \operatorname{var}(Y_{12}) = \theta_{12}/(1 - \lambda_{12})^3 \qquad (37.43)$$

and

$$\operatorname{var}(X_i) = \operatorname{var}(Y_i) + \operatorname{var}(Y_{12}) = \frac{\theta_i}{(1 - \lambda_i)^3} + \frac{\theta_{12}}{(1 - \lambda_{12})^3}, \quad i = 1, 2. \qquad (37.44)$$

In the special case when $\lambda_1 = \lambda_2 = \lambda_{12} = \lambda$, we observe from (37.43) and (37.44) that

$$\operatorname{corr}(X_1, X_2) = \frac{\theta_{12}}{\sqrt{(\theta_1 + \theta_{12})(\theta_2 + \theta_{12})}} \qquad (37.45)$$

as in the case of the bivariate Poisson distribution [see Eq. (37.11)]. In this case, the marginal distributions of X_1 and X_2 are generalized Poisson with parameters $(\theta_1 + \theta_{12}, \lambda)$ and $(\theta_2 + \theta_{12}, \lambda)$, respectively. Further, the regressions of X_2 on X_1 and of X_1 on X_2 are both linear with [Famoye and Consul (1995)]

$$E[X_2 \mid X_1 = x_1] = \frac{\theta_2}{1-\lambda} + \frac{\theta_{12}}{\theta_1 + \theta_{12}} x_1 \tag{37.46}$$

and

$$E[X_1 \mid X_2 = x_2] = \frac{\theta_1}{1-\lambda} + \frac{\theta_{12}}{\theta_2 + \theta_{12}} x_2 . \tag{37.47}$$

For the general distribution in (37.42), Famoye and Consul (1995) have presented a method of estimating the six parameters ($\theta_1, \theta_2, \theta_{12}; \lambda_1, \lambda_2, \lambda_{12}$) based on equating the observed frequency (f_{00}) of double zeros ($x_1 = x_2 = 0$) and the observed sample values of the first and second moments and the product moment to their population values in terms of the six parameters. They have obtained the formulas

$$\tilde{\lambda}_1 = 1 - \left\{ \frac{m'_{10} - m'_{01} + (1-\tilde{\lambda}_2)^2 (m_{02} - m_{11})}{m_{20} - m_{11}} \right\}^{1/2} , \tag{37.48}$$

$$\tilde{\lambda}_{12} = 1 - \left\{ \frac{m'_{01} - (1-\tilde{\lambda}_2)^2 (m_{02} - m_{11})}{m_{11}} \right\}^{1/2} , \tag{37.49}$$

$$\tilde{\theta}_1 = (1-\tilde{\lambda}_1)^3 (m_{20} - m_{11}) , \tag{37.50}$$

$$\tilde{\theta}_2 = (1-\tilde{\lambda}_2)^3 (m_{02} - m_{11}) , \tag{37.51}$$

$$\tilde{\theta}_{12} = (1-\tilde{\lambda}_{12})^3 m_{11} , \tag{37.52}$$

with $\tilde{\lambda}_2$ satisfying the equation

$$\begin{aligned} \log f_{00} &+ \left[\left\{ m'_{10} - m'_{01} + (1-\tilde{\lambda}_2)^2 (m_{02} - m_{11}) \right\}^{3/2} / \sqrt{m_{20} - m_{11}} \right] \\ &+ (1-\tilde{\lambda}_2)^3 (m_{02} - m_{11}) \\ &+ \left[\left\{ m'_{01} - (1-\tilde{\lambda}_2)^2 (m_{02} - m_{11}) \right\}^{3/2} / \sqrt{m_{11}} \right] = 0 . \end{aligned} \tag{37.53}$$

Equation (37.53) needs to be solved for $\tilde{\lambda}_2$ by numerical methods. Famoye and Consul (1995) have suggested taking the initial value of $\tilde{\lambda}_2$ to be zero. These authors have also provided formulas for determining the maximum likelihood estimates of the parameters. A simulation study for one special set of parameter values comparing the two methods of estimation, based on pseudorandom samples of sizes 200 and 500, has indicated that the standard deviations of the MLEs are smaller than the corresponding quantities for estimators based on moments and frequency of double zeros, though the biases of the MLEs are sometimes of larger magnitude. These values are summarized in Table 37.1. It has to be borne in mind that different comparisons may be found for some other sets of parameter values.

There are no bivariate distributions with Poisson marginals for which the conditional distribution of either variable, given the other, is Poisson. [In fact, if Z_2, given

Table 37.1. Comparison of Maximum Likelihood Estimators (MLEs) and Moment/ Double Zero Estimators[a]

Parameter Value	Sample Size	MLEs		Moment/Double Zero Estimators	
		Bias	Standard Deviation	Bias	Standard Deviation
$\theta_1 = 1.2$	200	−0.0102	0.169	−0.0325	0.239
	500	−0.0007	0.109	−0.0222	0.155
$\theta_2 = 1.0$	200	0.0372	0.153	0.0394	0.221
	500	0.0081	0.098	0.0012	0.150
$\theta_{12} = 0.8$	200	−0.0114	0.142	0.0052	0.203
	500	0.0006	0.090	0.0199	0.130
$\lambda_1 = 0.1$	200	0.0087	0.053	0.0214	0.069
	500	0.0025	0.037	0.0112	0.051
$\lambda_2 = 0.2$	200	−0.0252	0.056	−0.0299	0.076
	500	−0.0117	0.041	−0.0125	0.058
$\lambda_{12} = 0.3$	200	−0.0011	0.068	−0.0064	0.084
	500	−0.0018	0.043	−0.0094	0.053

[a] Some errors in the tables published have been brought to our attention by the authors. These are the corrected values.

$Z_1 = z_1$, has a Poisson distribution, then the overall distribution of Z_2 cannot be Poisson unless Z_1 and Z_2 are mutually independent or Z_1 can take on only a single value—because a mixture of Poisson distributions cannot be a Poisson distribution.]

Wesolowski (1994) has shown that if

(i) Z_1, Z_2 are random variables taking values $0, 1, 2, \ldots$,

(ii) the conditional distribution of Z_2, given $Z_1 = z_1$, is Poisson with mean $\lambda_2\lambda_{12}^{z_1}$, and

(iii) the regression of Z_1 on Z_2 is

$$E[Z_1 \mid Z_2 = z_2] = \lambda_1\lambda_{12}^{z_2},$$

then the joint probability mass function of Z_1 and Z_2 is

$$\Pr[Z_1 = z_1,\ Z_2 = z_2] = K(\lambda_1, \lambda_2, \lambda_{12})\lambda_1^{z_1}\lambda_2^{z_2}\lambda_{12}^{z_1 z_2}/(z_1!z_2!), \qquad (37.54)$$

where $K(\lambda_1, \lambda_2, \lambda_{12})$ is the norming constant of the distribution.

Wesolowski terms this a *bivariate Poisson conditionals distribution* and notes that a result of Arnold and Strauss (1991) shows that this is the *only* bivariate distribution for which both conditional distributions are Poisson. Further, the stated conditions also ensure that the conditional distribution of Z_1, given $Z_2 = z_2$, is Poisson. However, the marginal distributions of Z_1 and Z_2 are **not** Poisson. Wesolowski (1993) has given some other characterizations of this family, based on the conditional distributions and the properties

$$E\left[\theta_{12}^{\pm X_i} \mid X_{3-i} = x_{3-i}\right] = \exp\left\{\theta_1 \theta_{12}^{x_{3-i}}(\theta_{12}^{\pm 1} - 1)\right\}, \quad i = 1, 2.$$

Leiter and Hamdan (1973) introduced a bivariate distribution with one marginal as Poisson with parameter λ and the other marginal as Neyman Type A with parameters (λ, β). It has the structure

$$Y = Y_1 + Y_2 + \cdots + Y_X, \tag{37.55}$$

the conditional distribution of Y given $X = x$ as Poisson with parameter βx, and the joint probability mass function as

$$\Pr[X = x, Y = y] = e^{-(\lambda+\beta x)} \frac{\lambda^x}{x!} \frac{(\beta x)^y}{y!}, \quad \lambda, \beta > 0, \; x, y = 0, 1, \ldots .$$

This distribution has been discussed by Cacoullos and Papageorgiou (1980, 1981). This distribution has been referred to as the *bivariate Neyman Type A distribution* by Huang and Fung (1993b), which, as pointed out by Papageorgiou (1995), unfortunately is misleading since this name is used for bivariate distributions with both marginals being of Neyman Type A; see, for example, Holgate (1966), Gillings (1974), Papageorgiou and Kemp (1983), and Kocherlakota and Kocherlakota (1992). An appropriate name for the above distribution is the *bivariate Poisson-Poisson distribution* as suggested originally by Leiter and Hamdan (1973).

Huang and Fung (1993b) considered the structure in (37.55) with X as a Poisson random variable and $Y_1, Y_2, \ldots, Y_X$ as independent (also independently of X) and identically distributed doubly truncated Poisson (D-distributed) random variables. The distribution so derived, with the above structure and the conditional distribution of Y, given $X = x$, is a D-distribution with parameters (x, N, M, β) [see Huang and Fung (1993a)] and has the joint probability mass function

$$\Pr[X = x, Y = y] = e^{-\lambda} D_I(y, x; (N, M)) \beta^y \left\{\frac{\lambda}{e(N, M; \beta)}\right\}^x \Big/ (x!y!),$$

$$x = 0, 1, \ldots, \left[\frac{y}{N}\right]; \; y = xN, xN + 1, \ldots, xM;$$

$D_I(y, x; (N, M))$ are the D_I-numbers given by Huang and Fung (1993a), and $e(N, M; \beta)$ is the incomplete exponential function defined by

$$e(N, M; \beta) = \begin{cases} \sum_{i=N}^{M} \beta^i/i!, & \beta > 0, \; 0 \le N < M, \; N, M \text{ are integers} \\ \sum_{i=0}^{M} \beta^i/i!, & \beta > 0, \; N = -1, -2, \ldots \\ 0, & \text{otherwise.} \end{cases}$$

Huang and Fung (1993b) have referred to this bivariate distribution as the *D compound Poisson distribution*.

However, from the general result of Cacoullos and Papageorgiou (1981) that for distributions with structure (37.55) the joint probability generating function $G_{X,Y}(u, v)$

is

$$G_{X,Y}(u, v) = G_1\left(u\,G_2(v)\right),$$

where $G_1(\cdot)$ is the probability generating function of X and $G_2(\cdot)$ is the probability generating function of $Y_1, Y_2, \ldots, Y_X$, Papageorgiou (1995) has shown that many properties can be easily established for the D compound Poisson distribution. For example, noting that

$$G_1(u) = e^{\lambda(u-1)} \quad \text{and} \quad G_2(v) = e(N, M; v\beta)/e(N, M; \beta),$$

one readily obtains the joint probability generating function of the D compound Poisson distribution as

$$G_{X,Y}(u, v) = \exp\left\{\lambda\left(u\,\frac{e(N, M; v\beta)}{e(N, M; \beta)} - 1\right)\right\}$$

and the marginal probability generating function of Y as

$$G_Y(v) = G_{X,Y}(1, v) = \exp\left\{\lambda\left(\frac{e(N, M; v\beta)}{e(N, M; \beta)} - 1\right)\right\}.$$

This is the Poisson generalized by compounding with a D-distribution; see Papageorgiou (1995).

Papageorgiou (1986), Kocherlakota and Kocherlakota (1992), and Papageorgiou and Piperigou (1997) have all discussed various forms of *bivariate short distributions*. Since the univariate short distribution is the convolution of a Poisson random variable and a Neyman Type A random variable [see Section 9 of Chapter 9], Papageorgiou (1986) considered the structure

$$X = Z_1 + Z_3 \qquad \text{and} \qquad Y = Z_2 + Z_3,$$

where Z_1, Z_2, and Z_3 are independent univariate Neyman Type A or Poisson random variables, thus introducing two bivariate short distributions with joint probability generating functions

$$\exp\left[\lambda_1\left\{e^{\theta_1(u-1)} - 1\right\} + \lambda_2\left\{e^{\theta_2(v-1)} - 1\right\} + \phi(uv - 1)\right]$$

and

$$\exp\left[\phi_1(u - v) + \phi_2(v - 1) + \lambda\left\{e^{\theta(uv-1)} - 1\right\}\right].$$

It is clear that both these bivariate distributions have their univariate marginals to be univariate short distributions with probability generating function of the form

$$\exp\left[\lambda\left\{e^{\theta(u-1)} - 1\right\} + \phi(u - 1)\right].$$

Instead, by using the above structure of X and Y but with $(Z_1, Z_2)'$ having a bivariate Neyman Type A Type II distribution and Z_3 having independently a Poisson distribution, Kocherlakota and Kocherlakota (1992, p. 285) derived a more general bivariate short distribution with joint probability generating function

$$\exp\left[\lambda_1\left\{e^{\theta_1(u-1)}-1\right\}+\lambda_2\left\{e^{\theta_2(v-1)}-1\right\}+\lambda_3\left\{e^{\theta_1(u-1)+\theta_2(v-1)}-1\right\}+\lambda(uv-1)\right].$$

They also derived another general bivariate short distribution by assuming $(Z_1, Z_2)'$ to have a bivariate Poisson distribution and Z_3 to have an independent Neyman Type A distribution. The joint probability generating function of the resulting distribution is

$$\exp\left[\lambda_1(u-1)+\lambda_2(v-1)+\lambda_3(uv-1)+\lambda\left\{e^{\theta(uv-1)}-1\right\}\right].$$

It is of interest to mention here that this distribution may also be derived by taking Z_1 and Z_2 to be independent Poisson random variables and Z_3 having independently a short distribution.

Recently, Papageorgiou and Piperigou (1997) considered the more general structure

$$X = Z_1 + W_1 \qquad \text{and} \qquad Y = Z_2 + W_2$$

and derived three more general forms of bivariate short distributions. The joint probability generating functions of these three forms are

$$\exp\left[\lambda_1\left\{u\,e^{\lambda_2(v-1)}-1\right\}+\theta_1\left\{e^{\theta_2(u-1)}-1\right\}+\theta_3(v-1)\right],$$

$$\exp\left[\lambda_1\left\{u\,e^{\lambda_2(v-1)}-1\right\}+\theta_1\left\{v\,e^{\theta_2(u-1)}-1\right\}\right],$$

and

$$\exp\left[\lambda\left\{e^{\theta\,p(u-1)+\theta\,q(v-1)}-1\right\}+\phi\,p(u-1)+\phi\,q(v-1)\right]\quad(q=1-p).$$

The first form is derived by assuming $(Z_1, Z_2)'$ to have the bivariate Poisson–Poisson distribution [Leiter and Hamdan (1973); also see (37.55)] and W_1 and W_2 to be independent [and also independent of $(Z_1, Z_2)'$] Neyman Type A and Poisson random variables, respectively. The second form is derived by assuming $(Z_1, Z_2)'$ and $(W_1, W_2)'$ both to be distributed (independently) as bivariate Poisson–Poisson.

Further, this approach can also be used to derive several bivariate versions of discrete univariate distributions. For example, the *Delaporte distribution* is a convolution of a negative binomial and a Poisson [see Section 12.5 of Chapter 5; also see Willmot (1989) and Willmot and Sundt (1989)]. Then, several forms of bivariate Delaporte distributions, with marginal Delaporte distributions with probability generating function

$$\left(\frac{1-pu}{1-p}\right)^{-k} e^{\lambda(u-1)},$$

can be derived in this method. Papageorgiou and Piperigou (1997) have given four forms of bivariate Delaporte distributions with joint probability generating functions as follows:

$$\left(\frac{1-p_1u}{1-p_1}\right)^{-k_1}\left(\frac{1-p_2v}{1-p_2}\right)^{-k} e^{\lambda(uv-1)},$$

$$\left(\frac{1-puv}{1-p}\right)^{-k} e^{\lambda_1(u-1)+\lambda_2(v-1)},$$

$$\left(\frac{1-puv}{1-p}\right)^{-k} e^{\lambda_1(u-1)+\lambda_2(v-1)+\lambda_3(uv-1)},$$

and

$$\left(\frac{1-p_1u-p_2v-p_3uv}{1-p_1-p_2-p_3}\right)^{-k} e^{\lambda(u-1)}.$$

These distributions have been derived by assuming binomial, Poisson, negative binomial, bivariate Poisson, or bivariate negative binomial distributions for different random variables in the structure. Three other bivariate Delaporte distributions have also been derived by Papageorgiou and Piperigou (1997) by using the bivariate short distributions described earlier.

4 MULTIVARIATE FORMS

It is quite straightforward to generalize the definition in (37.2) of the bivariate Poisson distribution to the joint distribution of k (> 2) random variables $X_1, X_2, \ldots, X_k$ by setting

$$X_i = Y_i + Y, \qquad i = 1, 2, \ldots, k, \tag{37.56}$$

where $Y, Y_1, \ldots, Y_k$ are mutually independent Poisson random variables with means $\theta, \theta_1, \ldots, \theta_k$, respectively. [Note that Y takes the place of Y_{12} in (37.2).] Clearly, $X_1, \ldots, X_k$ marginally have Poisson distributions with means $\theta_1 + \theta, \ldots, \theta_k + \theta$, respectively, and the correlation coefficient between X_i and X_j is $\theta/\sqrt{(\theta_i + \theta)(\theta_j + \theta)}$. It is also clear that $X_i - X_j$ and $X_g - X_h$ are mutually independent if i, j, g, and h are distinct.

More elaborate structures of multivariate Poisson distributions than the one in (37.56) can easily be constructed, still retaining the expressions for each X_i as sums of independent Poisson variables. Teicher (1954) and Dwass and Teicher (1957) have shown that for all such multivariate distributions, the probability generating functions are of form

$$\exp\left\{\sum_{i=1}^{k} A_i t_i + \sum\sum_{1\le i<j\le k} A_{ij} t_i t_j + \cdots + A_{12\ldots k} t_1 t_2 \cdots t_k - A\right\}, \tag{37.57}$$

where

$$A = \sum_{i=1}^{k} A_i + \sum\sum_{1 \le i < j \le k} A_{ij} + \cdots + A_{12\ldots k}; \tag{37.58}$$

also see Mahamunulu (1967). These distributions are infinitely divisible.

Šidák (1972, 1973a,b) has presented several inequalities for some general multivariate discrete distributions, including the multivariate Poisson. Wang (1974) has established some characterization results in this case.

The special case ($k = 3$) of *trivariate Poisson distributions* has been discussed in detail by Mahamunulu (1967), Kawamura (1976), Loukas and Kemp (1983), and Loukas and Papageorgiou (1991). In the last-mentioned paper, there are extensions of the recurrence relations in (37.5) and their application in determining maximum likelihood estimates of the parameters [see Eqs. (37.16)–(37.21)].

In the bivariate case, as we have already seen in Section 2, the model [cf. (37.2)]

$$X_1 = Y_1 + Y_{12} \quad \text{and} \quad X_2 = Y_2 + Y_{12} \tag{37.59}$$

with Y_1, Y_2, and Y_{12} being independent Poisson random variables with parameters θ_1, θ_2 and θ_{12}, respectively, leads to a bivariate Poisson distribution. Similarly, if we consider a bivariate binomial experiment in which, in every trial, an item is declared conforming or nonconforming by two characteristics A and B with a corresponding 2×2 table of probabilities (see Section 2) and let $n \to \infty$, $p_{11}, p_{10}, p_{01} \to 0$ such that $np_{10} \to \theta_1$, $np_{01} \to \theta_2$, and $np_{11} \to \theta_{12}$, we obtain the same bivariate Poisson distribution. Then, in correspondence with the above model (37.59), this means we are taking N_{10} (number of nonconforming by A alone), N_{01} (number of nonconforming by B alone), and N_{11} (number of nonconforming by both A and B) to be independent Poisson random variables, and $X_1 = N_{10} + N_{11}$ (number of nonconforming by A) and $X_2 = N_{01} + N_{11}$ (number of nonconforming by B). Thus, both constructions lead to the same interpretation of the bivariate Poisson distribution. However, as Balakrishnan, Johnson, and Kotz (1995) recently pointed out, the two approaches lead to different interpretations of the resulting trivariate Poisson distribution. To see this, let us consider the following trivariate binomial experiment in which n trials are made with every item being declared conforming or nonconforming by three characteristics A, B, and C. We then have the following possibilities:

ABC	probability p_{111},	AB^cC	probability p_{101},
ABC^c	probability p_{110},	AB^cC^c	probability p_{100},
A^cBC	probability p_{011},	A^cB^cC	probability p_{001},
A^cBC^c	probability p_{010},	$A^cB^cC^c$	probability p_{000}.

Let us define the variables X_1 (number of nonconforming by A), X_2 (number of nonconforming by B), and X_3 (number of nonconforming by C). We then have

$$\Pr[X_1 = x_1,\ X_2 = x_2,\ X_3 = x_3]$$
$$= \sum_{a=0}^{\min(x_1,x_2)} \sum_{b=0}^{\min(x_2,x_3)} \sum_{c=0}^{\min(x_1,x_3)} \sum_{i=0}^{\min(a,b,c)} n!\big\{i!(a-i)!(c-i)!$$
$$\times (x_1 - a - c + i)!(b-i)!(x_2 - a - b + i)!(x_3 - b - c + i)!$$
$$\times (n - x_1 - x_2 - x_3 + a + b + c - i)!\big\}^{-1}$$
$$\times p_{111}^{i}\, p_{110}^{a-i}\, p_{101}^{c-i}\, p_{100}^{x_1-a-c+i}\, p_{011}^{b-i}\, p_{010}^{x_2-a-b+i}\, p_{001}^{x_3-b-c+i}$$
$$\times p_{000}^{n-x_1-x_2-x_3+a+b+c-i}\,. \qquad (37.60)$$

If we now let $n \to \infty$, $p_{111}, p_{110}, p_{101}, p_{100}, p_{011}, p_{010}$, and p_{001} all go to 0 such that $np_{111} \to \theta_{111}$, $np_{110} \to \theta_{110}$, $np_{101} \to \theta_{101}$, $np_{100} \to \theta_{100}$, $np_{011} \to \theta_{011}$, $np_{010} \to \theta_{010}$, and $np_{001} \to \theta_{001}$, we derive the limiting trivariate Poisson distribution as

$$\Pr[X_1 = x_1,\ X_2 = x_2,\ X_3 = x_3]$$
$$= \exp\{-(\theta_A + \theta_B + \theta_C) + (\theta_{AB} + \theta_{BC} + \theta_{CA}) - \theta_{ABC}\}$$
$$\times \sum_{a=0}^{\min(x_1,x_2)} \sum_{b=0}^{\min(x_2,x_3)} \sum_{c=0}^{\min(x_1,x_3)} \sum_{i=0}^{\min(a,b,c)} \big\{i!(a-i)!(c-i)!$$
$$\times (x_1 - a - c + i)!(b-i)!(x_2 - a - b + i)!(x_3 - b - c + i)!\big\}^{-1}$$
$$\times \theta_{ABC}^{i}(\theta_{AB} - \theta_{ABC})^{a-i}(\theta_{AC} - \theta_{ABC})^{c-i}$$
$$\times (\theta_A - \theta_{AB} - \theta_{AC} + \theta_{ABC})^{x_1-a-c+i}$$
$$\times (\theta_{BC} - \theta_{ABC})^{b-i}(\theta_B - \theta_{AB} - \theta_{BC} + \theta_{ABC})^{x_2-a-b+i}$$
$$\times (\theta_C - \theta_{BC} - \theta_{AC} + \theta_{ABC})^{x_3-b-c+i}. \qquad (37.61)$$

In (37.61), $\theta_{ABC} = \theta_{111}$, $\theta_{AB} = \theta_{110} + \theta_{111}$, $\theta_{BC} = \theta_{011} + \theta_{111}$, $\theta_{AC} = \theta_{101} + \theta_{111}$, $\theta_A = \theta_{111} + \theta_{110} + \theta_{101} + \theta_{100}$, and so on.

The trivariate Poisson distribution in (37.61), however, has a different interpretation than the trivariate Poisson distribution introduced by the model (37.56). In (37.56), we are essentially taking Y_1 (number of nonconforming by A and B but not C), Y_2 (number of nonconforming by B and C but not A), Y_3 (number of nonconforming by A and C but not B), and Y (number of nonconforming by A, B and C) all to be independent Poisson variables; consequently, the variables

$$X_1 = Y_1 + Y = \text{Number of nonconforming by } A \text{ and } B,$$
$$X_2 = Y_2 + Y = \text{Number of nonconforming by } B \text{ and } C,$$

and

$$X_3 = Y_3 + Y = \text{Number of nonconforming by } A \text{ and } C$$

form the trivariate Poisson distribution according to (37.56). In the trivariate Poisson distribution in (37.61), however, the variables X_1, X_2, and X_3 represent the number of nonconforming by A, B, and C, respectively.

Ong (1988) introduced the univariate Charlier series distribution with probability generating function

$$G(t) = e^{\theta p(t-1)}(1 - p + pt)^n, \qquad 0 < p < 1,\ \theta > 0 \tag{37.62}$$

as the conditional distribution of a bivariate Poisson distribution. Upton and Lampitt (1981) formed this distribution as the convolution of a Poisson with a binomial and used it to describe the interyear change in the size of bird populations. Similarly, by starting with the trivariate Poisson distribution with probability generating function [Loukas and Papageorgiou (1991) and Loukas (1993)]

$$\begin{aligned} G(t_1, t_2, t_3) = \exp\{&(a-d)(t_1-1) + (b-d)(t_2-1) + (c-d)(t_3-1) \\ &+ d(t_1t_2t_3 - 1)\}, \\ &\qquad a, b, c > 0,\ 0 < d < \min(a, b, c)\,, \end{aligned} \tag{37.63}$$

Papageorgiou and Loukas (1995) derived the bivariate Charlier series distribution with probability generating function

$$\exp\{\lambda_1 p(t_1 - 1) + \lambda_2 p(t_2 - 1)\}(q + pt_1t_2)^n,\ \lambda_1, \lambda_2 > 0 \tag{37.64}$$

as one of the conditionals. They also noted that this bivariate Charlier series distribution has the structure

$$X_1 = Y_1 + Y \qquad \text{and} \qquad X_2 = Y_2 + Y\,, \tag{37.65}$$

where Y_1, Y_2, and Y are independent random variables with Y_1 distributed as Poisson $(\lambda_1 p)$, Y_2 as Poisson $(\lambda_2 p)$, and Y as binomial (n, p). However, if we assume that Y_1 and Y_2 are independent binomials (n_i, p) $(i = 1, 2)$ and Y is Poisson (λp), then an alternative bivariate Charlier series distribution with probability generating function

$$(q + pt_1)^{n_1}(q + pt_2)^{n_2}\, e^{\lambda p(t_1t_2 - 1)} \tag{37.66}$$

is obtained. On the other hand, if we assume that Y_1 and Y_2 are jointly distributed as bivariate Poisson (bivariate binomial) and Y is independently distributed as binomial (Poisson), more general bivariate Charlier series distributions with probability generating functions

$$\exp\{\lambda_1 p(t_1 - 1) + \lambda_2 p(t_2 - 1) + \lambda_3 p(t_1t_2 - 1)\}(q + pt_1t_2)^n \tag{37.67}$$

and

$$(q + pt_1)^{n_1}(q + pt_2)^{n_2}(q + pt_1t_2)^{n_3}\, e^{\lambda p(t_1t_2 - 1)}\,, \tag{37.68}$$

respectively, can be derived; see, for example, Kocherlakota and Kocherlakota (1992, p. 286). Parameter estimation methods for the bivariate Charlier series distribution (37.64) have been discussed by Papageorgiou and Loukas (1995).

An explicit formula for the probability mass function corresponding to (37.57) is

$$\Pr\left[\bigcap_{i=1}^{k}(X_i = x_i)\right] = \left\{\prod_{i=1}^{k} e^{-\phi_i}\phi_i^{x_i}/x_i!\right\} \exp\left\{\sum_i \sum_j \phi_{ij} C(x_i)C(x_j) + \sum_i \sum_j \sum_g \phi_{ijg} C(x_i)C(x_j)C(x_g) + \cdots + \phi_{12\ldots k} C(x_1)\,C(x_2)\,\ldots\, C(x_k)\right\}, \tag{37.69}$$

where $C(\cdot)$ is a (Gram-) Charlier Type B polynomial (see Chapter 11, Section 8), and

$$\phi_{ijg\ldots} = E[X_i X_j X_g \cdots]\,;$$

in particular, $\phi_i = E[X_i]$. See also Krishnamoorthy (1951). Bahadur (1961) presented the approximation

$$\Pr\left[\bigcap_{i=1}^{k}(X_i = x_i)\right] \doteq \left\{\prod_{i=1}^{k} e^{-\phi_i}\phi_i^{x_i}/x_i!\right\}\left\{1 + \sum_{i<j=1}^{k}\sum \frac{\phi_{ij}}{\sqrt{\phi_i\phi_j}}\, \frac{x_i - \phi_i}{\sqrt{\phi_i}}\, \frac{x_j - \phi_j}{\sqrt{\phi_j}}\right\}. \tag{37.70}$$

Kawamura (1979) obtained this class of multivariate distributions as the limit of multivariate binomial distributions (see Chapter 36, Section 7) with probability generating functions

$$G_n(t_1, t_2, \ldots, t_k) = \left(\sum_{a_1=0}^{1}\sum_{a_2=0}^{1}\cdots\sum_{a_k=0}^{1} p_{a_1 a_2 \ldots a_k} t_1^{a_1} t_2^{a_2} \ldots t_k^{a_k}\right)^n \tag{37.71}$$

as $n \to \infty$ with $np_{a_1 a_2 \ldots a_k} \to w_{a_1 a_2 \ldots a_k} > 0$. He has shown, *inter alia*, that this class of distributions is closed under convolution and has Poisson marginals. Also, it is easy to see that

$$E[X_i] = \sum_{a_1=0}^{1}\cdots\sum_{a_{i-1}=0}^{1}\sum_{a_{i+1}=0}^{1}\cdots\sum_{a_k=0}^{1} w_{a_1\ldots a_{i-1}\,1\,a_{i+1}\ldots a_k}, \tag{37.72}$$

$$E[X_i X_j] = \sum_{a_1=0}^{1}\cdots\sum_{a_{i-1}=0}^{1}\sum_{a_{i+1}=0}^{1}\cdots\sum_{a_{j-1}=0}^{1}\sum_{a_{j+1}=0}^{1}\cdots\sum_{a_k=0}^{1} w_{a_1\ldots a_{i-1}\,1\,a_{i+1}\ldots a_{j-1}\,1\,a_{j+1}\ldots a_k}, \tag{37.73}$$

and

$$\text{cov}(X_i, X_j) = \text{sum of } w\text{'s over all sets of } a\text{'s for which } a_i = a_j = 1.$$

For the computation of the bivariate Poisson probability mass function (37.3), Kawamura (1985, 1987) presented the recurrence relations

$$a\,P(a,b) = \lambda_1\,P(a-1,b) + \lambda_{12}\,P(a-1,b-1), \quad a,b \geq 1,$$

$$b\,P(a,b) = \lambda_2\,P(a,b-1) + \lambda_{12}\,P(a-1,b-1), \quad a,b \geq 1,$$

$$a\,P(a,0) = \lambda_1\,P(a-1,0), \quad a \geq 1,$$

$$b\,P(0,b) = \lambda_2\,P(0,b-1), \quad b \geq 1,$$

and

$$P(0,0) = e^{-\lambda_1-\lambda_2-\lambda_{12}}\,. \tag{37.74}$$

Similar recurrence relations of the general form

$$\Pr[\boldsymbol{X} = \boldsymbol{a}] = \sum_{i=1}^{k} c_i\lambda_i\,\Pr[\boldsymbol{X} = \boldsymbol{a} - \boldsymbol{b}_i] \tag{37.75}$$

have been presented by Kano and Kawamura (1991) for the multivariate probability mass function.

These distributions are, in fact, *first-order generalized multivariate Hermite distributions* which are special cases of *mth-order* distributions of this kind. These have probability generating functions of form

$$G(t_1, t_2, \ldots, t_k) = \exp\left\{\sum_{0\leq \boldsymbol{\ell}'\mathbf{1}\leq m} a_{\boldsymbol{\ell}}\boldsymbol{t}^{\boldsymbol{\ell}}\right\}, \tag{37.76}$$

where m is a positive integer, $\boldsymbol{\ell}' = (\ell_1, \ell_2, \ldots, \ell_k)$ with ℓ_i as non-negative integers, and $\boldsymbol{t}^{\boldsymbol{\ell}} = t_1^{\ell_1} t_2^{\ell_2} \ldots t_k^{\ell_k}$. Milne and Westcott (1993) have proved that for (37.76) to be a proper probability generating function, in addition to $\sum_{\boldsymbol{\ell}} a_{\boldsymbol{\ell}} = 0$, it is necessary and sufficient that $a_{\boldsymbol{\ell}}$ be positive for all $\boldsymbol{\ell}'\mathbf{1} = 1$, or $\max_{1\leq i\leq k} \ell_i \geq m-1$, and that the magnitudes of every negative coefficient be "suitably small."

Kemp and Papageorgiou (1982) have described H_5 and H_8 distributions which are second-order and fourth-order Hermite distributions, respectively.

Among other definitions of classes of multivariate distributions with Poisson marginals, we note that of Lukacs and Beer (1977) which has characteristic function of form

$$\varphi(t_1, t_2, \ldots, t_k) = \exp\left\{\sum_{a_1=0}^{1} \cdots \sum_{a_k=0}^{1} \lambda_{a_1\ldots a_k} \exp\left(i\sum_{j=1}^{k} a_j t_j\right)\right\}. \tag{37.77}$$

In a note appended to a paper by Cuppens (1968), Lévy (1968) criticized this definition because he felt that "the logarithm of a Poisson distribution should have only one term" and then suggested

$$\varphi(t_1, t_2, \ldots, t_k) = \exp\left\{\lambda \exp\left(i \sum_{j=1}^{k} a_j t_j\right)\right\} \tag{37.78}$$

(with $\lambda > 0$) as a "suitable characteristic function." Motivated by this remark, Lukacs and Beer (1977) studied characteristic functions of form

$$\varphi(t_1, t_2, \ldots, t_k) = \exp\left\{\lambda \exp\left(i \sum_{j=1}^{k} t_j\right) - 1\right\}. \tag{37.79}$$

This is a special case of (37.77), and (37.77) corresponds to a convolution of multivariate distributions with characteristic functions of the same form. Lukacs and Beer (1977) then established the following characterization of this class of distributions:

> If $\boldsymbol{X}'_g \equiv (X_{1g}, X_{2g}, \ldots, X_{kg})$, $g = 1, 2, \ldots, n$, represent n independent and identically distributed $k \times 1$ vectors of random variables with a common distribution having finite moments up to and including order 2, then the expected values of
>
> $$S_j = \frac{1}{n-1} \sum_{g=1}^{n} (X_{jg} - \overline{X}_{j.})^2 - \overline{X}_{j.} \qquad \left(\text{with } \overline{X}_{j.} = \frac{1}{n} \sum_{g=1}^{n} X_{jg}\right), \tag{37.80}$$
>
> given $(\overline{X}_{1.}, \overline{X}_{2.}, \ldots, \overline{X}_{k.})$, are zero for all $j = 1, 2, \ldots, k$, if and only if the joint distribution has a characteristic function of form (37.77).

Lukacs and Beer (1977) also established a similar characterization result for distributions with characteristic function (37.79). Under the same conditions, this result requires that the expected values of

$$T_j = \frac{1}{n-1} \sum_{g=1}^{n} (X_{jg} - \overline{X}_{j.})(X_{j+1,g} - \overline{X}_{j+1,.}) - \overline{X}_{j.} \qquad (j = 1, 2, \ldots, k-1) \tag{37.81}$$

and

$$T_k = \frac{1}{n-1} \sum_{g=1}^{n} (X_{1g} - \overline{X}_{1.})(X_{kg} - \overline{X}_{k.}) - \overline{X}_{k.}\,, \tag{37.82}$$

given $(\overline{X}_{1.}, \overline{X}_{2.}, \ldots, \overline{X}_{k.})$, all be zero.

Liang (1989) has studied the structure of trivariate compound Poisson distributions with probability generating functions of form

$$G(t_1, t_2, t_3) = E\left[\exp\left\{\tau\left[\sum_{i=1}^{3} \lambda_i (t_i - 1) + \sum_{i<j=1}^{3} \lambda_{ij}(t_i t_j - 1) + \lambda_{123}(t_1 t_2 t_3 - 1)\right]\right\}\right], \tag{37.83}$$

where τ is a random variable with moment generating function $M_\tau[z]$. Then, (37.83) becomes

$$G(t_1, t_2, t_3) = M_\tau\left[\sum_{i=1}^{3} \lambda_i(t_i - 1) + \sum_{i<j=1}^{3}\sum \lambda_{ij}(t_i t_j - 1) + \lambda_{123}(t_1 t_2 t_3 - 1)\right]. \quad (37.84)$$

Extension of this family to general k-variate distributions can be written obviously as one with probability generating function

$$G(t_1, t_2, \ldots, t_k) = M_\tau\left[\sum_{i=1}^{k} \lambda_i(t_i - 1) + \sum_{i<j=1}^{k}\sum \lambda_{ij}(t_i t_j - 1) + \cdots + \lambda_{123\ldots k}(t_1 t_2 \ldots t_k - 1)\right], \quad (37.85)$$

where $M_\tau[z]$ is, as before, the moment generating function of the random variable τ.

Papageorgiou and David (1995a,b) have observed that, for the distribution with probability generating function (37.84), the conditional joint distribution of X_2 and X_3 (given $X_1 = x_1$) is the convolution of bivariate binomial distribution (see Chapter 36, Section 7) with parameters $\left(x_1; \frac{\lambda_{12}}{\lambda'}, \frac{\lambda_{13}}{\lambda'}, \frac{\lambda_{123}}{\lambda'}\right)$, where $\lambda' = \lambda_1 + \lambda_{12} + \lambda_{123}$, and a bivariate distribution with probability generating function

$$\frac{1}{M_\tau^{(x_1)}[-\lambda']} M_\tau^{(x_1)}\left[-\lambda' + \lambda_2(t_1 - 1) + \lambda_3(t_2 - 1) + \lambda_{23}(t_1 t_2 - 1)\right], \quad (37.86)$$

where $M_\tau^{(x)}[z] = \frac{d^x}{dz^x} M_\tau[z]$.

It is known in the univariate case that the mixture distribution

$$\text{Binomial}\,(N, p) \bigwedge_N \text{Poisson}\,(\lambda) \quad (37.87)$$

is in fact a Poisson(λp) distribution [see, for example, Chapter 8 (p. 336) for a more general result]. Sim (1993) has utilized an extension and generalization of this idea to construct multivariate Poisson models. The random variables $\boldsymbol{X}_i = (X_{ii}, \ldots, X_{ik})$, $k \geq i$, such that

$$\begin{pmatrix} X_{ii} \\ X_{i,i+1} \\ \vdots \\ X_{ik} \end{pmatrix} \rightsquigarrow \begin{pmatrix} \text{Binomial}\,(N, p_{ii}) \\ \text{Binomial}\,(N, p_{i,i+1}) \\ \vdots \\ \text{Binomial}\,(N, p_{ik}) \end{pmatrix} \bigwedge_N \text{Poisson}(\lambda_i) \quad (37.88)$$

have a joint multivariate Poisson distribution with

$$E[\boldsymbol{X}_i] = \lambda_i \boldsymbol{p}_i = \text{var}(\boldsymbol{X}_i) \quad (37.89)$$

and

$$\text{cov}(X_{ig}, X_{ih}) = \lambda_i p_{ig} p_{ih} \,. \tag{37.90}$$

Sim (1993) has then defined the variables

$$Z_1 = X_{11},\ Z_2 = X_{12} + X_{22}, \ldots, Z_k = X_{1k} + X_{2k} + \cdots + X_{kk}\,, \tag{37.91}$$

where $\boldsymbol{X}_1, \boldsymbol{X}_2, \ldots, \boldsymbol{X}_k$ are mutually independent sets and $p_{ii} = 1$ for all i. This is also a multivariate Poisson distribution, since $X_{1i}, X_{2i}, \ldots, X_{ii}$ are mutually independent Poisson random variables. Furthermore,

$$E[Z_i] = \sum_{j=1}^{i} \lambda_j p_{ji} = \text{var}(Z_i), \qquad i = 1, 2, \ldots, k, \tag{37.92}$$

and

$$\text{cov}(Z_i, Z_{i'}) = \sum_{j=1}^{i} \lambda_j p_{ji} p_{ji'}, \qquad 1 \le i < i' \le k. \tag{37.93}$$

Construction of a set of Z's having specified means and covariances can proceed sequentially, using in turn the equations

$$\begin{aligned} \lambda_1 &= E[Z_1], \qquad (\text{since } p_{11} = 1) \\ \lambda_1 p_{21} &= \text{cov}(Z_1, Z_2), \\ \lambda_2 + \lambda_1 p_{21} &= E[Z_2], \end{aligned} \tag{37.94}$$

and so on.

Radons *et al.* (1994) have used the multivariate Poisson distribution in the analysis, classification, and coding of neuronal spike patterns and have compared its performance with the analysis based on hidden Markov models.

BIBLIOGRAPHY

Ahmad, M. (1981). A bivariate hyper-Poisson distribution, in *Statistical Distributions in Scientific Work*, Vol. 4 (Eds. C. Taillie, G. P. Patil, and B. A. Baldessari), pp. 225–230, Dordrecht: D. Reidel.

Ahmed, M. S. (1961). On a locally most powerful boundary randomized similar test for the independence of two Poisson variables, *Annals of Mathematical Statistics*, **32**, 809–827.

Aitken, A. C. (1936). A further note on multivariate selection, *Proceedings of the Edinburgh Mathematical Society*, **5**, 37–40.

Aitken, A. C., and Gonin, H. T. (1935). On fourfold sampling with and without replacement, *Proceedings of the Royal Society of Edinburgh*, **55**, 114–125.

Arnold, B. C., and Strauss, D. J. (1991). Bivariate distributions with conditionals in prescribed exponential families, *Journal of the Royal Statistical Society, Series B*, **53**, 365–375.

Bahadur, R. R. (1961). A representation of the joint distribution of response of n dichotomous items, in *Studies of Item Analysis and Prediction* (Ed. H. Solomon), pp. 158–168, Stanford, CA: Stanford University Press.

Balakrishnan, N., Johnson, N. L., and Kotz, S. (1995). A note on an interpretational difference in trivariate Poisson distributions, *Report*, McMaster University.

Banerjee, D. P. (1959). On some theorems on Poisson distributions, *Proceedings of the National Academy of Science, India, Section A*, **28**, 30–33.

Belov, A. G. (1993). On uniqueness of maximum likelihood estimators of parameters of a bivariate Poisson distribution, *Vestnik MGU*, Series **15**, No. 1, 58–59 (in Russian).

Cacoullos, T., and Papageorgiou, H. (1980). On some bivariate probability models applicable to traffic accidents and fatalities, *International Statistical Review*, **48**, 345–356.

Cacoullos, T., and Papageorgiou, H. (1981). On bivariate discrete distributions generated by compounding, in *Statistical Distributions in Scientific Work*, Vol. 4 (Eds. C. Taillie, G. P. Patil, and B. A. Baldessari), pp. 197–212, Dordrecht: D. Reidel.

Campbell, J. T. (1938). The Poisson correlation function, *Proceedings of the Edinburgh Mathematical Society (Series 2)*, **4**, 18–26.

Consael, R. (1952). Sur les processus composés de Poisson a deux variables aléatoires, *Académie Royale de Belgique, Classe des Sciences, Mémoires*, **7**, 4–43.

Consul, P. C. (1989). *Generalized Poisson Distributions—Properties and Applications*, New York: Marcel Dekker.

Consul, P. C. (1994). Some bivariate families of Lagrangian probability distributions, *Communications in Statistics—Theory and Methods*, **23**, 2895–2906.

Consul, P. C., and Jain, G. C. (1973). A generalization of Poisson distribution, *Technometrics*, **15**, 791–799.

Consul, P. C., and Shoukri, M. M. (1985). The generalized Poisson distribution with the sample mean is larger than sample variance, *Communications in Statistics—Simulation and Computation*, **14**, 667–681.

Cuppens, R. (1968). Sur les produits finis de lois de Poisson, *Comptes Rendus de l'Académie des Sciences, Paris*, **266**, 726–728.

David, K. M., and Papageorgiou, H. (1994). On compounded bivariate Poisson distributions, *Naval Research Logistics*, **41**, 203–214.

Dwass, M., and Teicher, H. (1957). On infinitely divisible random vectors, *Annals of Mathematical Statistics*, **28**, 461–470.

Famoye, F., and Consul, P. C. (1995). Bivariate generalized Poisson distribution with some application, *Metrika*, **42**, 127–138.

Gillings, D. B. (1974). Some further results for bivariate generalizations of the Neyman type A distribution, *Biometrics*, **30**, 619–628.

Griffiths, R. C., Milne, R. K., and Wood, R. (1980). Aspects of correlation in bivariate Poisson distributions and processes, *Australian Journal of Statistics*, **22**, 238–255.

Hamdan, M. A., and Al-Bayyati, H. A. (1969). A note on the bivariate Poisson distribution, *The American Statistician*, **23**, No. 4, 32–33.

Hesselager, O. (1996a). A recursive procedure for calculation of some mixed compound Poisson distributions, *Scandinavian Actuarial Journal*, 54–63.

Hesselager, O. (1996b). Recursions for certain bivariate counting distributions and their compound distributions, *ASTIN Bulletin* (to appear).

Holgate, P. (1964). Estimation for the bivariate Poisson distribution, *Biometrika*, **51**, 241–245.

Holgate, P. (1966). Bivariate generalizations of Neyman's type A distribution, *Biometrika*, **53**, 241–245.

Huang, M. L., and Fung, K. Y. (1993a). D-distribution and its applications, *Statistical Papers*, **34**, 143–159.

Huang, M. L., and Fung, K. Y. (1993b). The D compound Poisson distribution, *Statistical Papers*, **34**, 319–338.

Johnson, M. E., and Brooks, D. G. (1985). Can we generate a bivariate Poisson distribution with a negative correlation? *Statistics Group G-1 Report*, Los Alamos National Laboratory, Los Alamos, NM.

Johnson, N. L., and Kotz, S. (1969). *Distributions in Statistics: Discrete Distributions*, New York: John Wiley & Sons.

Johnson, N. L., and Kotz, S. (1972). *Distributions in Statistics: Multivariate Continuous Distributions*, New York: John Wiley & Sons.

Kano, K., and Kawamura, K. (1991). On recurrence relations for the probability function of multivariate generalized Poisson distribution, *Communications in Statistics—Theory and Methods*, **20**, 165–178.

Kawamura, K. (1973). The diagonal distributions of the bivariate Poisson distributions, *Kodai Mathematical Seminar Reports*, **25**, 379–384.

Kawamura, K. (1976). The structure of trivariate Poisson distribution, *Kodai Mathematical Seminar Reports*, **28**, 1–8.

Kawamura, K. (1979). The structure of multivariate Poisson distribution, *Kodai Mathematical Journal*, **2**, 337–345.

Kawamura, K. (1985). A note on the recurrence relations for the bivariate Poisson distribution, *Kodai Mathematical Journal*, **8**, 70–78.

Kawamura, K. (1987). Calculation of density for the multivariate Poisson distribution, *Kodai Mathematical Journal*, **10**, 231–241.

Kemp, C. D., and Papageorgiou, H. (1982). Bivariate Hermite distributions, *Sankhyā, Series A*, **44**, 269–280.

Kocherlakota, S., and Kocherlakota, K. (1992). *Bivariate Discrete Distributions*, New York: Marcel Dekker.

Krishnamoorthy, A. S. (1951). Multivariate binomial and Poisson distributions, *Sankhyā*, **11**, 117–124.

Leiter, R. E., and Hamdan, M. A. (1973). Some bivariate probability models applicable to traffic accidents and fatalities, *International Statistical Review*, **41**, 81–100.

Lévy, P. (1968). Observations sur la note précedente, *Comptes Rendus de l'Académie des Sciences, Paris*, **266**, 728–729.

Liang, W. G. (1989). The structure of compounded trivariate Poisson distribution, *Kodai Mathematical Journal*, **12**, 30–40.

Loukas, S. (1993). Some methods of estimation for a trivariate Poisson distribution, *Zastosowania Mathematica*, **21**, 503–510.

Loukas, S., and Kemp, C. D. (1983). On computer sampling from trivariate and multivariate discrete distributions, *Journal of Statistical Computation and Simulation*, **17**, 113–123.

Loukas, S., and Kemp, C. D. (1986). The index of dispersion test for the bivariate Poisson distribution, *Biometrics*, **42**, 941–948.

Loukas, S., Kemp, C. D., and Papageorgiou, H. (1986). Even-point estimation for the bivariate Poisson distribution, *Biometrika*, **73**, 222–223.

Loukas, S., and Papageorgiou, H. (1991). On a trivariate Poisson distribution, *Applications of Mathematics*, **36**, 432–439.

Lukacs, E., and Beer, S. (1977). Characterization of the multivariate Poisson distribution, *Journal of Multivariate Analysis*, **7**, 1–12.

Mahamunulu, D. M. (1967). A note on regression in the multivariate Poisson distribution, *Journal of the American Statistical Association*, **62**, 251–258.

Mardia, K. V. (1970). *Families of Bivariate Distributions*, London: Griffin.

McKendrick, A. G. (1926). Application of mathematics to medical problems, *Proceedings of the Edinburgh Mathematical Society*, **4**, 106–130.

Milne, R. K., and Westcott, M. (1993). Generalized multivariate Hermite distributions and related point processes, *Annals of the Institute of Statistical Mathematics*, **45**, 367–381.

Morin, A. (1993). A propos de la loi de Poisson multivariée, *Journal de la Société de Statistique de Paris*, **134**, 3–12.

Ong, S. H. (1988). A discrete Charlier series distribution, *Biometrical Journal*, **30**, 1003–1009.

Ord, J. K. (1974). A characterization of a dependent bivariate Poisson distribution, in *Statistical Distributions in Scientific Work*, Vol. 3 (Eds. C. Taillie, G. P. Patil, and B. A. Baldessari), pp. 291–297, Dordrecht: D. Reidel.

Panjer, H. H. (1981). Recursive evaluation of a family of compound distributions, *ASTIN Bulletin*, **12**, 22–26.

Papageorgiou, H. (1986). Bivariate "short" distributions, *Communications in Statistics—Theory and Methods*, **15**, 893–905.

Papageorgiou, H. (1995). Some remarks on the D compound Poisson distribution, *Statistical Papers*, **36**, 371–375.

Papageorgiou, H., and David, K. M. (1995a). The structure of compounded trivariate binomial distributions, *Biometrical Journal*, **37**, 81–95.

Papageorgiou, H., and David, K. M. (1995b). On a class of bivariate compounded Poisson distributions, *Statistics & Probability Letters*, **23**, 93–104.

Papageorgiou, H., and Kemp, C. D. (1977). Even point estimation for bivariate generalized Poisson distributions, *Statistical Report No. 29*, School of Mathematics, University of Bradford, England.

Papageorgiou, H., and Kemp, C. D. (1983). Conditionality in bivariate generalized Poisson distributions, *Biometrical Journal*, **25**, 757–763.

Papageorgiou, H., and Kemp, C. D. (1988). A method of estimation for some bivariate discrete distributions, *Biometrical Journal*, **30**, 993–1001.

Papageorgiou, H., and Loukas, S. (1988). Conditional even point estimation for bivariate discrete distributions, *Communications in Statistics—Theory and Methods*, **17**, 3403–3412.

Papageorgiou, H., and Loukas, S. (1995). A bivariate discrete Charlier series distribution, *Biometrical Journal*, **37**, 105–117.

Papageorgiou, H., and Piperigou, V. E. (1997). On bivariate 'Short' and related distributions, in *Advances in the Theory and Practice of Statistics—A Volume in Honor of Samuel Kotz* (Eds. N. L. Johnson and N. Balakrishnan), New York: John Wiley & Sons (to appear).

Patil, G. P., and Bildikar, S. (1967). Multivariate logarithmic series distribution as a probability model in population and community ecology and some of its statistical properties, *Journal of the American Statistical Association*, **62**, 655–674.

Patil, S. A., Patel, D. I., and Kornes, J. L. (1977). On bivariate truncated Poisson distribution, *Journal of Statistical Computation and Simulation*, **6**, 49–66.

Paul, S. R., and Ho, N. I. (1989). Estimation in the bivariate Poisson distribution and hypothesis testing concerning independence, *Communications in Statistics—Theory and Methods*, **18**, 1123–1133.

Radons, G., Becker, J. D., Dülfer, B., and Krüger, J. (1994). Analysis, classification, and coding of multielectrode spike trains with hidden Markov models, *Biological Cybernetics*, **71**, 359–373.

Rayner, J. C. W., and Best, D. J. (1995). Smooth tests for the bivariate Poisson distribution, *Australian Journal of Statistics*, **37**, 233–245.

Renshaw, E. (1986). A survey of stepping-stone models in population dynamics, *Advances in Applied Probability*, **18**, 581–627.

Renshaw, E. (1991). *Modeling Biological Populations in Space and Time*, New York: Cambridge University Press.

Samaniego, F. J. (1976). A characterization of convoluted Poisson distributions with applications to estimation, *Journal of the American Statistical Association*, **71**, 475–479.

Shoukri, M. M., and Consul, P. C. (1982). On bivariate modified power series distribution; some properties, estimation and applications, *Biometrical Journal*, **8**, 787–799.

Sibuya, M., Yoshimura, I., and Shimizu, R. (1964). Negative multinomial distribution, *Annals of the Institute of Statistical Mathematics*, **16**, 409–426.

Šidák, Z. (1972). A chain of inequalities for some types of multivariate distributions, *Colloquium Mathematica Societatis Janos Bolyai*, 693–699.

Šidák, Z. (1973a). A chain of inequalities for some types of multivariate distributions, with nine special cases, *Aplikace Matematiky*, **18**, 110–118.

Šidák, Z. (1973b). On probabilities in certain multivariate distributions: Their dependence on correlations, *Aplikace Matematiky*, **18**, 128–135.

Sim, C. H. (1993). Generation of Poisson and gamma random vectors with given marginals and covariance matrix, *Journal of Statistical Computation and Simulation*, **47**, 1–10.

Stein, G. Z., and Juritz, J. M. (1987). Bivariate compound Poisson distributions, *Communications in Statistics—Theory and Methods*, **16**, 3591–3607.

Subrahmaniam, K. (1966). A test of "intrinsic correlation" in the theory of accident proneness, *Journal of the Royal Statistical Society, Series B*, **28**, 180–189.

Teicher, H. (1954). On the multivariate Poisson distribution, *Skandinavisk Aktuarietidskrift*, **37**, 1–9.

Upton, G. J. G., and Lampitt, G. A. (1981). A model for interyear change in the size of bird populations, *Biometrics*, **37**, 113–127.

Wang, Y. H. (1974). Characterizations of certain multivariate distributions, *Proceedings of the Cambridge Philosophical Society*, **75**, 219–234.

Wehrly, T. E., Matis, J. H., and Otis, J. W. (1993). Approximating multivariate distributions in stochastic models of insect population dynamics, in *Multivariate Environmental Statistics* (Eds. G. P. Patil and C. R. Rao), pp. 573–596, Amsterdam, Netherlands: Elsevier.

Wesolowski, J. (1993). Bivariate discrete measures via a power series conditional distribution and a regression function, *Technical Report*, Mathematical Institute, Warsaw University of Technology, Warsaw, Poland.

Wesolowski, J. (1994). A new conditional specification of the bivariate Poisson conditionals distribution, *Technical Report*, Mathematical Institute, Warsaw University of Technology, Warsaw, Poland.

Willmot, G. E. (1989). Limiting tail behaviour of some discrete compound distributions, *Insurance: Mathematics and Economics*, **8**, 175–185.

Willmot, G. E. (1993). On recursive evaluation of mixed Poisson probabilities and related quantities, *Scandinavian Actuarial Journal*, 114–133.

Willmot, G. E., and Sundt, B. (1989). On evaluation of the Delaporte distribution and related distributions, *Scandinavian Actuarial Journal*, 101–113.

Zheng, Q., and Matis, J. H. (1993). Approximating discrete multivariate distributions from known moments, *Communications in Statistics—Theory and Methods*, **22**, 3553–3567.

CHAPTER 38

Multivariate Power Series Distributions

1 INTRODUCTION

Several classical discrete multivariate distributions have probability mass functions of the form

$$P_{\boldsymbol{N}}(\boldsymbol{n}) = \Pr\left[\bigcap_{i=1}^{k}(N_i = n_i)\right] = \frac{1}{A(\boldsymbol{\theta})}\, a(\boldsymbol{n}) \prod_{i=1}^{k} \theta_i^{n_i}\,, \tag{38.1}$$

where n_i's are non-negative integers, θ_i's are positive parameters, and

$$A(\boldsymbol{\theta}) = A(\theta_1, \theta_2, \ldots, \theta_k) = \sum_{n_1=0}^{\infty} \cdots \sum_{n_k=0}^{\infty} a(n_1, n_2, \ldots, n_k) \prod_{i=1}^{k} \theta_i^{n_i}\,. \tag{38.2}$$

The family of distributions having probability mass functions of the form (38.1) is called the *multivariate power series distributions* family.

It may sometimes be the case that the functions $a(n_1, n_2, \ldots, n_k)$ depend on some other parameters (say, $\boldsymbol{\phi}$), but they *do not* depend on $\boldsymbol{\theta}$. The function $A(\boldsymbol{\theta})$ is called the *defining function* of the distribution while $a(\boldsymbol{n})$ is called the *coefficient function*.

The multivariate power series distribution family in (38.1) is a natural generalization of the univariate power series distribution family (see Chapter 2, Section 2). It is of interest to note that while the binomial (n, p) distribution belongs to the univariate power series distribution family since its probability mass function can be written as

$$\Pr[N_1 = n_1] = \binom{n}{n_1} \left(\frac{p}{1-p}\right)^{n_1} (1-p)^n, \qquad n_1 = 0, 1, \ldots, n, \tag{38.3}$$

the (nonsingular) multinomial distribution with probability mass function [see Eq. (35.1)]

$$\Pr\left[\bigcap_{i=1}^{k}(N_i = n_i)\right] = \binom{n}{n_1, n_2, \ldots, n_k} \left(1 - \sum_{i=1}^{k-1} p_i\right)^{n_k} \prod_{i=1}^{k-1} p_i^{n_i} \tag{38.4}$$

(with $n = \sum_{i=1}^{k} n_i$) is not a multivariate power series distribution in this sense since (38.4) cannot be written in the form (38.1).

Gerstenkorn (1981) has provided a simple way of testing whether a given multivariate probability mass function belongs to the multivariate power series distribution family. The conditions

$$\Pr\left[\bigcap_{i=1}^{k}(N_i = 0)\right] = a(\mathbf{0})/A(\boldsymbol{\theta}) \tag{38.5}$$

and

$$\frac{\Pr\left[\bigcap_{i=1}^{k}(N_i = n_i + m_i)\right]}{\Pr\left[\bigcap_{i=1}^{k}(N_i = n_i)\right]} = \frac{a(\boldsymbol{n} + \boldsymbol{m})}{a(\boldsymbol{n})}\prod_{i=1}^{k}\theta_i^{m_i} \tag{38.6}$$

are necessary and sufficient for the multivariate probability mass function to belong to the multivariate power series distribution family with defining function $A(\boldsymbol{\theta})$.

In the case of multivariate Poisson distribution with joint probability mass function (see Chapter 37)

$$P(x_1, x_2, \ldots, x_k) = e^{-\sum_{i=1}^{k}\lambda_i}\prod_{i=1}^{k}\left(\lambda_i^{x_i}/x_i!\right),$$

we have

$$\frac{\Pr\left[\bigcap_{i=1}^{k}(X_i = x_i + y_i)\right]}{\Pr\left[\bigcap_{i=1}^{k}(X_i = x_i)\right]} = \prod_{i=1}^{k}\left\{\frac{x_i!}{(x_i + y_i)!}\lambda_i^{y_i}\right\}$$

while

$$\Pr\left[\bigcap_{i=1}^{k}(X_i = 0)\right] = e^{-\sum_{i=1}^{k}\lambda_i},$$

and so the defining function of this family is $e^{\sum_{i=1}^{k}\lambda_i}$.

The multivariate power series distribution family has also been extended to the *generalized multivariate power series distribution* family by relaxing the limitation of possible values of n_i's being non-negative integers. In the majority of applications encountered so far, however, this limitation is satisfied and hence we will not pay special attention to this more general family of distributions.

The pioneering paper on multivariate power series distributions is by Patil (1965), which contains many details. Some relevant details are contained in an earlier paper by Khatri (1959), though this paper is mainly devoted to univariate power series distributions. A more general treatment, motivated in part by Khatri (1959), is presented by Gerstenkorn (1981). A concise review of the developments on multivariate power series distributions has been made by Patil (1985b).

2 MOMENTS AND OTHER PROPERTIES

If $N_1, N_2, \ldots, N_k$ have the joint probability mass function in (38.1), then the joint probability generating function is

$$E\left[\prod_{i=1}^{k} t_i^{N_i}\right] = \frac{1}{A(\boldsymbol{\theta})}\, a(\boldsymbol{n}) \sum_{n_1=0}^{\infty} \cdots \sum_{n_k=0}^{\infty} \prod_{i=1}^{k} (\theta_i t_i)^{n_i}$$
$$= A(\theta_1 t_1, \ldots, \theta_k t_k)/A(\theta_1, \ldots, \theta_k) \tag{38.7}$$

and the joint moment generating function is

$$E\left[\prod_{i=1}^{k} e^{t_i N_i}\right] = \frac{1}{A(\boldsymbol{\theta})}\, a(\boldsymbol{n}) \sum_{n_1=0}^{\infty} \cdots \sum_{n_k=0}^{\infty} \prod_{i=1}^{k} (\theta_i e^{t_i})^{n_i}$$
$$= A(\theta_1 e^{t_1}, \ldots, \theta_k e^{t_k})/A(\theta_1, \ldots, \theta_k). \tag{38.8}$$

In particular, we have

$$E[N_i] = \frac{\theta_i}{A(\boldsymbol{\theta})} \frac{\partial A(\boldsymbol{\theta})}{\partial \theta_i}, \tag{38.9}$$

$$\operatorname{var}(N_i) = \frac{\theta_i^2}{A^2(\boldsymbol{\theta})} \left\{ A(\boldsymbol{\theta}) \frac{\partial^2 A(\boldsymbol{\theta})}{\partial \theta_i^2} - \left(\frac{\partial A(\boldsymbol{\theta})}{\partial \theta_i}\right)^2 + \frac{A(\boldsymbol{\theta})}{\theta_i} \frac{\partial A(\boldsymbol{\theta})}{\partial \theta_i} \right\}, \tag{38.10}$$

$$\operatorname{cov}(N_i, N_j) = \frac{\theta_i \theta_j}{A^2(\boldsymbol{\theta})} \left\{ A(\boldsymbol{\theta}) \frac{\partial^2 A(\boldsymbol{\theta})}{\partial \theta_i \partial \theta_j} - \frac{\partial A(\boldsymbol{\theta})}{\partial \theta_i} \frac{\partial A(\boldsymbol{\theta})}{\partial \theta_j} \right\}, \tag{38.11}$$

and

$$\operatorname{corr}(N_i, N_j) = \frac{A(\boldsymbol{\theta}) \frac{\partial^2 A(\boldsymbol{\theta})}{\partial \theta_i \partial \theta_j} - \frac{\partial A(\boldsymbol{\theta})}{\partial \theta_i} \frac{\partial A(\boldsymbol{\theta})}{\partial \theta_j}}{\sqrt{\begin{array}{l}\left\{ A(\boldsymbol{\theta}) \frac{\partial^2 A(\boldsymbol{\theta})}{\partial \theta_i^2} - \left(\frac{\partial A(\boldsymbol{\theta})}{\partial \theta_i}\right)^2 + \frac{A(\boldsymbol{\theta})}{\theta_i} \frac{\partial A(\boldsymbol{\theta})}{\partial \theta_i} \right\} \\ \times \left\{ A(\boldsymbol{\theta}) \frac{\partial^2 A(\boldsymbol{\theta})}{\partial \theta_j^2} - \left(\frac{\partial A(\boldsymbol{\theta})}{\partial \theta_j}\right)^2 + \frac{A(\boldsymbol{\theta})}{\theta_j} \frac{\partial A(\boldsymbol{\theta})}{\partial \theta_j} \right\}\end{array}}}. \tag{38.12}$$

In general, the $\boldsymbol{r} = (r_1, r_2, \ldots, r_k)'$th decreasing factorial moment of $\boldsymbol{N}$ is

$$\mu'_{(\boldsymbol{r})}(\boldsymbol{N}) = E\left[\prod_{i=1}^{k} N_i^{(r_i)}\right]$$
$$= \frac{1}{A(\boldsymbol{\theta})} \prod_{i=1}^{k} \theta_i^{r_i} \left\{ \frac{\partial^{\sum_{i=1}^{k} r_i}}{\partial \theta_1^{r_1}\, \partial \theta_2^{r_2} \cdots \partial \theta_k^{r_k}} A(\boldsymbol{\theta}) \right\}. \tag{38.13}$$

We note the recurrence relation

$$\mu'_{(\boldsymbol{r}+\boldsymbol{e}_i)}(\boldsymbol{N}) = \theta_i \frac{\partial \mu'_{(\boldsymbol{r})}(\boldsymbol{N})}{\partial \theta_i} + (E[N_i] - r_i)\, \mu'_{(\boldsymbol{r})}(\boldsymbol{N}), \tag{38.14}$$

where $\boldsymbol{e}_i$ is the ith column of the $k \times k$ identity matrix.

Similarly, the following recurrence relations also hold [Patil (1985b)]:

$$\kappa_{\boldsymbol{r}+\boldsymbol{e}_i} = \theta_i \frac{\partial \kappa_{\boldsymbol{r}}}{\partial \theta_i} \tag{38.15}$$

among the cumulants;

$$\mu'_{\boldsymbol{r}+\boldsymbol{e}_i} = \theta_i \frac{\partial \mu'_{\boldsymbol{r}}}{\partial \theta_i} + \mu'_{\boldsymbol{r}}\, E[N_i] \tag{38.16}$$

among the moments about zero $\mu'_{\boldsymbol{r}} = E\left[\prod_{i=1}^{k} N_i^{r_i}\right]$; and

$$\mu_{\boldsymbol{r}+\boldsymbol{e}_i} = \theta_i \frac{\partial \mu_{\boldsymbol{r}}}{\partial \theta_i} + \sum_{j=1}^{k} r_j \operatorname{cov}(N_i, N_j) \mu_{\boldsymbol{r}-\boldsymbol{e}_i} \tag{38.17}$$

among the central moments $\mu_{\boldsymbol{r}} = E\left[\prod_{i=1}^{k} (N_i - E[N_i])^{r_i}\right]$.

The joint distribution of any subset $N_{b_1}, N_{b_2}, \ldots, N_{b_s}$ of random variables $N_1, N_2, \ldots, N_k$ is also a multivariate power series distribution with defining function $A(\theta_1, \theta_2, \ldots, \theta_k)$ regarded as a function of $\theta_{b_1}, \theta_{b_2}, \ldots, \theta_{b_s}$ with the other θ's being regarded as constants. Further, truncated multivariate power series distributions, with s_1 least and s_2 greatest values excluded, are also multivariate power series distributions [Patil (1985b)].

Some similar properties of multivariate factorial series distributions have been discussed by Berg (1977).

By considering the case when the support for $n_i = 0, 1, \ldots, m$ in (38.1), Kyriakoussis and Vamvakari (1996) have established the asymptotic ($m \to \infty$) normality of such a class of multivariate power series distribution.

3 ESTIMATION

Given m independent sets of variables $\boldsymbol{N}_h = (N_{1h}, N_{2h}, \ldots, N_{kh})$ $(h = 1, 2, \ldots, m)$, each having the same multivariate power series distribution with probability mass function (38.1), the maximum likelihood estimators of the expected values $E[N_i]$ $(i = 1, 2, \ldots, k)$ are simply the arithmetic means of the sample values, $\frac{1}{m}(N_{i1} + N_{i2} + \cdots + N_{im})$, respectively. The maximum likelihood estimators of the parameters θ_i [given by $\widehat{\boldsymbol{\theta}} = (\widehat{\theta}_1, \ldots, \widehat{\theta}_k)'$] are therefore obtained as solutions of the equations

$$\overline{N}_i = \frac{1}{m}(N_{i1} + N_{i2} + \cdots + N_{im}) = \theta_i \left. \frac{\partial A(\boldsymbol{\theta})}{\partial \theta_i} \right|_{\boldsymbol{\theta} = \widehat{\boldsymbol{\theta}}} \quad (i = 1, 2, \ldots, k). \tag{38.18}$$

This formula can also be obtained from the likelihood equations

$$\left.\frac{\partial \log L}{\partial \theta_i}\right|_{\boldsymbol{\theta}=\widehat{\boldsymbol{\theta}}} = 0\,, \tag{38.19}$$

where L is the likelihood function

$$L = \{A(\boldsymbol{\theta})\}^{-m} \left\{\prod_{h=1}^{m} a(\boldsymbol{N}_h)\right\} \prod_{i=1}^{k} \theta_i^{m\bar{N}_i} \tag{38.20}$$

with $\boldsymbol{N}_h = (N_{1h}, \ldots, N_{kh})$. The elements of the inverse of the Fisher information matrix are

$$-E\left[\frac{\partial^2 \log L}{\partial\theta_i \partial\theta_j}\right] = m\left\{\frac{\partial^2 \log A(\boldsymbol{\theta})}{\partial\theta_i \partial\theta_j} - \delta_{ij}\,\frac{E[N_i]}{\theta_i^2}\right\} \tag{38.21}$$

where $\delta_{ij} = 0$ if $i \neq j$ and $\delta_{ii} = 1$. Equivalently,

$$-E\left[\frac{\partial^2 \log L}{\partial\theta_i \partial\theta_j}\right] = \begin{cases} \dfrac{m}{\theta_i\theta_j}\,\mathrm{cov}(N_i, N_j) & \text{if } i \neq j \\[2ex] \dfrac{m}{\theta_i^2}\,\mathrm{var}(N_i) & \text{if } i = j \end{cases} \tag{38.22}$$

since [see Eqs. (38.9)–(38.11)]

$$\frac{\partial E[N_j]}{\partial\theta_i} = \begin{cases} \theta_j\,\dfrac{\partial^2 A(\boldsymbol{\theta})}{\partial\theta_i \partial\theta_j} & \text{if } i \neq j \\[2ex] \dfrac{\partial A(\boldsymbol{\theta})}{\partial\theta_i} + \theta_i\,\dfrac{\partial^2 A(\boldsymbol{\theta})}{\partial\theta_i^2} & \text{if } i = j. \end{cases} \tag{38.23}$$

These properties are shared by all members of the multivariate power series distribution family. We now turn to study a specific, but important, member of this family, with its own characteristic properties.

4 MULTIVARIATE LOGARITHMIC SERIES DISTRIBUTIONS

It is known that (see Chapter 7, Section 2) univariate logarithmic series distributions can be derived as limiting forms of truncated (by omission of the zero class) negative binomial distributions. The limit is taken as the value of the index parameter n tends to 0. A similar limiting operation can be performed on multivariate negative binomial (negative multinomial) distributions. From Chapter 36, we find the probability mass function, truncated by omission of $n_1 = n_2 = \cdots = n_k = 0$, to be

$$P(n_1, n_2, \ldots, n_k \mid n) = \frac{\Gamma\left(n + \sum_{i=1}^{k} n_i\right)}{\left(\prod_{i=1}^{k} n_i!\right)\Gamma(n)(1 - Q^{-n})}\, Q^{-n} \prod_{i=1}^{k} \left(\frac{P_i}{Q}\right)^{n_i}. \tag{38.24}$$

As n tends to 0, we have

$$\lim_{n\to 0} \frac{\Gamma\left(n+\sum_{i=1}^{k} n_i\right)}{\Gamma(n)(1-Q^{-n})} = \lim_{n\to 0}\left(n+\sum_{i=1}^{k} n_i - 1\right)\cdots(n+1)n(1-Q^{-n})^{-1}$$

$$= \left(\sum_{i=1}^{k} n_i - 1\right)!\ \lim_{n\to 0} n(1-Q^{-n})^{-1}$$

$$= \left(\sum_{i=1}^{k} n_i - 1\right)!/\log Q\,. \qquad (38.25)$$

Using this in (38.24), we obtain

$$\lim_{n\to 0} P(n_1, n_2, \ldots, n_k \mid n) = \frac{\left(\sum_{i=1}^{k} n_i - 1\right)!}{\left(\prod_{i=1}^{k} n_i!\right)\log Q} \prod_{i=1}^{k}\left(\frac{P_i}{Q}\right)^{n_i}. \qquad (38.26)$$

Setting $P_i/Q = \theta_i$ $(0 < \sum_{i=1}^{k} \theta_i < 1)$ and noting that $Q = (1 - \sum_{i=1}^{k} \theta_i)^{-1}$, the right-hand side of Eq. (38.26) can be rearranged to give a formula defining the *multivariate logarithmic series distribution with parameters* $\theta_1, \theta_2, \ldots, \theta_k$:

$$P(n_1, n_2, \ldots, n_k) = \frac{\left(\sum_{i=1}^{k} n_i - 1\right)!}{\left(\prod_{i=1}^{k} n_i!\right)\left\{-\log\left(1-\sum_{i=1}^{k}\theta_i\right)\right\}} \prod_{i=1}^{k} \theta_i^{n_i},$$

$$n \geq 0,\ \sum_{i=1}^{k} n_i > 0\,. \qquad (38.27)$$

This distribution has been studied by Patil and Bildikar (1967). They have also utilized the bivariate logarithmic series distribution to model the observed joint distribution of numbers of males and females in different occupations in a city. Applications of the multivariate logarithmic series distribution have also been considered by Chatfield, Ehrenberg, and Goodhardt (1966). They concluded that multivariate negative binomial distributions have some general advantages in terms of flexibility though multivariate logarithmic series distributions can be valuable when marginal distributions are very skewed. Patil (1985a) has provided a concise review of the developments on the multivariate logarithmic series distribution.

By summing (38.27), keeping $\sum_{i=1}^{k} n_i = n > 0$, we have

$$\sum_{n_1,\ldots,n_k}\cdots\sum \prod_{i=1}^{k} (\theta_i^{n_i}/n_i!) = \left(\sum_{i=1}^{k}\theta_i\right)^n \Big/ n! \qquad (38.28)$$

and, hence, it follows that $\sum_{i=1}^{k} N_i$ has a logarithmic series distribution with parameter $\sum_{i=1}^{k} \theta_i$.

One point of special importance is to note that the marginal distributions of (38.27) are *not* logarithmic series distributions. This is clearly evident, since the value of $n_1 = 0$ is not excluded (all that is excluded is the *combination* of values $n_1 = n_2 = \cdots = n_k = 0$). In fact, the joint probability generating function is [see Chapter 36]

$$\lim_{n\to 0} \frac{\left(Q - \sum_{i=1}^k P_i t_i\right)^{-n} - Q^{-n}}{1 - Q^{-n}} = \frac{-\log\left\{1 - \sum_{i=1}^k P_i t_i / Q\right\}}{\log Q}$$

$$= \frac{\log\left(1 - \sum_{i=1}^k \theta_i t_i\right)}{\log\left(1 - \sum_{i=1}^k \theta_i\right)}. \tag{38.29}$$

Setting $t_2 = t_3 = \cdots = t_k = 1$ in (38.29), we obtain the probability generating function of N_1 as

$$\frac{\log\left(1 - \sum_{i=2}^k \theta_i - \theta_1 t_1\right)}{\log\left(1 - \sum_{i=1}^k \theta_i\right)}, \tag{38.30}$$

whence

$$\Pr[N_1 = 0] = \frac{\log\left(1 - \sum_{i=2}^k \theta_i\right)}{\log\left(1 - \sum_{i=1}^k \theta_i\right)} \tag{38.31}$$

and

$$\Pr[N_1 = n_1] = \frac{1}{n_1\left\{-\log\left(1 - \sum_{i=1}^k \theta_i\right)\right\}} \left(\frac{\theta_1}{1 - \sum_{i=2}^k \theta_i}\right)^{n_1}, \quad n_1 \geq 1. \tag{38.32}$$

This is, in fact, the *zero-modified logarithmic series distribution* (see Chapter 8, Section 2.2). Note that (38.32) can be written as

$$\Pr[N_1 = n_1] =$$

$$\left\{1 - \frac{\log\left(1 - \sum_{i=2}^k \theta_i\right)}{\log\left(1 - \sum_{i=1}^k \theta_i\right)}\right\} \frac{1}{n_1\left\{\log\left(1 + \frac{\theta_1}{1 - \sum_{i=1}^k \theta_i}\right)\right\}} \left(\frac{\theta_1}{1 - \sum_{i=2}^k \theta_i}\right)^{n_1}. \tag{38.33}$$

Philippou and Roussas (1974) have shown that the distribution in (38.33), generalized (in the sense of random sum) by a Poisson distribution, is a multivariate negative binomial distribution (see Chapter 36).

The marginal distribution of any N_i is similarly a zero-modified logarithmic series distribution. Also the joint distribution of any subset $N_{b_1}, N_{b_2}, \ldots, N_{b_s}$ $(1 < s < k)$ is a modified multivariate logarithmic series distribution with joint probability generating function

$$
\begin{aligned}
&\frac{\log\left(1 - \sum_{j \neq b_i} \theta_j - \sum_{i=1}^{s} \theta_{b_i} t_{b_i}\right)}{\log\left(1 - \sum_{i=1}^{k} \theta_i\right)} \\
&\quad = \frac{\log\left(1 - \sum_{j \neq b_i} \theta_j\right) + \log\left(1 - \frac{\sum_{i=1}^{s} \theta_{b_i} t_{b_i}}{1 - \sum_{j \neq b_i} \theta_j}\right)}{\log\left(1 - \sum_{i=1}^{k} \theta_i\right)} \\
&\quad = \frac{\log\left(1 - \sum_{j \neq b_i} \theta_j\right)}{\log\left(1 - \sum_{i=1}^{k} \theta_i\right)} \\
&\qquad + \frac{\log\left\{1 - \sum_{i=1}^{s} \frac{\theta_{b_i}}{1 - \sum_{i=1}^{k} \theta_i} t_{b_i}\right\}}{\log\left\{1 - \sum_{i=1}^{s} \frac{\theta_{b_i}}{1 - \sum_{i=1}^{k} \theta_i}\right\}} \cdot \left\{1 - \frac{\log\left(1 - \sum_{j \neq b_i} \theta_j\right)}{\log\left(1 - \sum_{i=1}^{k} \theta_i\right)}\right\}. \qquad (38.34)
\end{aligned}
$$

The parameters of the unmodified multivariate logarithmic series distribution are $\theta_{b_i}/(1 - \sum_{j \neq b_i} \theta_j)$. The modification consists of assigning a probability of $\log(1 - \sum_{j \neq b_i} \theta_j)/\log(1 - \sum_{i=1}^{k} \theta_i)$ to the event $N_{b_1} = N_{b_2} = \cdots = N_{b_s} = 0$.

As is to be expected from the nature of the joint distribution of $N_{b_1}, N_{b_2}, \ldots, N_{b_s}$, the conditional distribution of the remaining N_j's, given $N_{b_i} = n_{b_i}$ $(i = 1, 2, \ldots, s)$, takes a different form when $n_{b_1} = n_{b_2} = \cdots = n_{b_s} = 0$. If this event occurs, then the conditional distribution of the remaining N_j's is a multivariate logarithmic series distribution with parameters $\{\theta_j\}$, $j \neq b_1, b_2, \ldots, b_s$. In all other cases (that is, when $\sum_{i=1}^{s} n_{b_i} > 0$), the conditional distribution is not of logarithmic series type at all, but is a *multivariate negative binomial distribution* with parameters $\sum_{i=1}^{s} n_{b_i}$, $\{\theta_j\}$ $(j \neq b_1, b_2, \ldots, b_s)$. Chatfield, Ehrenberg, and Goodhardt (1966) seemed to regard this as implying some kind of lack of homogeneity in the multivariate logarithmic series distribution. While there may not always be good reason to expect that all conditional distributions should be of the same form as the overall distribution, it is undoubtedly true that it is easier to visualize a distribution with conditional distributions of common form. It makes theoretical analysis tidier, and may correspond to practical requirements if the different variables represent similar kinds of physical quantities. From these points of view, the multivariate logarithmic series distribution is at a disadvantage compared with distributions considered in Chapters 35–37.

Remembering that the mode of the univariate logarithmic series distribution is at 1, it is not surprising that the mode of the multivariate logarithmic series distribution is at

$$n_i = 1,\ n_j = 0\ (j \neq i), \quad \text{where } \theta_i = \max(\theta_1, \theta_2, \ldots, \theta_k)\,. \qquad (38.35)$$

If two or more θ's are equal maxima, then there is a corresponding number of equal modal values.

From (38.27), we derive the $\boldsymbol{r} = (r_1, r_2, \ldots, r_k)'$th descending factorial moment as

$$\mu'_{(\boldsymbol{r})} = E\left[\prod_{i=1}^{k} N_i^{(r_i)}\right]$$

$$= \left\{\frac{\left(\sum_{i=1}^k r_i - 1\right)!}{-\log\left(1-\sum_{i=1}^k \theta_i\right)} + E\left[\left(\sum_{i=1}^k N_i\right)^{[\sum_{i=1}^k r_i]}\right]\right\}\prod_{i=1}^k \theta_i^{r_i} \qquad (38.36)$$

[the first term inside the braces on the right-hand side of (38.36) is necessary because the summation formally includes $\sum_{i=1}^k n_i = 0$]. Since, as already observed, $\sum_{i=1}^k N_i$ has a logarithmic series distribution with parameter $\sum_{i=1}^k \theta_i$, we have

$$\mu'_{(\boldsymbol{r})} = \frac{\left(\sum_{i=1}^k r_i - 1\right)!\prod_{i=1}^k \theta_i^{r_i}}{\left\{-\log\left(1-\sum_{i=1}^k \theta_i\right)\right\}\left(1-\sum_{i=1}^k \theta_i\right)^{\sum_{i=1}^k r_i}} \qquad (r_i\text{'s not all 0})\,. \qquad (38.37)$$

Hence,

$$E[N_i] = \frac{\theta_i}{\left\{-\log\left(1-\sum_{h=1}^k \theta_h\right)\right\}\left(1-\sum_{h=1}^k \theta_h\right)}\,, \qquad (38.38)$$

$$\begin{aligned}\operatorname{var}(N_i) = {} & \frac{\theta_i^2}{\left\{-\log\left(1-\sum_{h=1}^k \theta_h\right)\right\}^2\left(1-\sum_{h=1}^k \theta_h\right)^2} \\ & \cdot\left[-\log\left(1-\sum_{h=1}^k \theta_h\right) - 1\right. \\ & \left. + \frac{\left\{-\log\left(1-\sum_{h=1}^k \theta_h\right)\right\}\left(1-\sum_{h=1}^k \theta_h\right)}{\theta_i}\right],\end{aligned} \qquad (38.39)$$

$$\begin{aligned}\operatorname{cov}(N_i, N_j) = {} & \frac{\theta_i\theta_j}{\left\{-\log\left(1-\sum_{h=1}^k \theta_h\right)\right\}^2\left(1-\sum_{h=1}^k \theta_h\right)^2} \\ & \cdot\left\{-\log\left(1-\sum_{h=1}^k \theta_h\right) - 1\right\},\end{aligned} \qquad (38.40)$$

and

$$\operatorname{corr}(N_i, N_j) = \left[\left\{ 1 + \frac{\left\{\log\left(1 - \sum_{h=1}^k \theta_h\right)\right\}\left(1 - \sum_{h=1}^k \theta_h\right)}{\left\{\log\left(1 - \sum_{h=1}^k \theta_h\right) + 1\right\}\theta_i} \right\} \right.$$

$$\left. \cdot \left\{ 1 + \frac{\left\{\log\left(1 - \sum_{h=1}^k \theta_h\right)\right\}\left(1 - \sum_{h=1}^k \theta_h\right)}{\left\{\log\left(1 - \sum_{h=1}^k \theta_h\right) + 1\right\}\theta_j} \right\} \right]^{-1/2} . \quad (38.41)$$

In Gupta and Jain's (1976) generalized bivariate logarithmic series distribution, however, the term *generalized* is used in the Lagrangian sense. The distribution obtained is a limit distribution of generalized (Lagrangian) bivariate negative binomial distribution [see Chapter 36] as $N \to 0$ and is

$$P_{X_1,X_2}(x_1, x_2) = \{-\log(1 - \theta_1 - \theta_2)\}^{-1} \frac{\theta_1^{x_1}\theta_2^{x_2}}{x_1!x_2!}(1 - \theta_1\theta_2)^{(\gamma_1-1)x_1+(\gamma_2-1)x_2}$$
$$\times \frac{\Gamma(\gamma_1 x_1 + \gamma_2 x_2)}{\Gamma\left((\gamma_1 - 1)x_1 + (\gamma_2 - 1)x_2 + 1\right)},$$

where $\theta_i > 0$, $\theta_1 + \theta_2 < 1$, $\gamma_i > 1$.

5 MIXTURES OF MULTIVARIATE POWER SERIES DISTRIBUTIONS

A compound multivariate power series distribution can be formed from (38.1) by ascribing a joint distribution to the parameter $\boldsymbol{\theta} = (\theta_1, \theta_2, \ldots, \theta_k)$; see Chapter 34 for details. The probability mass function of the resulting distribution is

$$P_{\mathbf{X}}(\mathbf{x}) = a(\mathbf{x}) \int_0^\infty \cdots \int_0^\infty \{A(\boldsymbol{\theta})\}^{-1} \left(\prod_{i=1}^k \theta_i^{x_i}\right) dF(\theta_1, \theta_2, \ldots, \theta_k), \quad (38.42)$$

where $F(\theta_1, \theta_2, \ldots, \theta_k)$ is the ascribed joint cumulative distribution function of $\boldsymbol{\theta}$. Equivalently,

$$P_{\mathbf{X}}(\mathbf{x}) = a(\mathbf{x})\, E_{\boldsymbol{\theta}} \left[\{A(\boldsymbol{\theta})\}^{-1} \prod_{i=1}^k \theta_i^{x_i} \right], \quad (38.43)$$

where $E_{\boldsymbol{\theta}}$ denotes expectation with respect to $\boldsymbol{\theta}$.

If $F(\boldsymbol{\theta})$ can be determined from the knowledge of the values of

$$\{a(\mathbf{x})\}^{-1} P_{\mathbf{X}}(\mathbf{x}) = E_{\boldsymbol{\theta}} \left[\{A(\boldsymbol{\theta})\}^{-1} \prod_{i=1}^k \theta_i^{x_i} \right] \quad (38.44)$$

for all $\boldsymbol{x} = (x_1, x_2, \ldots, x_k)'$, the mixture is *identifiable*. Sapatinas (1995) has shown that the condition

$$\sum_{j=1}^{\infty} \left[\sum_{i=1}^{k} \{a(\mathbf{0}'_{j-1}, i, \mathbf{0}'_{n-j})\}^{-1} P_{\boldsymbol{X}}(\mathbf{0}'_{j-1}, \boldsymbol{x}, \mathbf{0}'_{n-j}) \right]^{-1/(2j)} \quad \text{diverges}, \qquad (38.45)$$

where $\mathbf{0}'_i = (0, 0, \ldots, 0)_{1\times i}$ is sufficient to ensure identifiability. The similarity of the condition (38.45) to Carleman's condition for a distribution to be determined by its moments [see, for example, Feller (1957) or Shohat and Tamarkin (1963, pp. 19–20)] is noteworthy. The proof is based on the Cramér–Wold condition for the k-dimensional Stieltjes moment problem which implies that the distribution P in (38.42) is uniquely determined by the sequence $\{P(0, \ldots, 0, x, 0, \ldots, 0), x = 0, 1, 2, \ldots\}$.

It may be easily shown that mixtures of multivariate Poisson distributions [see Chapter 37], multivariate negative binomial distributions [see Chapter 36], or multivariate logarithmic series distributions [as in Eq. (38.27)] are all identifiable. The restricted class of mixtures with probability mass function of type

$$P_{\boldsymbol{X}} = \int_0^{\infty} \frac{a(\boldsymbol{x}) \prod_{i=1}^{k} h_i^{x_i}}{A(h_1\theta, h_2\theta, \ldots, h_k\theta)} \, \theta^{\sum_{i=1}^{k} x_i} \, dF(\theta)\,, \qquad (38.46)$$

that is, as in (38.42) but with $\theta_1 = \theta_2 = \cdots = \theta_k = \theta$ and θ having cumulative distribution function $F(\theta)$, is identifiable if

$$\sum_{j=1}^{\infty} \left\{ \frac{a(x, \mathbf{0}'_{k-1})}{P_{\boldsymbol{X}}(x, \mathbf{0}'_{k-1})} \right\}^{1/(2j)} \quad \text{diverges}. \qquad (38.47)$$

An alternative sufficient condition to (38.44) for identifiability is that the joint distribution has infinitely divisible (possibly shifted) univariate marginal distributions. [*Infinite divisibility* is defined in Chapter 1, p. 48]. Sapatinas (1995) has pointed out that the assumption of infinite divisibility for all univariate marginal distributions is weaker than the assumption that the joint distribution is itself infinitely divisible. As an example, he has presented a bivariate power series distribution defined as in (38.1) with $k = 2$ and

$$a(x_1, x_2) = \begin{cases} \prod_{i=1}^{2} (e^{-\phi_i} \phi_i^{x_i}/x_i!) + \alpha & \text{if } (x_1, x_2) = (0, 1) \text{ or } (1, 0) \\ \prod_{i=1}^{2} (e^{-\phi_i} \phi_i^{x_i}/x_i!) - \alpha & \text{if } (x_1, x_2) = (0, 0) \text{ or } (1, 1) \\ \prod_{i=1}^{2} (e^{-\phi_i} \phi_i^{x_i}/x_i!) & \text{otherwise,} \end{cases} \qquad (38.48)$$

with $0 < \alpha < e^{-(\phi_1+\phi_2)} \min(1, \phi_1\phi_2)$ $(\phi_1, \phi_2 > 0)$. For $\theta_1 = \theta_2 = 1$, the probability generating function $G(t_1, t_2)$ of this distribution is not $\exp\{-\sum_{i=1}^{2} \phi_i(1 - t_i)\}$, but as $t_1, t_2 \to \infty$ its ratio to this value tends to 1. Since the probability generating function is positive at $t_1 = t_2 = 0$, it follows that the distribution is *not* infinitely divisible even though the marginal distributions are Poisson and so are infinitely divisible.

Characterization of multivariate power series distribution mixtures can be achieved in terms of the regression functions

$$\mu_1'(\theta_i \mid x_1, \ldots, x_i, \mathbf{0}_{k-i}') = E\left[\theta_i \left| \bigcap_{j=1}^{i}(X_j = x_j) \bigcap_{j=i+1}^{k} (X_j = 0)\right.\right], \quad (i = 1, 2, \ldots, k). \tag{38.49}$$

A characterization of multivariate power series distribution mixtures in (38.42) in terms of regression functions [generalizing the result of Johnson and Kotz (1989)] is as follows:

Let $\boldsymbol{X}$ be a k-dimensional random vector with the multivariate power series distribution mixture as defined in (38.42). Let the regression functions $\mu_1'(\theta_i \mid x_1, \ldots, x_i, \mathbf{0}_{k-i}')$ given in (38.49), also denoted by $m_i(x_1, \ldots, x_i, 0, \ldots, 0)$, be such that

$$\sum_{h=1}^{\infty} \left\{ \prod_{i=0}^{h-1} m_1(i, 0, \ldots, 0) + \prod_{i=0}^{h-1} m_2(0, i, 0, \ldots, 0) + \cdots + \prod_{i=0}^{h-1} m_k(0, \ldots, 0, i) \right\}^{-1/(2h)} = \infty . \tag{38.50}$$

Then, the distribution of $\boldsymbol{\theta}$ is uniquely determined by m_i, $i = 1, 2, \ldots, k$.

Sapatinas (1995) showed that the condition (38.50) is met if the condition

$$\limsup_{h} \left\{ \frac{m_1(h, 0, \ldots, 0)}{h} + \frac{m_2(0, h, 0, \ldots, 0)}{h} + \cdots + \frac{m_k(0, \ldots, 0, h)}{h} \right\} < \infty \tag{38.51}$$

is valid.

As an example, if the conditional joint distribution of $\boldsymbol{X}$, given $\boldsymbol{\theta}$, is negative multinomial with parameters $(r; \boldsymbol{\theta})$ [see Chapter 36], we obtain

$$\mu_1'(\theta_i \mid x_1, \ldots, x_i, \mathbf{0}_{k-i}') = (x_i + \alpha_i)/(x + r + \alpha) \tag{38.52}$$

(where $x = \sum_{i=1}^{k} x_i$ and $\alpha = \sum_{i=1}^{k} \alpha_i$) if and only if $\boldsymbol{X}$ has an unconditional joint negative multinomial distribution with parameters $(r; \alpha_1, \ldots, \alpha_k)$; then $\boldsymbol{\theta}$ must have a joint Dirichlet $(\alpha_1, \alpha_2, \ldots, \alpha_k)$ distribution, as defined in Johnson and Kotz (1972).

This result is a generalization of univariate results established by Johnson (1957, 1967) for a Poisson (θ) mixture, which show that if $E[\theta \mid X = x]$ is a linear function of x, the prior distribution of θ must be a gamma distribution. More generally, Ericson (1969) showed that if the regression of θ on x is linear, then the first two moments of the prior distribution are determined by the regression function and the variance of X. [See also Diaconis and Ylvisaker (1979) for a more general result on exponential families.]

Khatri (1971) termed all distributions (univariate or multivariate) of the form

$$F_1(\boldsymbol{\theta}) \underset{\boldsymbol{\theta}}{\wedge} F_2$$

contagious distributions. He also referred to these as $F_1 - F_2$ distributions, while $F_2 - F_1$ distributions is a more conventional usage. Observing that "the general class of multivariate contagious distributions is too wide," Khatri proceeded to discuss some cases of special importance. These include the following:

(a) *Multivariate Neyman Type A distribution* with probability generating function

$$G(\boldsymbol{t}) = \exp\left[\lambda\left\{e^{\sum_{i=1}^k \theta_i(t_i-1)+\sum\sum_{i<j=1}^k \theta_{ij}(t_i-1)(t_j-1)} - 1\right\}\right], \qquad \lambda,\ \theta_i,\ \theta_{ij} \geq 0\,. \quad (38.53)$$

Particular attention is given to the case when all θ_{ij}'s are zero.

(b) *Poisson general multinomial distribution* with probability generating function

$$G(\boldsymbol{t}) = \left\{1 + \sum_{i=1}^k \theta_i(t_i - 1) + \sum\sum_{i<j=1}^k \theta_{ij}(t_i - 1)(t_j - 1)\right\}^N \quad (38.54)$$

with either (i) N is a positive integer, θ's are all non-negative and their sum is less than 1 or (ii) $N < 0$ and θ's are all nonpositive.

(c) *General binomial–multivariate power series distribution* with probability generating function

$$G(\boldsymbol{t}) = \{1 - p + p\,G_2(\boldsymbol{t})\}^N\,, \quad (38.55)$$

where $G_2(\boldsymbol{t})$ is the probability generating function of a multivariate power series distribution, and either (i) N is a positive integer and $0 < p \leq 1$ or (ii) $N < 0$ and $p \leq 0$.

(d) *Logarithmic–multivariate power series distribution* with probability generating function

$$G(\boldsymbol{t}) = 1 - \omega + \omega\,\frac{\log\{1 - \theta\,G_2(\boldsymbol{t})\}}{\log(1 - \theta)}, \qquad 0 \leq \omega \leq 1,\ \theta > 0\,. \quad (38.56)$$

This is an "added-zeros" distribution.

6 MULTIVARIATE SUM-SYMMETRIC POWER SERIES DISTRIBUTIONS

A special case of the multivariate power series distribution (38.1) is the *multivariate sum-symmetric power series distribution*. This is the case when $A(\boldsymbol{\theta}) = B(\theta)$ with $\theta = \sum_{i=1}^k \theta_i$ and $B(\theta)$ admits a power series expansion in powers of θ for $0 \leq \theta < \rho$,

ρ being the radius of convergence. In other words, $B(\theta) = \sum b(n)\, \theta^n$ with summation over S for which $S = \{n : n = \sum_{i=1}^{k} n_i,\ \boldsymbol{n} \in T\}$, where T is the support of the multivariate power series distribution (38.1), so that $b(n) > 0$ for $n \in S$ and is independent of θ. This distribution, denoted by MSSPSD($\boldsymbol{\theta}$, $B(\theta)$), was introduced by Joshi and Patil (1972). Note that if $\boldsymbol{N} \rightsquigarrow$ MSSPSD($\boldsymbol{\theta}$, $B(\theta)$), then $N = \sum_{i=1}^{k} N_i$ has a power series distribution with parameters (θ, $B(\theta)$). We also observe that $n \in S$ iff $\boldsymbol{n} \in T$ and that

$$a(\boldsymbol{n}) = \binom{n}{\boldsymbol{n}} b(n) = \frac{n!}{\prod_{i=1}^{k}(n_i!)}\, b(n)\,. \tag{38.57}$$

Furthermore, it is easy to establish that if $\boldsymbol{N}_1, \boldsymbol{N}_2, \ldots, \boldsymbol{N}_\ell$ are independent and identically distributed as MSSPSD($\boldsymbol{\theta}$, $B(\theta)$), then $\boldsymbol{N} = (N_1, \ldots, N_k)' = \sum_{i=1}^{\ell} \boldsymbol{N}_i$ and $N = \sum_{j=1}^{k} N_j$ are distributed as MSSPSD($\boldsymbol{\theta}$, $(B(\theta))^\ell$) and power series distributions with parameters (θ, $(B(\theta))^\ell$), respectively, with the support of $\boldsymbol{N}$ being the ℓ-fold sum of T and that of N being the ℓ-fold sum of S.

For a multivariate sum-symmetric power series distribution, a polynomial of degree d in $\pi_1, \pi_2, \ldots, \pi_k$ (with $\pi_i = \theta_i/\theta$) is MVU (minimum variance unbiased) estimable on the basis of a random sample of size ℓ iff $\min\{\ell[S]\} \geq d$. More precisely, the only parametric functions $\beta(\boldsymbol{\pi})$ that are MVU estimable on the basis of a random sample of size ℓ from a MSSPSD with finite range not containing the origin are polynomials in $\pi_1, \pi_2, \ldots, \pi_k$ of degree not exceeding $\min\{\ell[S]\}$.

In addition, if $\alpha(\theta)$ and $\beta(\boldsymbol{\pi})$ (a polynomial in $\pi_1, \pi_2, \ldots, \pi_k$) are MVU estimable on the basis of a random sample of size ℓ from a MSSPSD, then their product is also MVU estimable for the same sample size ℓ and the MVU estimator of the product is the product of the individual MVU estimators.

Joshi and Patil (1972) established the following remarkable characterizations of MSSPSD via MVU estimators in the class of multivariate power series distributions:

- **(i)** A k-variate power series distribution having a basis of I_k as its range is a MSSPSD if the MVU estimators of the components of its parameter vector $\boldsymbol{\theta}$ are proportional to the components of the sample totals vector $\boldsymbol{N}$.
- **(ii)** A k-variate power series distribution having a basis of I_k as its range is a MSSPSD if the ratios of the MVU estimators of θ_i and θ are equal to the ratios of the corresponding sample totals N_i and N, that is,

$$\frac{\tilde{\theta}_i}{\tilde{\theta}} = \frac{N_i}{N} \qquad \text{for } i = 1, 2, \ldots, k\,. \tag{38.58}$$

Note that for a MSSPSD with finite range, the components of $\boldsymbol{\theta}$ are not MVU estimable and thus the above characterizations are inapplicable for multivariate power series distributions having finite range, which includes an important distribution such as the multinomial distribution. For the case $k = 2$, Joshi and Patil (1972) provided a more general result as follows: A bivariate power series distribution admitting θ_i/θ to be MVU estimable with $\tilde{\theta}_i/\theta = N_i/N$ for some i is a bivariate sum-symmetric power series distribution.

Joshi and Patil (1972) also established the following results:

(i) Let $u(\boldsymbol{\theta})$ be a function of the parameter vector $\boldsymbol{\theta}$ of a MSSPSD admitting a power series expansion in the components θ_i of $\boldsymbol{\theta}$. Then a sufficient condition for $u(\boldsymbol{\theta})$ to be MVU estimable is that $\theta = \sum_{i=1}^{k} \theta_i$ be MVU estimable.

(ii) For $u(\boldsymbol{\theta})$ as above, a sufficient condition for $u(\boldsymbol{\theta})$ to be MVU estimable is that the range of the MSSPSD be a Cartesian product.

(iii) $\boldsymbol{\theta}$ is componentwise estimable iff $\theta = \sum_{i=1}^{k} \theta_i$ is MVU estimable.

(iv) The probability mass function P of the MSSPSD is MVU estimable for every sample size ℓ and the MVU estimator is

$$\tilde{P} = \tilde{p} \prod_{i=1}^{k} \binom{N_i}{n_i} \Big/ \binom{N}{n}, \tag{38.59}$$

where $\tilde{}$ denotes the MVU estimator, and p is the probability mass function of the component sum of the MSSPSD.

The result in (**iv**) is quite interesting in that on the right-hand side appears two factors of which the first factor is the MVU estimator for the univariate distribution of the component sum of the MSSPSD and the second factor is a simple multiplying factor of multinomial coefficients. This factorization reduces the estimation problem from the multivariate to the univariate case and tabulations for the univariate case can be convenient and useful for the multivariate case.

Note that the hypergeometric factor $\prod_{i=1}^{k} \binom{N_i}{n_i} / \binom{N}{n}$ in (38.59) is the MVU estimator of the probability mass function of a singular multinomial distribution at $\boldsymbol{n}$. Also the conditional distribution of $\boldsymbol{N}$, given N, for a MSSPSD is a singular multinomial with parameters N and $\pi_i = \theta_i/\theta$ $(i = 1, 2, \ldots, k)$. Therefore, in (38.59), we have for an MSSPSD the MVU estimator for the probability mass function of $\boldsymbol{N}$ as the MVU estimator for the probability mass function of N times the MVU estimator for the probability mass function of the conditional distribution of $\boldsymbol{N}$ given N. Thus the property of the probability mass function of $\boldsymbol{N}$ is inherited by its MVU estimator.

7 EQUIVALENT FAMILIES OF DISTRIBUTIONS

Gerstenkorn (1981) introduced the concept of *equivalent families of distributions*, which is described briefly in this section.

Two families $\{\boldsymbol{X}(\boldsymbol{z}),\ \boldsymbol{z} = (z_1, \ldots, z_k)' \in Z\}$ and $\{\boldsymbol{Y}(\boldsymbol{u}),\ \boldsymbol{u} = (u_1, \ldots, u_k)' \in U\}$, where $\boldsymbol{u} = \boldsymbol{u}(\boldsymbol{z}) = (u_1(z_1, \ldots, z_k), \ldots, u_k(z_1, \ldots, z_k))$ is an invertible function, are said to be *equivalent* if $\boldsymbol{X}(\boldsymbol{z}) = \boldsymbol{Y}(\boldsymbol{u}(\boldsymbol{z}))$ for every $\boldsymbol{z} \in Z$.

A sufficient condition for the equivalence of two families of random vectors is as follows:

If the families $\{\boldsymbol{X}(\boldsymbol{z}), \boldsymbol{z} \in Z\}$ and $\{\boldsymbol{Y}(\boldsymbol{u}), \boldsymbol{u} \in U\}$ with the defining functions $A_1(z_1, \ldots, z_k)$ and $A_2(u_1, \ldots, u_k)$, respectively, contain a common random vector

$$\boldsymbol{X} = \boldsymbol{X}(\boldsymbol{z}_0) = \boldsymbol{Y}(\boldsymbol{u}_0)\,, \tag{38.60}$$

where $z_0 = (z_{10}, \ldots, z_{k0})' \neq (0, 0, \ldots, 0)' \neq (u_{10}, \ldots, u_{k0})' = \boldsymbol{u}_0$, then they are equivalent and their equivalence is determined by

$$z_i = \frac{z_{i0}}{u_{i0}}\, u_i \qquad \text{or} \qquad u_i = \frac{u_{i0}}{z_{i0}}\, z_i, \qquad i = 1, 2, \ldots, k\,. \tag{38.61}$$

Moreover, there exists a positive constant c such that

$$A_1(z_1, \ldots, z_k) = c\, A_2\left(\frac{u_{10}}{z_{10}}\, z_1, \ldots, \frac{u_{k0}}{z_{k0}}\, z_k\right) \tag{38.62}$$

and

$$A_2(u_1, \ldots, u_k) = \frac{1}{c}\, A_1\left(\frac{z_{10}}{u_{10}}\, u_1, \ldots, \frac{z_{k0}}{u_{k0}}\, u_k\right)\,. \tag{38.63}$$

Determination of the distribution of a sum of families of distributions of the multivariate power series distribution type is facilitated by the following result:

Let $\{\boldsymbol{X}(\boldsymbol{z}), \boldsymbol{z} \in Z_1\}$ and $\{\boldsymbol{Y}(\boldsymbol{z}), \boldsymbol{z} \in Z_2\}$ be families of random vectors of the multivariate power series distribution type with the defining functions $A_1(z_1, \ldots, z_k)$ and $A_2(z_1, \ldots, z_k)$, respectively. Let the random vectors $\boldsymbol{X}(\boldsymbol{z})$ and $\boldsymbol{Y}(\boldsymbol{z})$ be independent. Then, for any $(z_1, \ldots, z_k)' \in Z = Z_1 \cap Z_2$, the family of random vectors $\{\boldsymbol{W}(\boldsymbol{z}), \boldsymbol{z} \in Z\}$ is of the multivariate power series distribution type with the defining function $A(\boldsymbol{z})$ being the product of the two defining functions $A_1(\boldsymbol{z})$ and $A_2(\boldsymbol{z})$ if and only if $\boldsymbol{W}(\boldsymbol{z}) = \boldsymbol{X}(\boldsymbol{z}) + \boldsymbol{Y}(\boldsymbol{z})$.

This leads to the following definition:

Let a family of probability distributions $\{P(z_1, \ldots, z_k)\}$ of a family of random vectors $\{\boldsymbol{X}(\boldsymbol{z}), \boldsymbol{z} \in Z\}$ be given. If, for any values $\boldsymbol{z}'$ and $\boldsymbol{z}''$ such that $\boldsymbol{z}' + \boldsymbol{z}'' \in Z$, the random vector $\boldsymbol{Y}$ being a sum of independent vectors $\boldsymbol{X}(\boldsymbol{z}')$ and $\boldsymbol{X}(\boldsymbol{z}'')$ has a distribution being an element of the family once again, then we say that for the family of vectors $\{\boldsymbol{X}(\boldsymbol{z}), \boldsymbol{z} \in Z\}$ or the family of probability distributions $\{P(\boldsymbol{z})\}$ *the theorem on addition with respect to the parameter* $\boldsymbol{z}$ *holds*. Recall from Section 1 that this property holds for the multivariate Poisson distribution with respect to the parameters $\lambda_1, \ldots, \lambda_k$.

This is also valid for the class of multivariate power series distributions through the defining functions. In fact, if the defining function $A(z_1, \ldots, z_k)$ of a family $\{P(z_1, \ldots, z_k)\}$ of distributions, for any $\boldsymbol{z}', \boldsymbol{z}'' \in Z$ such that $\boldsymbol{z}' + \boldsymbol{z}'' \in Z$, satisfies the condition

$$A(z_1', \ldots, z_k')\, A(z_1'', \ldots, z_k'') = A(z_1' + z'', \ldots, z_k' + z_k''), \tag{38.64}$$

then the theorem on addition with respect to the parameter $\boldsymbol{z}$ holds for this family. Conversely, if for the family of vectors $\{\boldsymbol{X}(\boldsymbol{z}), \boldsymbol{z} \in Z\}$ of the multivariate power series distribution type the theorem on addition with respect to the parameter $\boldsymbol{z}$ holds, then the defining function of this family satisfies the condition

$$A(z_1', \ldots, z_k')\, A(z_1'', \ldots, z_k'') = c\, A(z_1' + z_1'', \ldots, z_k' + z_k'')\,, \tag{38.65}$$

where c is some positive constant.

Furthermore, if the defining function $A(z_1, \ldots, z_k)$ depends on some other parameter $\boldsymbol{w}$ on which the coefficients $a(n_1, \ldots, n_k)$ depend, then the following result holds:

If for the family of random vectors $\{\boldsymbol{X}(z; \boldsymbol{w}), \boldsymbol{w} \in W\}$ of the multivariate power series distribution type (with respect to the parameter z) the theorem on addition with respect to the parameter $\boldsymbol{w}$ holds, then the defining function $A(z; \boldsymbol{w})$ of this family satisfies the condition

$$A(z; \boldsymbol{w}_1)A(z; \boldsymbol{w}_2) = A(z; \boldsymbol{w}_1 + \boldsymbol{w}_2) \tag{38.66}$$

and conversely.

BIBLIOGRAPHY

Berg, S. (1977). Certain properties of the multivariate factorial series distributions, *Scandinavian Journal of Statistics*, **4**, 25–30.

Chatfield, C., Ehrenberg, A. S. C., and Goodhardt, G. J. (1966). Progress on a simplified model of stationary purchasing behaviour, *Journal of the Royal Statistical Society, Series A*, **129**, 317–367.

Diaconis, P., and Ylvisaker, D. (1979). Conjugate priors for exponential families, *Annals of Statistics*, **7**, 269–281.

Ericson, W. A. (1969). A note on the posterior mean of a population mean, *Journal of the Royal Statistical Society, Series B*, **31**, 332–334.

Feller, W. (1957). *An Introduction to Probability Theory and Its Applications*, second edition, New York: John Wiley & Sons.

Gerstenkorn, T. (1981). On multivariate power series distributions, *Revue Roumaine de Mathématiques Pures et Appliquées*, **26**, 247–266.

Gupta, R. P., and Jain, G. C. (1976). A generalized bivariate logarithmic series distribution, *Biometrische Zeitschrift*, **18**, 169–173.

Johnson, N. L. (1957). Uniqueness of a result in the theory of accident proneness, *Biometrika*, **44**, 530–531.

Johnson, N. L. (1967). Note on the uniqueness of a result in a certain accident proneness model, *Journal of the American Statistical Association*, **62**, 288–289.

Johnson, N. L., and Kotz, S. (1972). *Multivariate Continuous Distributions*, New York: John Wiley & Sons.

Johnson, N. L., and Kotz, S. (1989). Characterization based on conditional distributions, *Annals of the Institute of Statistical Mathematics*, **41**, 13–17.

Joshi, S. W., and Patil, G. P. (1972). Sum-symmetric power series distributions and minimum variance unbiased estimation, *Sankhyā, Series A*, **34**, 377–386.

Khatri, C. G. (1959). On certain properties of power series distributions, *Biometrika*, **46**, 486–490.

Khatri, C. G. (1971). On multivariate contagious distributions, *Sankhyā, Series B*, **33**, 197–216.

Kyriakoussis, A. G., and Vamvakari, M. G. (1996). Asymptotic normality of a class of bivariate-multivariate discrete power series distributions, *Statistics & Probability Letters*, **27**, 207–216.

Patil, G. P. (1965). On multivariate generalized power series distribution and its applications to the multinomial and negative multinomial, in *Classical and Contagious Discrete Distributions* (Ed. G. P. Patil), Proceedings of International Symposium at McGill University on August 15–20, 1963, pp. 183–194, Calcutta: Statistical Publishing Society; Oxford: Pergamon Press.

Patil, G. P. (1985a). Multivariate logarithmic series distribution, in *Encyclopedia of Statistical Sciences* (Eds. S. Kotz, N. L. Johnson, and C. B. Read), Vol. 6, pp. 82–85, New York: John Wiley & Sons.

Patil, G. P. (1985b). Multivariate power series distributions, in *Encyclopedia of Statistical Sciences* (Eds. S. Kotz, N. L. Johnson, and C. B. Read), Vol. 6, pp. 104–108, New York: John Wiley & Sons.

Patil, G. P., and Bildikar, S. (1967). Multivariate logarithmic series distribution as a probability model in population and community ecology and some of its statistical properties, *Journal of the American Statistical Association*, **62**, 655–674.

Philippou, A. N., and Roussas, G. G. (1974). A note on the multivariate logarithmic series distribution, *Communications in Statistics*, **3**, 469–472.

Sapatinas, T. (1995). Identifiability of mixtures of power series distributions and related characterizations, *Annals of the Institute of Statistical Mathematics*, **47**, 447–459.

Shohat, J., and Tamarkin, J. D. (1963). *The Problem of Moments*, second edition, New York: American Mathematical Society.

CHAPTER 39

Multivariate Hypergeometric and Related Distributions

1 INTRODUCTION AND GENESIS

In Chapters 3 and 6 we have already seen that, while binomial distributions can arise in repeated sampling from a finite population provided that each chosen item is replaced before the next choice is made (*sampling with replacement*), univariate hypergeometric distributions arise under *sampling without replacement*.

There exists a similar relationship between multinomial distributions (see Chapter 35, Section 1) and multivariate hypergeometric distributions. While the former can arise in repeated *sampling with replacement* from a population divided into k (> 2) categories, the latter will arise under *sampling without replacement* from a population comprising k (> 2) categories.

According to Hald (1990), multivariate hypergeometric distribution was introduced in a different notation (using the symbol $\underset{m}{\overset{n}{\square}}$ for the binomial coefficient) by de Montmort in his famous essay in 1708. In this essay, multinomial distribution is also discussed.

2 DISTRIBUTION AND MOMENTS

Consider a population of m individuals, of which m_1 are of type C_1, m_2 of type $C_2, \ldots,$ and m_k of type C_k with $\sum_{i=1}^{k} m_i = m$. Suppose a random sample of size n is chosen by sampling without replacement from among these m individuals; that is, we just choose a set of n from the m individuals available. There are $\binom{m}{n}$ possible sets amongst which $\prod_{i=1}^{k} \binom{m_i}{n_i}$ will contain n_1 individuals of type C_1, n_2 of type $C_2, \ldots,$ and n_k of type C_k. The joint distribution of $\boldsymbol{N} = (N_1, N_2, \ldots, N_k)'$, representing the numbers of individuals of types $C_1, C_2, \ldots, C_k$, respectively, has probability mass function

$$P(n_1, n_2, \ldots, n_k) = \Pr[\boldsymbol{N} = \boldsymbol{n}] = \Pr\left[\bigcap_{i=1}^{k}(N_i = n_i)\right]$$

$$= \left\{\prod_{i=1}^{k}\binom{m_i}{n_i}\right\}\Big/\binom{m}{n} \tag{39.1}$$

with $\sum_{i=1}^{k} n_i = n$, $0 \le n_i \le m_i$ $(i = 1, 2, \ldots, k)$.

The distribution in (39.1) is called a *multivariate hypergeometric distribution* with parameters $(n; m_1, m_2, \ldots, m_k)$. We will denote it by the symbols

$$\text{Mult. Hypg.}(n; m_1, m_2, \ldots, m_k)$$

or

$$\text{Mult. Hypg.}(n; \boldsymbol{m}).$$

When $k = 2$, it reduces to the ordinary (univariate) hypergeometric distribution described in Chapter 6. [Note that there are really only $k - 1$ distinct variables since $n_k = n - \sum_{i=1}^{k-1} n_i$.]

Several of the relationships among multinomial variables also hold, with minor modifications, for variables with joint multivariate hypergeometric distributions. This is not surprising in view of the similarity of geneses of the two classes of distributions. As a matter of fact, as m tends to infinity with $m_i/m = p_i$ for $i = 1, 2, \ldots, k$, the multivariate hypergeometric distribution in (39.1) tends to the multinomial distribution with parameters $(n; p_1, p_2, \ldots, p_k)$ [see Eq. (35.13)].

From (39.1), the marginal distribution of N_i is univariate hypergeometric $(n; m_i, m)$; that is,

$$\Pr[N_i = n_i] = \binom{m_i}{n_i}\binom{m - m_i}{n - n_i}\Big/\binom{m}{n}. \tag{39.2}$$

Similarly, the joint distribution of $N_{a_1}, N_{a_2}, \ldots, N_{a_s}, n - \sum_{i=1}^{s} N_{a_i}$ $(s \le k - 1)$ is

$$\text{Mult. Hypg.}\left(n; m_{a_1}, m_{a_2}, \ldots, m_{a_s}, n - \sum_{i=1}^{s} m_{a_i}\right). \tag{39.3}$$

Also, the conditional joint distribution of the remaining elements of $\boldsymbol{N}$, given $N_{a_i} = n_{a_i}$ for $i = 1, 2, \ldots, s$, is

$$\text{Mult. Hypg.}\left(n - \sum_{i=1}^{s} n_{a_i}; m_{b_1}, m_{b_2}, \ldots, m_{b_{k-s}}\right), \tag{39.4}$$

where $(b_1, b_2, \ldots, b_{k-s})$ is the complement of $(a_1, a_2, \ldots, a_s)$ in $(1, 2, \ldots, k)$. In particular, the conditional distribution of N_j, given $N_{a_i} = n_{a_i}$ $(i = 1, 2, \ldots, s)$, is

$$\text{Hypg.}\left(n - \sum_{i=1}^{s} n_{a_i};\ m_j, \sum_{i=1}^{k-s} b_i\right) \quad \text{for } j \in (b_1, b_2, \ldots, b_{k-s}). \tag{39.5}$$

The joint descending $\boldsymbol{r} = (r_1, r_2, \ldots, r_k)'$th factorial moment of $\boldsymbol{N}$ is

$$\begin{aligned}\mu'_{(\boldsymbol{r})}(\boldsymbol{N}) = \mu'_{(r_1, r_2, \ldots, r_k)}(N_1, N_2, \ldots, N_k) &= E\left[\prod_{i=1}^{k} N_i^{(r_i)}\right] \\ &= \frac{n^{(\sum_{i=1}^{k} r_i)}}{m^{(\sum_{i=1}^{k} r_i)}} \prod_{i=1}^{k} m_i^{(r_i)}, \end{aligned} \tag{39.6}$$

where, as before [Chapter 34, Eq. (34.12)], $a^{(b)} = a(a-1)\ldots(a-b+1)$ is the descending factorial.

From (39.6), we have in particular

$$E[N_i] = \frac{n}{m}\, m_i\,, \tag{39.7}$$

$$\begin{aligned}\text{var}(N_i) &= \frac{n(n-1)}{m(m-1)}\, m_i(m_i - 1) + \frac{n}{m}\, m_i - \left(\frac{n}{m}\, m_i\right)^2 \\ &= \frac{n(m-n)}{m^2(m-1)}\, m_i(m - m_i), \end{aligned} \tag{39.8}$$

and

$$\begin{aligned}\text{cov}(N_i, N_j) &= \frac{n(n-1)}{m(m-1)}\, m_i m_j - \left(\frac{n}{m}\, m_i\right)\left(\frac{n}{m}\, m_j\right) \\ &= -\frac{n(m-n)}{m^2(m-1)}\, m_i m_j\,, \end{aligned} \tag{39.9}$$

whence

$$\text{corr}(N_i, N_j) = -\sqrt{\frac{m_i m_j}{(m - m_i)(m - m_j)}}\,. \tag{39.10}$$

Note that the covariance and the correlation coefficient are both negative. The similarity of this result with the corresponding formula in (35.9) for the multinomial distribution is quite apparent.

In fact, with n being fixed, and letting $m \to \infty$, Eqs. (39.7)–(39.10) are readily reduced to the limiting results

$$E[N_i] = np_i, \tag{39.11}$$

$$\text{var}(N_i) = np_i(1 - p_i), \tag{39.12}$$

$$\text{cov}(N_i, N_j) = -np_i p_j\,, \tag{39.13}$$

and

$$\text{corr}(N_i, N_j) = -\sqrt{\frac{p_i p_j}{(1-p_i)(1-p_j)}}\,, \tag{39.14}$$

which, not surprisingly, agree with the corresponding results for the multinomial distribution as presented in Chapter 35.

From Eqs. (39.3) and (39.7), we obtain the regression of N_j on $N_{a_1}, N_{a_2}, \ldots, N_{a_s}$ to be

$$E\left[N_j \,\middle|\, \bigcap_{i=1}^{s}(N_{a_i} = n_{a_i})\right] = \left(n - \sum_{i=1}^{s} n_{a_i}\right) m_j \Big/ \left(m - \sum_{i=1}^{s} m_{a_i}\right). \tag{39.15}$$

Note that $m - \sum_{i=1}^{s} m_{a_i} = \sum_{i=1}^{k-s} m_{b_i}$. Thus, the regression of N_j on $N_{a_1}, N_{a_2}, \ldots, N_{a_s}$ is linear. Also, from (39.8) we obtain

$$\begin{aligned} &\text{var}\left(N_j \,\middle|\, \bigcap_{i=1}^{s}(N_{a_i} = n_{a_i})\right) \\ &\quad = \frac{\left(n - \sum_{i=1}^{s} n_{a_i}\right)\left\{m - n - \sum_{i=1}^{s}(m_{a_i} - n_{a_i})\right\}}{\left(m - \sum_{i=1}^{s} m_{a_i}\right)^2 \left(m - \sum_{i=1}^{s} m_{a_i} - 1\right)}\, m_j \left(m - \sum_{i=1}^{s} m_{a_i} - m_j\right). \end{aligned} \tag{39.16}$$

Equation (39.16) shows that the regression is heteroscedastic.

Multivariate hypergeometric distributions can also arise when sampling in sequence without replacement, from a finite population of size m, where each individual is classified according to a *single* characteristic, $\mathcal{D}$ (say). If there are d individuals with this characteristic, and the numbers of individuals in successive samples of sizes $n_1, n_2, \ldots, n_w$ possessing the characteristic $\mathcal{D}$ are $Y_1, Y_2, \ldots, Y_w$, then the joint probability mass function of the Y's is

$$\Pr\left[\bigcap_{i=1}^{w}(Y_i = y_i)\right] = \binom{m - \sum_{i=1}^{w} n_i}{d - \sum_{i=1}^{w} y_i} \prod_{i=1}^{w} \binom{n_i}{y_i} \Big/ \binom{m}{d}; \tag{39.17}$$

that is, $(Y_1, Y_2, \ldots, Y_w, d - \sum_{i=1}^{w} Y_i)$ is distributed as

$$\text{Mult. Hypg.}\left(d; n_1, n_2, \ldots, n_w, m - \sum_{i=1}^{w} n_i\right). \tag{39.18}$$

The w samples, together with the remainder $m - \sum_{i=1}^{w} n_i$, now constitute the k $(= w+1)$ types $C_1, C_2, \ldots, C_k$ into which the population is divided, and d *plays the role of sample size*. An application of this genesis of multivariate hypergeometric distribution is presented in Section 5. In order to indicate that only $\boldsymbol{Y} = (Y_1, Y_2, \ldots, Y_w)'$ is to be considered, we shall write

$$Y \rightsquigarrow \text{Mult. Hypg.}(d; n_1, n_2, \ldots, n_w). \tag{39.19}$$

Panaretos (1983a) presented the following interesting characterization of the multivariate hypergeometric distributions. Suppose $\boldsymbol{X} = (X_1, \ldots, X_k)'$ and $\boldsymbol{Y} = (Y_1, \ldots, Y_k)'$ are two random vectors, where X_i and Y_i $(i = 1, 2, \ldots, k)$ are non-negative integer-valued random variables with $\boldsymbol{X} \geq \boldsymbol{Y}$. Suppose $\boldsymbol{X}$ has a multinomial distribution with probability mass function

$$\Pr[\boldsymbol{X} = \boldsymbol{n}] = \frac{n!}{n_0!n_1! \cdots n_k!} p_0^{n_0} p_1^{n_1} \cdots p_k^{n_k},$$

where $n_0 = n - \sum_{i=1}^k n_i$ and $p_0 = 1 - \sum_{i=1}^k p_i$. Suppose the conditional distribution of $\boldsymbol{Y}$ given $\boldsymbol{X}$ is of the form

$$\Pr[\boldsymbol{Y} = \boldsymbol{m} | \boldsymbol{X} = \boldsymbol{n}] = \frac{a_{\boldsymbol{m}} b_{\boldsymbol{n}-\boldsymbol{m}}}{c_{\boldsymbol{n}}},$$

where $\{(a_{\boldsymbol{m}}, b_{\boldsymbol{m}}) : \boldsymbol{m} = (m_1, \ldots, m_k),\ m_i = 0, 1, \ldots;\ i = 1, 2, \ldots, k\}$ is a sequence of real vectors satisfying some conditions [see Panaretos (1983a)] and $c_{\boldsymbol{n}} = \sum_{\boldsymbol{m}=\boldsymbol{0}}^{\boldsymbol{n}} a_{\boldsymbol{m}} b_{\boldsymbol{n}-\boldsymbol{m}}$. Then, with $\boldsymbol{Z} = \boldsymbol{X} - \boldsymbol{Y}$, the condition

$$\begin{aligned}\Pr[\boldsymbol{Y} = \boldsymbol{m} | \boldsymbol{Z} = \boldsymbol{0}] = \Pr[\boldsymbol{Y} = \boldsymbol{m} | Z_{h_1} &= \ell_{h_1} r_{h_1} + 1, \ldots, \\ Z_{h_j} &= \ell_{h_j} r_{h_j} + 1,\ Z_{h_{j+1}} = \cdots = Z_{h_k} = 0]\end{aligned}$$

for all $j = 1, 2, \ldots, k$, where ℓ_i $(i = 1, 2, \ldots, k)$ are non-negative integers such that $\sum_{i=1}^k \ell_i r_i \leq n - m - k$ [here, $\{r_1, r_2, \ldots, r_k\}$ denotes an integer partition of the integer m; that is, $r_1 + \cdots + r_k = m$, $\{h_1, h_2, \ldots, h_w\}$ $(w \leq k)$ denotes an arbitrary subset of size w of $\{1, 2, \ldots, k\}$, $\sum_{i=1}^k n_i \leq n$, and $\sum_{i=1}^k m_i \leq m$], is necessary and sufficient for the conditional distribution of $\boldsymbol{Y} \mid (\boldsymbol{X} = \boldsymbol{n})$ to be multivariate hypergeometric with probability mass function

$$\binom{n_0}{m_0}\binom{n_1}{m_1} \cdots \binom{n_k}{m_k} \Big/ \binom{n}{m},$$

$$n_0 = n - \sum_{i=1}^k n_i \quad \text{and} \quad m_0 = m - \sum_{i=1}^k m_i;$$

reference may also be made to a related work by Panaretos (1983b).

Boland and Proschan (1987) have established the Schur convexity of the maximum likelihood function for the multivariate hypergeometric distribution.

Childs and Balakrishnan (1996) have examined some approximations to the multivariate hypergeometric distribution by continuous random variables which are chosen such that they have the same range of variation, means, variances, and covariances as their discrete counterparts. Approximations of this type for the multinomial distribution have already been described in Chapter 35. For example, their Dirichlet approximation is by regarding the relative frequencies $Y_i = \frac{N_i}{n}$ $(i = 1, 2, \ldots, k)$

as continuous random variables having the Dirichlet distribution with parameters $\frac{m_i}{m}\left(\frac{n(m-1)}{m-n}-1\right)$, $i = 1, 2, \ldots, k$—that is, with joint probability density function

$$f(y_1, \ldots, y_k) = \Gamma\left(\frac{n(m-1)}{m-n} - 1\right) \prod_{i=1}^{k} \left\{ \frac{y_i^{\frac{m_i}{m}\left(\frac{n(m-1)}{m-n}-1\right)-1}}{\Gamma\left(\frac{m_i}{m}\left(\frac{n(m-1)}{m-n}-1\right)\right)} \right\},$$

$$0 \le y_i \le 1, \quad \sum_{i=1}^{k} y_i = 1 .$$

It can be easily shown that the random variables nY_i $(i = 1, 2, \ldots, k)$ have the same means, variances, and covariances as those of N_i $(i = 1, 2, \ldots, k)$ presented in Eqs. (39.7)–(39.9). Furthermore, it can also be established by a simple change of variables that $Y_1, Y_2, \ldots, Y_k$ have the same joint distribution as the random variables

$$\frac{Z_1}{\sum_{i=1}^{k} Z_i}, \frac{Z_2}{\sum_{i=1}^{k} Z_i}, \ldots, \frac{Z_k}{\sum_{i=1}^{k} Z_i},$$

where Z_is are independent chi-square random variables with $\frac{2m_i}{m}\left(\frac{n(m-1)}{m-n}-1\right)$ degrees of freedom, $i = 1, 2, \ldots, k$.

The normal approximation discussed by Childs and Balakrishnan (1996) is by considering the standardized multivariate hypergeometric random variables

$$W_i = \frac{N_i - \frac{nm_i}{m}}{\sqrt{\frac{nm_i}{m}\left(1 - \frac{m_i}{m}\right)\left(\frac{m-n}{m-1}\right)}}$$

with

$$\mathrm{Cov}(W_i, W_j) = -\frac{1}{m}\sqrt{\frac{m_i m_j}{\left(1 - \frac{m_i}{m}\right)\left(1 - \frac{m_j}{m}\right)}}.$$

As noted by Childs and Balakrishnan (1996), under the hypothesis that $m_1 = m_2 = \cdots = m_k$, we have

$$\mathrm{Cov}(W_i, W_j) = -1/(k-1) .$$

Consequently, under the above hypothesis, the joint distribution of $W_1, \ldots, W_k$ can be approximated by a multivariate normal distribution with means as 0, variances as 1, and equal covariances as $-1/(k-1)$. Recall that such a multivariate normal distribution is the same as the joint distribution of the random variables

$$\sqrt{\frac{k}{k-1}}\,(X_1 - \overline{X}), \sqrt{\frac{k}{k-1}}\,(X_2 - \overline{X}), \ldots, \sqrt{\frac{k}{k-1}}\,(X_k - \overline{X}),$$

where X_is are independent standard normal variables.

Childs and Balakrishnan (1996) have made use of these approximations to develop some tests for the hypothesis $H_0 : m_1 = m_2 = \cdots = m_k = m^*$ (with $m = km^*$)

based on the test statistics

$$T_1 = \frac{N_{(1)}}{N_{(k)}}, \quad T_2 = \frac{N_{(k)} - N_{(1)}}{n} \quad \text{and} \quad T_3 = \frac{N_{(k)}}{n},$$

where $N_{(1)} \le N_{(2)} \le \cdots \le N_{(k)}$ are the order statistics from the multivariate hypergeometric distribution in (39.1).

3 TRUNCATED MULTIVARIATE HYPERGEOMETRIC DISTRIBUTIONS

Applying the general results of Gerstenkorn (1977) described earlier in Chapters 34 and 35 to the multivariate hypergeometric distribution (39.1), we have the probability mass function of the truncated multivariate hypergeometric distribution (with the variable N_j being restricted in the interval $[b+1, c]$) as

$$\begin{aligned} P^T(n_1, n_2, \ldots, n_k) &= \frac{1}{\sum_{\ell=b+1}^{c} \binom{m_j}{\ell}\binom{m-m_j}{n-\ell}} \prod_{i=1}^{k} \binom{m_i}{n_i} \\ &= \frac{1}{\sum_{\ell=b+1}^{c} \frac{m_j^{(\ell)}(m-m_j)^{(n-\ell)}}{\ell!(n-\ell)!}} \prod_{i=1}^{k} \frac{m_i^{(n_i)}}{n_i!} \end{aligned} \qquad (39.20)$$

with $\sum_{i=1}^k n_i = n$, $\sum_{i=1}^k m_i = m$, $0 \le n_i \le m_i$ $(i \ne j)$, and $b < n_j \le c$. Similarly, the probability mass function of the truncated multivariate hypergeometric distribution (with the variables $N_1, N_2, \ldots, N_s$ all doubly truncated, where N_j is restricted in the interval $[b_j + 1, c_j]$ for $j = 1, 2, \ldots, s$) is

$$\begin{aligned} &P^T(n_1, n_2, \ldots, n_k) \\ &= \frac{1}{\sum_{\ell_1=b_1+1}^{c_1} \cdots \sum_{\ell_s=b_s+1}^{c_s} \left\{\prod_{i=1}^{s} \binom{m_i}{\ell_i}\right\} \binom{m-\sum_{i=1}^s m_i}{n-\sum_{i=1}^s \ell_i}} \prod_{i=1}^{k} \binom{m_i}{n_i} \\ &= \frac{1}{\sum_{\ell_1=b_1+1}^{c_1} \cdots \sum_{\ell_s=b_s+1}^{c_s} \left\{\prod_{i=1}^{s} \frac{m_i^{(\ell_i)}}{\ell_i!}\right\} \frac{(m-\sum_{i=1}^s m_i)^{(n-\sum_{i=1}^s \ell_i)}}{(n-\sum_{i=1}^s \ell_i)!}} \prod_{i=1}^{k} \frac{m_i^{(n_i)}}{n_i!}, \end{aligned} \qquad (39.21)$$

where $b_i < n_i \le c_i$ $(i = 1, 2, \ldots, s)$ and $0 \le n_i \le m_i$ $(i = s+1, s+2, \ldots, k)$.

The ***r***th descending factorial moment of the truncated multivariate hypergeometric distribution (39.20) is

$$\begin{aligned} \mu'_{(\boldsymbol{r})T} &= E[N_i^{(r_i)}] \\ &= \sum_{n_1=0}^{n} \cdots \sum_{n_{j-1}=0}^{n} \sum_{n_j=b+1}^{c} \sum_{n_{j+1}=0}^{n} \cdots \sum_{n_k=0}^{n} n_i^{(r_i)} P^T(n_1, n_2, \ldots, n_k) \end{aligned}$$

$$= \sum_{n_1=0}^{n} \cdots \sum_{n_{j-1}=0}^{n} \sum_{n_j=b+1}^{c} \sum_{n_{j+1}=0}^{n} \cdots \sum_{n_k=0}^{n} \frac{1}{\sum_{\ell=b+1}^{c} \frac{m_j^{(\ell)}(m-m_j)^{(n-\ell)}}{\ell!(n-\ell)!}} \prod_{i=1}^{k} \frac{m_i^{(n_i)}}{n_i!} n_i^{(r_i)}$$

$$= \frac{1}{\sum_{\ell=b+1}^{c} \frac{m_j^{(\ell)}(m-m_j)^{(n-\ell)}}{\ell!(n-\ell)!}} \sum_{\ell=b+1}^{c} \frac{(m_j - r_j)^{(\ell-r_j)}(m - r - m_j + r_j)^{(n-r-\ell+r_j)}}{(\ell - r_j)!(n - r - \ell + r_j)!}, \tag{39.22}$$

where $r = \sum_{i=1}^{k} r_i$. Similarly, the $\boldsymbol{r}$th descending factorial moment of the truncated multivariate hypergeometric distribution (with the variables $N_1, N_2, \ldots, N_s$ all doubly truncated) in (39.21) is

$$\begin{aligned} \mu'_{(\boldsymbol{r})T} &= \sum_{n_1=b_1+1}^{c_1} \cdots \sum_{n_s=b_s+1}^{c_s} \sum_{n_{s+1}=0}^{n} \cdots \sum_{n_k=0}^{n} n_i^{(r_i)} P^T(n_1, n_2, \ldots, n_k) \\ &= \frac{1}{\sum_{\ell_1=b_1+1}^{c_1} \cdots \sum_{\ell_s=b_s+1}^{c_s} \left\{ \prod_{i=1}^{s} \frac{m_i^{(\ell_i)}}{\ell_i!} \right\} \frac{(m-\sum_{i=1}^{s} m_i)^{(n-\sum_{i=1}^{s} \ell_i)}}{(n-\sum_{i=1}^{s} \ell_i)!}} \\ &\quad \times \sum_{n_1=b_1+1}^{c_1} \cdots \sum_{n_s=b_s+1}^{c_s} \sum_{n_{s+1}=0}^{n} \cdots \sum_{n_k=0}^{n} \prod_{i=1}^{k} \frac{m_i^{(n_i)}}{n_i!} n_i^{(r_i)} \\ &= \frac{1}{\sum_{\ell_1=b_1+1}^{c_1} \cdots \sum_{\ell_s=b_s+1}^{c_s} \left\{ \prod_{i=1}^{s} \frac{m_i^{(\ell_i)}}{\ell_i!} \right\} \frac{(m-\sum_{i=1}^{s} m_i)^{(n-\sum_{i=1}^{s} \ell_i)}}{(n-\sum_{i=1}^{s} \ell_i)!}} \\ &\quad \times \sum_{\ell_1=b_1+1}^{c_1} \cdots \sum_{\ell_s=b_s+1}^{c_s} \left\{ \prod_{i=1}^{s} \frac{(m_i - r_i)^{(\ell_i - r_i)}}{(\ell_i - r_i)!} \right\} \\ &\quad \times \frac{(m - r - \sum_{i=1}^{s} m_i + \sum_{i=1}^{s} r_i)^{(n-r-\sum_{i=1}^{s} \ell_i + \sum_{i=1}^{s} r_i)}}{(n - r - \sum_{i=1}^{s} \ell_i + \sum_{i=1}^{s} r_i)!}. \end{aligned} \tag{39.23}$$

The results for the left-truncated case are obtained from the above formulas by taking the c's to be corresponding m's. For example, we have the probability mass function of the left-truncated multivariate hypergeometric distribution (with the variable N_j being at least $b + 1$) as

$$P^{LT}(n_1, n_2, \ldots, n_k) = \frac{1}{\sum_{\ell=b+1}^{n} \frac{m_j^{(\ell)}(m-m_j)^{(n-\ell)}}{\ell!(n-\ell)!}} \prod_{i=1}^{k} \frac{m_i^{(n_i)}}{n_i!}, \tag{39.24}$$

where, once again, $\sum_{i=1}^{k} n_i = n$, $\sum_{i=1}^{k} m_i = m$, $0 \le n_i \le m_i$ $(i \ne j)$, and $b < n_j \le m_j$. Similarly, the probability mass function of the left-truncated multivariate hypergeometric distribution (with the variables $N_1, N_2, \ldots, N_s$ all truncated on the left) is

$$P^{LT}(n_1, n_2, \ldots, n_k)$$

$$= \frac{1}{\sum_{\ell_1=b_1+1}^{m_1} \cdots \sum_{\ell_s=b_s+1}^{m_s} \left\{\prod_{i=1}^{s} \binom{m_i}{\ell_i}\right\} \binom{m-\sum_{i=1}^{s} m_i}{n-\sum_{i=1}^{s} \ell_i}} \prod_{i=1}^{k} \frac{m_i^{(n_i)}}{n_i!}, \qquad (39.25)$$

where $b_i < n_i \leq m_i$ $(i = 1, 2, \ldots, s)$ and $0 \leq n_i \leq m_i$ $(i = s + 1, s + 2, \ldots, k)$.

4 RELATED DISTRIBUTIONS

4.1 Multivariate Inverse Hypergeometric and Negative Hypergeometric Distributions

Janardan and Patil (1972) have pointed out that the parameters in (39.1) need not all be non-negative integers. Using the conventions that (if θ is a positive integer)

$$(-\theta)! = (-1)^{\theta-1}/(\theta - 1)!$$

and

$$(-\theta)!/(-\theta - \phi)! = (-1)^{\phi}(\theta + \phi - 1)!/(\theta - 1)!,$$

they have defined

(a) *multivariate inverse hypergeometric distributions* with probability mass function

$$\Pr\left[\bigcap_{i=1}^{k}(X_i = x_i)\right]$$

$$= \frac{\binom{m_0}{d-1}\prod_{i=1}^{k}\binom{m_i}{x_i}}{\binom{m_0+\sum_{i=1}^{k} m_i}{d+\sum_{i=1}^{k} x_i - 1}} \; \frac{m_0 - d + 1}{m_0 - d + \sum_{i=1}^{k}(m_i - x_i) + 1}, \qquad (39.26)$$

(b) *multivariate negative hypergeometric distributions* with probability mass function

$$\Pr\left[\bigcap_{i=0}^{k}(X_i = x_i)\right] = \frac{n!\,\Gamma(m)}{\Gamma(n + m)} \prod_{i=0}^{k} \frac{\Gamma(m_i + x_i)}{\Gamma(m_i)\, x_i!}, \qquad (39.27)$$

and

(c) *multivariate negative inverse hypergeometric distributions* with probability mass function

$$\Pr\left[\bigcap_{i=0}^{k}(X_i = x_i)\right]$$

$$= \frac{\Gamma(d + \sum_{i=0}^{k} x_i)\,\Gamma(\sum_{i=0}^{k} m_i)}{\Gamma(d + \sum_{i=0}^{k} x_i + \sum_{i=0}^{k} m_i)} \frac{\Gamma(d + m_0)}{\Gamma(d)\,\Gamma(m_0)} \prod_{i=0}^{k} \frac{\Gamma(m_i + x_i)}{\Gamma(m_i)x_i!}. \qquad (39.28)$$

4.2 Multivariate Generalized Hypergeometric and Unified Multivariate Hypergeometric Distributions

Sibuya and Shimizu (1981) have defined a *multivariate generalized hypergeometric family* of distributions (multivariate GHg distributions) through the joint probability mass function

$$\Pr\left[\bigcap_{i=1}^{k}(X_i = x_i)\right] = K\left\{\prod_{i=1}^{k} \frac{\alpha_i^{[x_i]}}{x_i!}\right\} \frac{\lambda^{[\sum x_i]}}{\omega^{[\sum x_i]}}, \qquad x_i \geq 0, \qquad (39.29)$$

where K is a normalizing factor [determined below in (39.31)]. Clearly,

$$\Pr\left[\sum_{i=1}^{k} X_i = s\right] = K \frac{\lambda^{[s]}}{\omega^{[s]}} \sum_{\sum x_i = s}\cdots\sum \left\{\prod_{i=1}^{k} \frac{\alpha_i^{[x_i]}}{x_i!}\right\}$$

$$= K \frac{\lambda^{[s]}}{\omega^{[s]} s!} \left(\sum_{i=1}^{k} \alpha_i\right)^{[s]}. \qquad (39.30)$$

Thus, the distribution of $S = \sum_{i=1}^{k} X_i$ is a univariate generalized hypergeometric distribution with parameters $(\lambda, \sum_{i=1}^{k} \alpha_i; \omega)$ (see Chapter 6), which shows that

$$K = \frac{\Gamma(\omega - \sum_{i=1}^{k} \alpha_i)\Gamma(\omega - \lambda)}{\Gamma(\omega - \sum_{i=1}^{k} \alpha_i - \lambda)\Gamma(\omega)}. \qquad (39.31)$$

Special cases of (39.29) correspond to special values of the parameters. If $\omega = \lambda = 0$, we have a *singular multivariate hypergeometric distribution* with probability mass function

$$\Pr\left[\bigcap_{i=1}^{k}(X_i = x_i)\right] = \frac{\left(\sum_{i=1}^{k} x_i\right)!}{\left(\sum_{i=1}^{k} \alpha_i\right)^{[\sum_{i=1}^{k} x_i]}} \prod_{i=1}^{k} \left(\frac{\alpha_i^{[x_i]}}{x_i!}\right),$$

$$x_i \geq 0\ (i = 1, \ldots, k). \qquad (39.32)$$

Sibuya and Shimizu (1981) have also given some representations of the bivariate generalized hypergeometric distributions in terms of urn models.

A considerable part of the increased attention devoted to discrete multivariate distributions in recent years has been directed to generalizations of multivariate hypergeometric distributions. Note that the term *generalized* is not used in the limited *random sum* sense. Although this work has been done fairly recently, most of the distributions proposed and discussed are included in a very broad class described by Steyn (1951) about 45 years ago.

Steyn (1951, 1955) introduced a probability generating function for a class of probability distributions which have linear regression equations and called them *multivariate probability distributions of the hypergeometric type*. He also demonstrated that multivariate hypergeometric, multivariate inverse hypergeometric, and multivariate Pólya distributions belong to this class. The probability mass function of this general class is

$$P(\boldsymbol{x}) = \frac{(c - \sum_{i=1}^{k} b_i - a)^{[a]}}{c^{[a]}} \frac{a^{[\sum_{i=1}^{k} x_i]}}{c^{[\sum_{i=1}^{k} x_i]}} \prod_{i=1}^{k} \frac{b_i^{[x_i]}}{x_i!}, \tag{39.33}$$

where the values of a, b_i, and c are such that $P(\boldsymbol{x}) \geq 0$. Ishii and Hayakawa (1960) and Mosimann (1962) derived independently the multivariate negative hypergeometric distribution in (39.27) by treating the parameter vector of a multinomial distribution as a random variable having a Dirichlet distribution. Mosimann (1963) obtained the multivariate negative inverse hypergeometric distribution in (39.28) as a Dirichlet mixture of the negative multinomial distribution. Särndal (1964) derived the multivariate negative hypergeometric distribution in (39.27) as a posterior distribution when the parameter vector of the multivariate hypergeometric distribution is assumed to have a Bose–Einstein uniform prior distribution (see Chapter 26) or any prior of the multivariate negative hypergeometric type.

Janardan and Patil (1972) showed that Steyn's class also contains multivariate negative hypergeometric, multivariate negative inverse hypergeometric and multivariate inverse Pólya distributions, and then introduced a subclass called *unified multivariate hypergeometric distributions* [see (39.34)]. This subclass includes many known distributions such as the multivariate hypergeometric, inverse hypergeometric, negative hypergeometric, negative inverse hypergeometric, Pólya–Eggenberger, and inverse Pólya–Eggenberger distributions.

Janardan and Patil (1972) also summarized some possible geneses of the distributions as follows:

1. If $\boldsymbol{X}$ and $\boldsymbol{Y}$ are independent and each distributed as Multinomial $(n, \boldsymbol{p})$, then

$$\boldsymbol{X} \mid (\boldsymbol{X} + \boldsymbol{Y} = \boldsymbol{N}) \rightsquigarrow \text{Multivariate Hypergeometric } (n, \boldsymbol{N}).$$

2. If $\boldsymbol{X}$ and $\boldsymbol{Y}$ are independent and distributed as Negative Multinomial with parameters $(n, \boldsymbol{p})$ and $(n + 1, \boldsymbol{p})$, respectively, then

$$\boldsymbol{X} \mid (\boldsymbol{X} + \boldsymbol{Y} = \boldsymbol{N}) \rightsquigarrow \text{Multivariate Inverse Hypergeometric } (n, \boldsymbol{N}).$$

3. If $X_1, \ldots, X_k$ are independent Binomial (N_i, p) variables and p is distributed as Beta $(n, N_0 - n + 1)$, then

$$\boldsymbol{X} \rightsquigarrow \text{Multivariate Inverse Hypergeometric } (n, \boldsymbol{N}).$$

4. If $X_1, \ldots, X_k$ are independent Negative Binomial (N_i, p) variables and $p/(1+p)$ is distributed as Beta (N_0, n), then

$$\boldsymbol{X} \rightsquigarrow \text{Multivariate Negative Inverse Hypergeometric } (n, \boldsymbol{N}).$$

5. Multivariate Inverse Hypergeometric $(n, \boldsymbol{N}) \bigwedge_{\boldsymbol{N}}$ Multivariate

$$\text{Negative Inverse Hypergeometric } (N_0 + 1, \boldsymbol{\theta})$$
$$\rightsquigarrow \text{Multivariate Negative Inverse Hypergeometric } (n, \boldsymbol{\theta})\,.$$

Furthermore, the following property of the multivariate negative inverse hypergeometric distribution is of special importance in the Bayesian estimation of parameters of a finite population in terms of inverse sampling:

6. Let the conditional distribution of $\boldsymbol{X}$ given $\boldsymbol{N}$ be multivariate inverse hypergeometric with parameters d and $\boldsymbol{N}$. Let the distribution of $\boldsymbol{N}$ be multivariate negative inverse hypergeometric with parameters $N_0 + 1$ and $\boldsymbol{\alpha}$. Then, (i) the marginal distribution of $\boldsymbol{X}$ is multivariate negative inverse hypergeometric with parameters d and $\boldsymbol{\alpha}$, and (ii) the posterior distribution of $\boldsymbol{N} - \boldsymbol{X}$ given $\boldsymbol{X}$ is multivariate negative inverse hypergeometric with parameters $N_0 - d + 1$ and $\boldsymbol{\alpha} + \boldsymbol{X}$.

The parameters $m_1, m_2, \ldots, m_k$ and $n_1, n_2, \ldots, n_k$ in (39.1) are, from the nature of the genesis presented in this chapter (see Sections 1 and 2), essentially non-negative integers. It is possible to obtain proper distributions by relaxing this condition, interpreting $\binom{a}{b}$ in (39.1) as $\frac{\Gamma(a+1)}{\Gamma(b+1)\Gamma(a-b+1)}$. Such distributions were termed *unified multivariate hypergeometric distributions* by Janardan and Patil (1972). They defined these distributions by the probability mass function

$$P(\boldsymbol{n}) = \Pr[\boldsymbol{N} = \boldsymbol{n}] = \Pr\left[\bigcap_{i=1}^{k} (N_i = n_i)\right]$$
$$= \binom{a_0}{n - \sum_{i=1}^k n_i} \prod_{i=1}^{k} \binom{a_i}{n_i} \Big/ \binom{a}{n}, \qquad (39.34)$$

where $a = \sum_{i=0}^k a_i$. Note that this definition is valid for not necessarily a positive n. When $a_i = m_i$ $(i = 0, 1, \ldots, k)$ with positive integer m_i's, we have the regular multivariate hypergeometric distribution described earlier in Sections 2 and 3, but using the additional (C_0) type described at the beginning of Section 2. Similarly, the parameter relations for various other distributions in terms of a_i's in the unified multivariate hypergeometric distributions (39.34) are as follows:

Multivariate inverse hypergeometric distributions:

$$a_i = m_i \quad \text{for } i = 1, 2, \ldots, k\,,$$

$$a = \sum_{i=0}^{k} a_i = -m_0 - 1, \quad a_0 = -m_0 - \sum_{i=1}^{k} m_i - 1, \quad n = -d\,.$$

Multivariate negative hypergeometric distributions:

$$a_i = -m_i \quad \text{for } i = 0, 1, \ldots, k\,,$$

$$a = \sum_{i=0}^{k} a_i = -m, \quad n = n\,.$$

Multivariate negative inverse hypergeometric distributions:

$$a_i = -m_i \quad \text{for } i = 1, 2, \ldots, k\,,$$

$$a = \sum_{i=0}^{k} a_i = m_0 - 1, \quad a_0 = m_0 - 1 - \sum_{i=1}^{k} a_i, \quad n = -d\,.$$

Multivariate Pólya–Eggenberger distributions (see Chapter 40):

$$a_i = -n_i/c \quad \text{for } i = 0, 1, 2, \ldots, k\,,$$

$$a = \sum_{i=0}^{k} a_i = -n/c, \quad n = n\,.$$

Multivariate inverse Pólya–Eggenberger distributions (see Chapter 40):

$$a_i = -n_i/c \quad \text{for } i = 1, 2, \ldots, k\,,$$

$$a = \sum_{i=0}^{k} a_i = (n_0/c) - 1, \quad n = -d\,.$$

Janardan and Patil (1972) used the notation UMH$(a; a_1, \ldots, a_k; a_0; n)$ for the unified multivariate hypergeometric distribution (39.34) of $(N_1, \ldots, N_k)'$, while the distribution of $(N_0, N_1, \ldots, N_k)'$ has been termed by them *singular unified multivariate hypergeometric* and denoted by SUMH$(a_0, a_1, \ldots, a_k; n)$. The probability generating function of the distribution (39.34) is

$$G_N(\boldsymbol{t}) = \frac{a_0^{(n)}}{a^{(n)}}\, F(-n; -a_1, -a_2, \cdots, -a_k; a_0 - n + 1; t_1, t_2, \ldots, t_k)\,,$$

where $F(\cdots)$ is the multivariate hypergeometric series defined by

$$F(\alpha; \beta_1, \beta_2, \ldots, \beta_k; \gamma; t_1, t_2, \ldots, t_k) = \sum_{\substack{x_i \ge 0 \\ \sum_{i=1}^{k} x_i = x}} \cdots \sum \frac{\alpha^{(x)} \prod_{i=1}^{k} \beta_i^{(x_i)}}{\gamma^{(x)}} \prod_{j=1}^{k} \frac{t_j^{x_j}}{x_j!}\,.$$

An alternative expression, given by Janardan and Patil (1972), is

$$G_{\boldsymbol{N}}(\boldsymbol{t}) = \sum_m C_m \left[\frac{\partial^m}{\partial u^m} H(\boldsymbol{t}; u) \right]_{u=0},$$

where

$$C_m = \frac{\binom{a_0}{n-m}}{m!\binom{a}{n}}$$

and

$$H(\boldsymbol{t}; u) = \prod_{i=1}^{k} (1 + ut_i)^{a_i} \qquad \text{with } u \text{ having the same sign as } a_i.$$

We then note that

$$\left[\frac{\partial^m}{\partial u^m} H(\boldsymbol{t}; u) \right]_{u=0} = m! \underset{\sum_{i=1}^{k} x_i = m}{\sum \cdots \sum_{x_i \geq 0}} \left\{ \prod_{i=1}^{k} \binom{a_i}{x_i} t_i^{x_i} \right\}.$$

The moment generating function of the unified multivariate hypergeometric distribution (39.34) can be similarly written as

$$\sum_m C_m \left[\frac{\partial^m}{\partial u^m} H^*(\boldsymbol{t}; u) \right]_{u=0},$$

where C_m is as defined above and

$$H^*(\boldsymbol{t}; u) = \prod_{i=1}^{k} (1 + u\, e^{t_i})^{a_i} \qquad \text{with } u \text{ having the same sign as } a_i.$$

The unified multivariate hypergeometric family enjoys a number of closure properties as listed by Steyn (1955) and Janardan and Patil (1972). For example, if $(N_1, N_2, \ldots, N_k)' \rightsquigarrow \text{UMH}(a; a_1, a_2, \ldots, a_k; a_0; n)$, then:

(i) $(N_{i_1}, N_{i_2}, \ldots, N_{i_k})' \rightsquigarrow \text{UMH}(a; a_{i_1}, a_{i_2}, \ldots, a_{i_k}; a_0; n)$, where $(i_1, i_2, \ldots, i_k)$ is a permutation of $(1, 2, \ldots, k)$;

(ii) $(N_1, N_2, \ldots, N_\ell)' \rightsquigarrow \text{UMH}(a; a_1, a_2, \ldots, a_\ell; a_0^*; n)$, where $a_0^* = a - \sum_{i=1}^{\ell} a_i$, $1 \leq \ell \leq k$;

(iii) $N_1 + N_2 + \cdots + N_k \rightsquigarrow \text{UMH}_{(1)}(a; \sum_{i=1}^{k} a_i; a_0; n)$, where $\text{UMH}_{(1)}$ denotes the univariate case of UMH; and

(iv) $(N_1, N_2, \ldots, N_\ell)' \mid (N_{\ell+1}, \ldots, N_k)' \rightsquigarrow \text{UMH}(a^*; a_1, \ldots, a_\ell; a_0; n^*)$, where $a^* = a - \sum_{i=\ell+1}^{k} a_i$ and $n^* = n - \sum_{i=\ell+1}^{k} n_i$.

As was shown by Steyn (1951), the linear regression $E[N_k \mid N_1, \ldots, N_{k-1}]$ is

$$E\left[N_k \mid N_1, \ldots, N_{k-1}\right] = \frac{a_k}{a_0 + a_k}\left(n - \sum_{i=1}^{k-1} N_i\right).$$

Thus, the regression function depends only on the sum of the conditioning variables and not on their individual values. Also, direct calculations show that [see Janardan and Patil (1972)]

$$\begin{aligned}&\operatorname{var}(N_k \mid N_1, \ldots, N_{k-1})\\&\quad = \left(n - \sum_{i=1}^{k-1} N_i\right)\frac{a_0 a_k}{(a_0 + a_k)^2}\,\frac{\left(a_0 + a_k - n + \sum_{i=1}^{k-1} N_i\right)}{(a_0 + a_k - 1)},\end{aligned}$$

which reveals that the regression is heteroscedastic.

For positive values of n, the expression for the moment generating function can be somewhat simplified as

$$\frac{(a-n)!}{a!}\left[\frac{\partial^n}{\partial u^n} H^*(\boldsymbol{t}; u)\right]_{u=0},$$

where

$$H^*(\boldsymbol{t}; u) = \prod_{i=1}^{k}(1 + u\,e^{t_i})^{a_i} \qquad \text{with } u \text{ having the same sign as } a_i.$$

From this expression of the moment generating function, an explicit (but rather cumbersome) formula for the $\boldsymbol{r}$th raw moment of $\boldsymbol{N}$ can be obtained as [Rao and Janardan (1981)]

$$\mu_{\boldsymbol{r}}'(\boldsymbol{N}) = E\left[\prod_{i=1}^{k} N_i^{r_i}\right] = \frac{1}{\binom{a}{n}}\sum_{j=k}^{r} C_j \binom{a-r}{n-j},$$

where

$$C_j = \sum_{1 \le q_i \le r_i}\cdots\sum \prod_{i=1}^{k} J_{r_i,q_i}(a_i), \quad r = \sum_{i=1}^{k} r_i, \ \sum_{i=1}^{k} q_i = j,$$

and

$$J_{p,q}(a) = \sum_i \binom{p-i}{q-i}\binom{a}{i}\Delta^i O^p.$$

$J_{p,q}$ can be determined recursively as

$$\begin{aligned}&J_{p+1,1}(a) = J_{p,1}(a),\\&J_{p+1,q}(a) = q\,J_{p,q}(a) + (a - p + q - 1)J_{p,q-1}(a)\\&\qquad\qquad\qquad\qquad\qquad\qquad\qquad\qquad \text{for } q = 2, 3, \ldots, p,\end{aligned}$$

and

$$J_{p+1,p+1}(a) = a\, J_{p,p}(a)\,.$$

Hence, the ***r***th raw moment of ***N*** can be written as

$$\mu'_{\boldsymbol{r}}(\boldsymbol{N}) = \frac{1}{\binom{a}{n}} \prod_{i=1}^{m} \sum_{j=1}^{r_i} \binom{a_i}{j} \binom{a-j}{n-j} \Delta_i^j O^{r_i}\,,$$

and, in particular, we have

$$E[N_i N_j] = \frac{n\, a_i a_j (a-n)}{a(a-1)} \qquad \text{for } 1 \le i \ne j \le k\,.$$

Corresponding results for various specific distributions can be deduced from these formulas by choosing a_0, a_i, n_i, and n as presented earlier.

The ***r***th descending factorial moment of the unified multivariate hypergeometric distribution (39.34) is easily seen to be

$$\mu'_{(\boldsymbol{r})} = E\left[\prod_{i=1}^{k} N_i^{(r_i)}\right] = \frac{n^{(\sum_{i=1}^{k} r_i)}}{a^{(\sum_{i=1}^{k} r_i)}} \prod_{i=1}^{k} a_i^{(r_i)}\,, \tag{39.35}$$

which is the same as the corresponding expression in (39.6) for the multivariate hypergeometric distribution, with m's replaced by a's.

4.3 Restricted Occupancy Distributions (Equiprobable Multivariate Hypergeometric Distributions)

Charalambides (1981) studied the distribution of the numbers (Y_h) of N_i's equal to h (say)—that is, the numbers of different types among $C_1, C_2, \ldots, C_k$ represented by just h individuals in the sample. For the special case of *equiprobable classes* ($m_1 = m_2 = \cdots = m_k = m/k$), he obtained the joint probability mass function as

$$\Pr\left[\bigcap_{i=0}^{n} (Y_i = y_i)\right] = \binom{k}{y_0, y_1, \ldots, y_n} \left\{\prod_{i=0}^{n} \binom{m/k}{i}^{y_i}\right\} \Big/ \binom{m}{n},$$

$$\sum_{i=0}^{n} y_i = k, \ \sum_{i=1}^{n} i y_i = n. \tag{39.36}$$

An alternative form of (39.36) is

$$\Pr\left[\bigcap_{i=0}^{n} (Y_i = y_i)\right] = \frac{k!}{m^{(n)}}\, n! \prod_{i=0}^{n} \left\{ \binom{m/k}{i}^{y_i} \Big/ y_i! \right\}. \tag{39.37}$$

The probability mass function of

$$Z = \sum_{i=1}^{n} Y_i = k - Y_0\,,$$

the number of different classes represented in the sample, is

$$\Pr[Z = z] = \frac{(m/k)^{(z)}}{m^{(n)}} \sum_{y_1} \cdots \sum_{y_n} \left\{ \frac{n!}{\prod_{i=1}^{n} y_i!} \prod_{i=1}^{n} \binom{m/k}{i}^{y_i} \right\},$$

$$y_i > 0, \ \sum_{i=1}^{n} y_i = z, \ \sum_{i=1}^{n} i y_i = n\,. \tag{39.38}$$

An alternative expression to (39.38) is

$$\Pr[Z = z] = \frac{(m/k)^{(z)}}{m^{(n)}} \cdot \frac{1}{n!} \left[\Delta^z \left(\frac{m}{k} - x \right)^{(n)} \right] \Bigg|_{x=0}. \tag{39.39}$$

Note that $S(n, z) = \frac{1}{n!} \left[\Delta^z x^{(n)}\right]\Big|_{x=0}$ is a Stirling number of the second kind [see Chapter 34, Section 2.3].

4.4 "Noncentral" Multivariate Hypergeometric Distributions

Wallenius (1963) defined a *"noncentral" univariate hypergeometric distribution* as one representing the number of times (N_1) that a white ball is chosen in n successive drawings under sampling without replacement from an urn initially containing m_1 white balls and m_2 ($= m - m_1$) black balls, and the probability of drawing a white ball, when there are x_1 white and x_2 black balls in the urn, is

$$\omega_1 x_1 / (\omega_1 x_1 + \omega_2 x_2), \qquad \omega_1, \omega_2 > 0. \tag{39.40}$$

See Chapter 6, Section 12 for details.

Clearly, if $\omega_1 = \omega_2$ we obtain a simple univariate hypergeometric distribution with parameters $(n; m_1, m_2)$, while if sampling is done with replacement, N_1 has a binomial distribution with parameters $(n; \frac{\omega_1 m_1}{\omega_1 m_1 + \omega_2 m_2})$.

Chesson (1976) extended this idea to define *'noncentral' multivariate hypergeometric distributions* arising from a similar sampling scheme with k different colors $C_1, C_2, \ldots, C_k$ of balls, there being initially m_j balls of color C_j with $\sum_{i=1}^{m} m_i = m$. If there are $x_1, x_2, \ldots, x_k$ balls of colors $C_1, C_2, \ldots, C_k$, respectively, the probability of drawing a ball of color C_i is

$$\omega_i x_i \Big/ \sum_{j=1}^{k} \omega_j x_j, \qquad \omega_1, \omega_2, \ldots, \omega_k > 0. \tag{39.41}$$

Introducing the notation $\boldsymbol{N}(n) = (N_1(n), N_2(n), \ldots, N_k(n))'$ for the numbers of balls of colors $C_1, C_2, \ldots, C_k$, respectively, chosen in n drawings, Chesson (1976) obtained the recurrence relation

$$\begin{aligned}\Pr[\boldsymbol{N}(n) = \boldsymbol{n}] &= \Pr\left[\bigcap_{i=1}^{k}\{N_i(n) = n_i\}\right] \\ &= \frac{1}{\sum_{i=1}^{k} \omega_i n_i} \cdot \sum_{i=1}^{k} \omega_i n_i \; \Pr[\boldsymbol{N}(n-1) = \boldsymbol{n} - \boldsymbol{e}_i] \end{aligned} \tag{39.42}$$

with $\Pr[\boldsymbol{N}(0) = \boldsymbol{0}] = 1$, where $\boldsymbol{e}_i'$ is the ith row of the $k \times k$ identity matrix; that is, $\boldsymbol{e}_i' = (\underbrace{0, \ldots, 0}_{i-1}, 1, \underbrace{0, \ldots, 0}_{k-i})$. Chesson then showed that the solution of (39.42) is

$$\Pr[\boldsymbol{N}(n) = \boldsymbol{n}] = \prod_{i=1}^{k} \binom{m_i}{n_i} \int_0^1 \prod_{i=1}^{k} (1 - t^{\omega_i c})^{n_i}\, dt, \tag{39.43}$$

where

$$c = 1 \Big/ \sum_{i=1}^{k} \omega_i (m_i - n_i), \qquad n_i \le m_i\,.$$

Applying Lyons' (1980) method of expanding the integrand, we obtain

$$\begin{aligned}&\Pr[\boldsymbol{N}(n) = \boldsymbol{n}] \\ &= \left\{\prod_{i=1}^{k} \binom{m_i}{n_i}\right\} \sum_{\ell_1=0}^{n_1} \cdots \sum_{\ell_k=0}^{n_k} \left\{\prod_{i=1}^{k} \binom{n_i}{\ell_i}\right\} \Big/ \left(c \sum_{i=1}^{k} \omega_i \ell_i + 1\right).\end{aligned} \tag{39.44}$$

Chesson's (1976) work arose from consideration of models for achieved catches by predators when the chance of being caught varies with the species of prey, as well as relative numbers thereof, as described by Manly (1974). Macdonald, Power, and Fuller (1994) have applied the same distribution to some problems arising in oil exploration.

4.5 Multivariate Hypergeometric-Multinomial Model

As in Section 2, let $\boldsymbol{N} = (N_1, N_2, \ldots, N_k)'$ denote the numbers of individuals of types $C_1, C_2, \ldots, C_k$, respectively, drawn by sampling without replacement n individuals from a population of m out of which m_i are of type C_i (with $m = \sum_{i=1}^{k} m_i$). Then, the joint probability mass function of $\boldsymbol{N}$ (of course, given $\boldsymbol{m} = (m_1, m_2, \ldots, m_k)'$) is [see (39.1)]

$$P(\boldsymbol{n} \mid \boldsymbol{m}) = \left\{\prod_{i=1}^{k} \binom{m_i}{n_i}\right\} \Big/ \binom{m}{n}, \tag{39.45}$$

where $\boldsymbol{n} = (n_1, n_2, \ldots, n_k)'$, $\sum_{i=1}^{k} n_i = n$, and $0 \le n_i \le m_i$ for $i = 1, 2, \ldots, k$. In many situations (like acceptance sampling problems), the number of individuals of type C_i (m_i), $i = 1, 2, \ldots, k$, in the population will be unknown. Then, if $\boldsymbol{m}$ is considered as a random vector and if we further assign a multinomial *prior distribution* with (unknown) hyperparameter $\boldsymbol{p} = (p_1, \ldots, p_k)'$, where $0 < p_i < 1$ and $\sum_{i=1}^k p_i = 1$, and joint probability mass function [see Chapter 35]

$$P^*(\boldsymbol{m}) = \frac{m!}{\prod_{i=1}^k (m_i!)} \prod_{i=1}^k p_i^{m_i}, \quad m_i \geq 0, \ \sum_{i=1}^k m_i = m, \tag{39.46}$$

then the *marginal* joint probability mass function of $\boldsymbol{N}$ is obtained from (39.45) and (39.46) to be

$$\begin{aligned}
P(\boldsymbol{n}) &= \sum_{\boldsymbol{m}} P(\boldsymbol{n} \mid \boldsymbol{m}) P^*(\boldsymbol{m}) \\
&= \sum_{m_1=n_1}^{m} \cdots \sum_{m_k=n_k}^{m} \frac{\prod_{i=1}^k \binom{m_i}{n_i}}{\binom{m}{n}} \frac{m!}{\prod_{i=1}^k (m_i!)} \prod_{i=1}^k p_i^{m_i} \\
&= \frac{n!(m-n)!}{\prod_{i=1}^k (n_i!)} \sum_{m_1=n_1}^{m} \cdots \sum_{m_k=n_k}^{m} \prod_{i=1}^k \left\{ \frac{p_i^{m_i}}{(m_i - n_i)!} \right\} \\
&= \frac{n!}{\prod_{i=1}^k (n_i!)} \prod_{i=1}^k p_i^{n_i},
\end{aligned} \tag{39.47}$$

where $n_i \geq 0$ $(i = 1, 2, \ldots, k)$, $0 < p_i < 1$, $\sum_{i=1}^k n_i = n$, and $\sum_{i=1}^k p_i = 1$.

It may be noted that the marginal distribution of $\boldsymbol{N}$ in (39.47) is simply a multinomial distribution with parameters $(n; p_1, p_2, \ldots, p_k)$. Referring to this setup as a *multivariate hypergeometric-multinomial model*, Balakrishnan and Ma (1996) have discussed empirical Bayes estimators as well as empirical Bayes rules for selecting the most (category with $\max_i m_i$) and the least (category with $\min_i m_i$) probable multivariate hypergeometric event. They have further shown that the Bayes risk of these empirical Bayes selection rules tends to the minimum Bayes risk with a rate of convergence of order $O(e^{-c\ell})$, where c is some positive constant and ℓ is the number of accumulated past observations. It should be mentioned here that Steyn (1990) has similarly discussed some tests of hypotheses for the parameters of the multivariate hypergeometric distribution, but by assuming the prior distributions under the null and alternative hypotheses to be both uniform.

Dwass (1979) defined a class of multivariate distributions by

$$P(\boldsymbol{n}) = \binom{n}{n_1, \ldots, n_k} \prod_{i=1}^k A_i^{(n_i)} \Big/ \left(\sum_{i=1}^k A_i \right)^{(n)}, \tag{39.48}$$

where, again $\sum_{i=1}^k n_i = n$. This is based on the identity

$$\left(\sum_{i=1}^k A_i \right)^{(n)} = \sum \cdots \sum_{\sum_{i=1}^k n_i = n} \binom{n}{n_1, \ldots, n_k} \prod_{i=1}^k A_i^{(n_i)}. \tag{39.49}$$

4.6 Multivariate Lagrange Distributions

Consul and Shenton (1972), utilizing a multivariate extension of Lagrange's expansion due to Good (1960), described a general approach for constructing multivariate distributions with univariate Lagrange distributions (see Chapter 2, Section 5.2, pp. 96–100) as marginals. They showed that such distributions appear naturally in queueing theory, presented general formulas for variances and covariances, and also constructed a *multivariate Borel–Tanner* (double Poisson) distribution with probability mass function

$$P(\boldsymbol{x}) = \left[\prod_{j=1}^{k} \frac{(\sum_{i=1}^{k} \lambda_{ij} x_i)^{x_j - r_j} \exp\left\{-(\sum_{i=1}^{k} \lambda_{ji}) x_j\right\}}{(x_j - r_j)!}\right] |I - A(\boldsymbol{x})|\,, \qquad (39.50)$$

where $A(\boldsymbol{x})$ is a $(k \times k)$ matrix with (p, q)th element given by $\lambda_{pq}(x_p - r_p)/\sum_{i=1}^{k} \lambda_{ij} x_i$, and the λ's are positive constants for $x_j \geq r_j$. [For a discussion on the univariate Borel–Tanner distribution, see Chapter 9, Section 11.] A queueing theory interpretation has also been given by Consul and Shenton (1972). The bivariate case has been considered in more detail by Shenton and Consul (1973).

Jain (1974) extended his treatment of univariate Lagrange power series distributions to multivariate distributions. In particular, Jain and Singh (1975) obtained the *Lagrangian bivariate negative binomial distribution* with probability mass function

$$P(x_1, x_2) = \frac{N\,\Gamma(N + \gamma_1 x_1 + \gamma_2 x_2)}{\Gamma(N + (\gamma_1 - 1)x_1 + (\gamma_2 - 1)x_2 + 1)} \times \frac{(q - p_1 - p_2)^{N + (\gamma_1 - 1)x_1 + (\gamma_2 - 1)x_2}\, p_1^{x_1} p_2^{x_2}}{x_1! x_2! q^{N + \gamma_1 x_1 + \gamma_2 x_2}}\,, \qquad (39.51)$$

where $0 < p_1, p_2$ with $p_1 + p_2 < q$, $N > 0$, and $\gamma_1, \gamma_2 > 1$. They also obtained a *bivariate Borel–Tanner distribution* with probability mass function

$$P(x_1, x_2) = \frac{\lambda(\lambda + \lambda_1 x_1 + \lambda_2 x_2)^{x_1 + x_2 - 1}\, e^{-(\lambda - \lambda_1 x_1 - \lambda_2 x_2)(\theta_1 + \theta_2)}\, \theta_1^{x_1} \theta_2^{x_2}}{x_1! x_2!}\,, \qquad (39.52)$$

where $\lambda, \lambda_1, \lambda_2, \theta_1, \theta_2 > 0$, which differs from the form in (39.50).

Bivariate Lagrange distributions have been described by Shenton and Consul (1973) and Jain and Singh (1975). In the bivariate power series distribution with probability mass function (also see Chapter 38)

$$P(x_1, x_2) = \frac{a_{x_1, x_2} \theta_1^{x_1} \theta_2^{x_2}}{f(\theta_1, \theta_2)}\,, \qquad x_1, x_2 = 0, 1, 2, \ldots, \qquad (39.53)$$

which is a particular case of (38.1), the a's are derived from the expansion of $f(\theta_1, \theta_2)$ as a power series in θ_1, θ_2. Instead, if the Lagrange expansion in powers of $\phi_1(\theta_1, \theta_2)$

and $\phi_2(\theta_1, \theta_2)$ given by

$$f(\theta_1, \theta_2) = \sum_{i_1=0}^{\infty} \sum_{i_2=0}^{\infty} b_{i_1,i_2} \phi_1^{i_1} \phi_2^{i_2} \tag{39.54}$$

is used, and if none of the b's is negative, we obtain a *bivariate Lagrange distribution* with probability mass function

$$P(x_1, x_2) = \frac{b_{x_1,x_2}\{\phi_1(\theta_1, \theta_2)\}^{x_1}\{\phi_2(\theta_1, \theta_2)\}^{x_2}}{f(\theta_1, \theta_2)}, \tag{39.55}$$

where

$$\begin{aligned} b_{x_1,x_2} &= \frac{1}{x_1!x_2!} \Big[D_1^{x_1-1} D_2^{x_2-1} \left\{ \phi_1^{x_1} \phi_2^{x_2} D_1 D_2 + \phi_1^{x_1} (D_1 \phi_2^{x_2}) D_2 \right. \\ &\qquad \left. + \phi_2^{x_2} (D_2 \phi_1^{x_1}) D_1 \right\} - f(\theta_1, \theta_2) \Big] \Big|_{\theta_1=\theta_2=0} . \end{aligned} \tag{39.56}$$

Writing

$$f(\theta_1, \theta_2) \equiv g(\phi_1(\theta_1, \theta_2), \phi_2(\theta_1, \theta_2)) = g(\phi_1, \phi_2),$$

the probability generating function of the distribution (39.55) is

$$E\left[t_1^{X_1} t_2^{X_2}\right] = g(t_1\phi_1, t_2\phi_2)/g(\phi_1, \phi_2). \tag{39.57}$$

Generalization to k-variate distributions is straightforward.

Jain and Singh (1975) constructed a *generalized bivariate negative binomial distribution* by taking

$$\begin{aligned} f(\theta_1, \theta_2) = (1 - \theta_1 - \theta_2)^n, \ \phi_i(x_1, x_2) = (1 - \theta_1 - \theta_2)^{1-\gamma_i}, \\ i = 1, 2, \ \theta_i \geq 0, \ \theta_1 + \theta_2 < 1 . \end{aligned} \tag{39.58}$$

For this distribution,

$$g(\phi_1, \phi_2) = 1 - \sum_{i_1,i_2} \frac{\phi_1^{i_1} \phi_2^{i_2}}{i_1!i_2!} (\gamma_1 i_1 + \gamma_2 i_2 - 2)^{i_1+i_2} \tag{39.59}$$

and

$$P(x_1, x_2) = \frac{\phi_1^{x_1} \phi_2^{x_2}}{x_1!x_2!} \{1 - g(\phi_1, \phi_2)\}^n (n + \gamma_1 x_1 + \gamma_2 x_2 - 1)^{x_1+x_2} . \tag{39.60}$$

4.7 Iterated Hypergeometric Distributions

For $n = 0, 1, \ldots, m$, define matrices $H(n)$, called *hypergeometric* matrices, by $H(n) = ((h_{s,t}(n)))$, $s, t = 0, 1, \ldots, m$, where

$$h_{s,t}(n) = \frac{\binom{t}{s}\binom{m-t}{n-s}}{\binom{m}{n}} .$$

Then $H(n)$ is the transition matrix of the following urn sampling scheme: An urn contains m balls, some of which are marked with a certain color. From this urn, n balls are drawn randomly by without replacement scheme, marked with a second color and put back into the urn. Clearly, $h_{s,t}(n)$ is the conditional probability of having s balls marked with both colors if initially t balls were marked.

Röhmel, Streitberg, and Tismer (1996) have derived a formula for the probability of having s balls marked with all colors, after performing the experiment k times with k different colors and $\boldsymbol{n} = (n_1, \ldots, n_k)$ on an urn with m balls, where initially all balls are marked. These authors called this distribution as the *iterated hypergeometric distribution* which has the probability mass function as

$$H(n_k)H(n_{k-1}) \cdots H(n_2)e(n_1) ,$$

where $e(n_1)$ is a unit vector with the 1 in position n_1. The support of this univariate distribution, related to the multivariate hypergeometric experiment, is clearly $0, 1, \ldots, \min(n_1, \ldots, n_k)$.

4.8 Generalized Waring Distributions

The bivariate generalized Waring distribution with parameters $a > 0$, $k > 0$, $m > 0$, and $\rho > 0$, studied by Xekalaki (1977, 1984a,b), has the joint probability mass function

$$\Pr[X = x,\ Y = y] = \frac{\rho^{[k+m]} a^{[x+y]} k^{[x]} m^{[y]}}{(a+\rho)^{[k+m]} (a+k+m+\rho)^{[x+y]} x! y!} , \qquad x, y = 0, 1, 2, \ldots .$$

This probability mass function has been derived as the general term of a bivariate series of ascending factorials defining a two-dimensional extension of Waring's series expansion of a function of the form $1/(x-a)^{[k+m]}$, $x > a > 0$, $k, m > 0$. The joint probability generating function of this distribution is

$$G_{X,Y}(t_1, t_2) = \frac{\rho^{[k+m]}}{(a+\rho)^{[k+m]}} F_1(a; k, m; a+k+m+\rho; t_1, t_2) , \qquad -1 \le t_1, t_2 \le 1 ,$$

where F_1 is the Appell hypergeometric function of the first type defined by

$$F_1(a; b, b'; c; u, v) = \sum_{i=0}^{\infty} \sum_{j=0}^{\infty} \frac{a^{[i+j]} b^{[i]} b'^{[j]}}{c^{[i+j]}} \frac{u^i v^j}{i! j!} , \qquad a, b, b', c-a-b-b' > 0,\ -1 \le u, v \le 1 .$$

The marginal as well as the conditional distributions are all univariate generalized Waring distributions. Xekalaki (1984a) has illustrated some applications of these distributions in the analysis of accident data, while Xekalaki (1984b) has discussed different models that generate these distributions.

5 APPLICATIONS TO FAULTY INSPECTION

In the last decade, numerous applications of multivariate hypergeometric distributions and their mixtures have been found in faulty inspection procedures and, in particular, in models dealing with consequences of errors in sampling inspection of lots for attributes. These results have been summarized in a monograph by Johnson, Kotz, and Wu (1991), and interested readers may refer to this monograph and the references cited therein for more details.

Consider a lot containing m items of which d are nonconforming. If inspection is not perfect, then some nonconforming (NC) items may not be detected while some of the conforming (C) items may be falsely designated as nonconforming. Let us introduce the notation

p—probability that a NC item is correctly classified as NC

and

p^*—probability that a C item is wrongly classified as NC.

Possible variation of these probabilities with sample, inspectors, and so on, can be introduced by use of appropriate suffices.

Now suppose we have a sequence of w samples of sizes $n_1, n_2, \ldots, n_w$ chosen without replacement and $Z_1, Z_2, \ldots, Z_w$ represent the numbers of items *classified as NC* in these samples. The numbers of items *actually NC*, say, $Y_1, Y_2, \ldots, Y_w$, in the sequence of samples jointly have a multivariate hypergeometric distribution with parameters $(m; n_1, n_2, \ldots, n_w)$ and with joint probability mass function

$$\Pr\left[\bigcap_{i=1}^{w}(Y_i = y_i)\right] = \left\{\prod_{i=1}^{w}\binom{n_i}{y_i}\right\}\binom{m - \sum_{i=1}^{w} n_i}{d - \sum_{i=1}^{w} y_i}\Big/\binom{m}{d},$$

$$0 \le y_i \le n_i,\ m - d + \sum_{i=1}^{w} y_i \ge \sum_{i=1}^{w} n_i. \qquad (39.61)$$

Given $Y_i = y_i$, the number Z_i of items classified as NC in the ith sample is distributed as the sum (convolution) of two independent univariate binomial variables with parameters (y_i, p_i) and $(n_i - y_i, p_i^*)$, respectively. Formally, we write

$$\boldsymbol{Z} \rightsquigarrow \text{Bin}(\boldsymbol{Y}, \boldsymbol{p}) \star \text{Bin}(\boldsymbol{n} - \boldsymbol{Y}, \boldsymbol{p}^*) \bigwedge_{\boldsymbol{Y}} \text{Mult. Hypg.}\,(d; n_1, n_2, \ldots, n_w). \qquad (39.62)$$

In explicit terms, we have

$$\Pr\left[\bigcap_{i=1}^{w}(Z_i = z_i) \mid \boldsymbol{Y} = \boldsymbol{y}\right]$$

$$= \prod_{i=1}^{w}\left\{\sum_{j=0}^{z_i}\binom{y_i}{j}\binom{n_i - y_i}{z_i - j}p_i^j p_i^{*z_i - j}(1 - p_i)^{y_i - j}(1 - p_i^*)^{n_i - y_i - z_i + j}\right\}. \qquad (39.63)$$

This distribution may be called *Multivariate Hypergeometric–Convoluted Binomial(s)*.

Factorial moments of $\boldsymbol{Z}$ of order $\boldsymbol{r} = (r_1, r_2, \ldots, r_w)'$ are easily derived from the structure in (39.62). We have

$$\mu'_{(\boldsymbol{r})}(\boldsymbol{Z}) = E\left[\prod_{i=1}^{w} Z_i^{(r_i)}\right]$$

$$= \left\{\prod_{i=1}^{w} n_i^{(r_i)}\right\}\left\{\sum_{j=0}^{\sum_{i=1}^{w} r_i} d^{(j)}(m - d)^{(\sum_{i=1}^{w} r_i - j)}g_j\right\}\Big/ m^{(\sum_{i=1}^{w} r_i)}, \qquad (39.64)$$

where g_j is the coefficient of x^j in $\prod_{i=1}^{w}(p_i^* + p_i x)^{r_i}$.

[Note that

(i) If sampling were *with replacement*, (39.64) would be

$$\mu'_{(\boldsymbol{r})}(\boldsymbol{Z}) = \left[\prod_{i=1}^{w} n_i^{(r_i)}\{dp_i + (m - d)p_i^*\}^{r_i}\right]\Big/ m^{\sum_{i=1}^{w} r_i}; \qquad (39.65)$$

(ii) If $p_i = p$ and $p_i^* = p^*$ for all $i = 1, 2, \ldots, w$, corresponding to constant quality of inspection throughout, then

$$g_j = \binom{\sum_{i=1}^{w} r_i}{j} p^j \, p^{*\sum_{i=1}^{w} r_i - j}.] \qquad (39.66)$$

From Eq. (39.64), we obtain in particular

$$E[Z_i] = n_i\left\{\frac{d}{m}p_i + \left(1 - \frac{d}{m}\right)p_i^*\right\} = n_i\bar{p}_i, \qquad (39.67)$$

$$\operatorname{var}(Z_i) = n_i\bar{p}_i(1 - \bar{p}_i) - n_i(n_i - 1)(m - 1)\frac{d}{m}\left(1 - \frac{d}{m}\right)(p_i - p_i^*)^2, \qquad (39.68)$$

and

$$\operatorname{cov}(Z_i, Z_j) = -\frac{n_i n_j}{m - 1}\frac{d}{m}\left(1 - \frac{d}{m}\right)(p_i - p_i^*)(p_j - p_j^*), \qquad (39.69)$$

where

$$\bar{p}_i = \frac{d}{m} p_i + \left(1 - \frac{d}{m}\right) p_i^* \tag{39.70}$$

is the probability that an item chosen at random from the ith sample will be *classified as NC* irrespective of whether it is truly NC or not.

Next, noting that

$$E[Z_i \mid Y_i = y_i] = y_i p_i + (n_i - y_i) p_i^*, \tag{39.71}$$

$$E[Y_i \mid Y_j = y_j] = n_i(d - y_j)/(m - n_j), \;\; j \neq i, \tag{39.72}$$

and

$$E[Y_i \mid Z_j = z_j] = n_i\{d - E[Y_j \mid Z_j = z_j]\}/(m - n_j), \tag{39.73}$$

we obtain the regression function of Z_i on Z_j as

$$\begin{aligned} E[Z_i \mid Z_j = z_j] &\\ &= n_i p_i^* + (p_i - p_i^*) E[Y_i \mid Z_j = z_j] \\ &= n_i p_i^* + n_i(p_i - p_i^*)\{d - E[Y_j \mid Z_j = z_j]\}/(m - n_j) \\ &= n_i p_i^* + n_i(p_i - p_i^*) \frac{d}{m} \left\{1 - \frac{(1 - \frac{d}{m})(p_j - p_j^*)}{(m - n_j)\bar{p}_j(1 - \bar{p}_j)} (z_j - n_j \bar{p}_j)\right\}. \end{aligned} \tag{39.74}$$

Thus, the regression is linear. Comparing (39.74) and (39.69), we observe that the slope of the regression and the correlation have the opposite sign to $(p_i - p_i^*)(p_j - p_j^*)$. For further details, interested readers may refer to Johnson and Kotz (1985).

In the special case of a fully sequential sampling procedure with $n_1 = n_2 = \cdots = n_w = 1$, the Y's and Z's have only 0 and 1 as possible values. In this case, we have

$$\begin{aligned} \Pr[\mathbf{Z} = \mathbf{z}] &\\ &= \frac{1}{\binom{m}{d}} \sum_{y_1} \cdots \sum_{y_w} \binom{m - w}{d - \sum_{i=1}^{w} y_i} \\ &\quad \times \prod_{i=1}^{w} \left\{ y_i p_i^{z_i} (1 - p_i)^{1 - z_i} + (1 - y_i) p_i^{* z_i} (1 - p_i^*)^{1 - z_i} \right\} \\ &= \frac{1}{\binom{m}{d}} \sum_{y} \binom{m - w}{d - y} \sum \cdots \sum_{\boldsymbol{\alpha}, \boldsymbol{\alpha}^*, \boldsymbol{\beta}, \boldsymbol{\beta}^*} \prod_{i=1}^{w} p_i^{\alpha_i} p_i^{* \alpha_i^*} (1 - p_i)^{\beta_i} (1 - p_i^*)^{\beta_i^*}, \end{aligned} \tag{39.75}$$

where the summations with respect to $\boldsymbol{\alpha} = (\alpha_1, \ldots, \alpha_w)'$, $\boldsymbol{\beta} = (\beta_1, \ldots, \beta_w)'$, $\boldsymbol{\alpha}^* = (\alpha_1^*, \ldots, \alpha_w^*)'$, and $\boldsymbol{\beta}^* = (\beta_1^*, \ldots, \beta_w^*)'$ are constrained by

$$\sum_{i=1}^{w} (\alpha_i + \beta_i) = \sum_{i=1}^{w} y_i = y \text{ and } \sum_{i=1}^{w} (\alpha_i^* + \beta_i^*) = \sum_{i=1}^{w} z_i = z. \tag{39.76}$$

Now

$$\sum_{\boldsymbol{\alpha},\boldsymbol{\alpha}^*,\boldsymbol{\beta},\boldsymbol{\beta}^*}\cdots\sum \prod_{i=1}^{w} p_i^{\alpha_i} p_i^{*\alpha_i^*}(1-p_i)^{\beta_i}(1-p_i^*)^{\beta_i^*}$$

$$= \text{Coefficient of } t^z u^y \text{ in } \prod_{i=1}^{w}\{p_i tu + p_i^* t + (1-p_i)t + (1-p_i^*)\}. \qquad (39.77)$$

This immediately implies that if m, p_i and p_i^* $(i = 1, \ldots, w)$ are all known, then $Z = \sum_{i=1}^{w} Z_i$ is a sufficient statistic for d.

An extension of the above discussed model is obtained if we consider a situation with several kinds of nonconformity, say $NC_1, NC_2, \ldots, NC_a$. Denoting the numbers of individuals in the population with combinations $\boldsymbol{g} = (g_1, g_2, \ldots, g_a)$ of types of nonconformity (with $g_i = 0, 1$ denoting absence or presence of NC_i, respectively) by $d_{\boldsymbol{g}}$, and the number of individuals in a single sample (without replacement) of size n by $Y_{\boldsymbol{g}}$, the joint distribution of $Y_{\boldsymbol{g}}$'s has probability mass function

$$\Pr\left[\bigcap_{g_1=0}^{1}\cdots\bigcap_{g_a=0}^{1}(Y_{\boldsymbol{g}} = y_{\boldsymbol{g}})\right] = \left\{\prod_{g_1=0}^{1}\cdots\prod_{g_a=0}^{1}\binom{d_{\boldsymbol{g}}}{y_{\boldsymbol{g}}}\right\}\bigg/\binom{m}{n},$$

$$\sum_{g_1=0}^{1}\cdots\sum_{g_a=0}^{1} d_{\boldsymbol{g}} = m, \quad \sum_{g_1=0}^{1}\cdots\sum_{g_a=0}^{1} y_{\boldsymbol{g}} = n; \qquad (39.78)$$

more concisely,

$$\Pr\left[\bigcap_{\boldsymbol{g}}(Y_{\boldsymbol{g}} = y_{\boldsymbol{g}})\right] = \left\{\prod_{\boldsymbol{g}}\binom{d_{\boldsymbol{g}}}{y_{\boldsymbol{g}}}\right\}\bigg/\binom{m}{n}. \qquad (39.79)$$

The joint descending $\boldsymbol{r}$th factorial moment is

$$\mu'_{(\boldsymbol{r})}(\{Y_{\boldsymbol{g}}\}) = E\left[\prod_{\boldsymbol{g}} Y_{\boldsymbol{g}}^{(r_{\boldsymbol{g}})}\right] = n^{(r)}\left\{\prod_{\boldsymbol{g}} d_{\boldsymbol{g}}^{(r_{\boldsymbol{g}})}\right\}\bigg/ m^{(r)}, \qquad (39.80)$$

where $r = \sum_{\boldsymbol{g}} r_{\boldsymbol{g}}$.

For the multivariate hypergeometric distribution, the minimax estimation of $\left(\frac{m_1}{m}, \frac{m_2}{m}, \ldots, \frac{m_k}{m}\right)'$ has been discussed by Trybula (1958), Hodges and Lehmann (1950), and Amrhein (1995). For fixed n, Trybula (1958) showed that the estimator $\boldsymbol{\delta}^* = (\delta_1^*, \ldots, \delta_k^*)'$ defined by

$$\delta_i^*(\boldsymbol{N}, n) = \begin{cases} \dfrac{1}{k} & \text{if } n = 0 \\[2ex] \dfrac{N_i + \frac{1}{k}\sqrt{\frac{n(m-n)}{m-1}}}{n + \sqrt{\frac{n(m-n)}{m-1}}} & \text{if } n > 0 \end{cases}$$

is minimax and admissible with respect to the squared error loss function

$$L(\boldsymbol{m}, \boldsymbol{a}) = \sum_{i=1}^{k} \left(\frac{m_i}{m} - a_i \right)^2 ,$$

where $\boldsymbol{a} = (a_1, \ldots, a_k)'$ is an estimate of $\left(\frac{m_1}{m}, \frac{m_2}{m}, \ldots, \frac{m_k}{m}\right)'$. The special case $k = 2$ was solved earlier by Hodges and Lehmann (1950); see also Wilczynski (1985) for a discussion on a more general loss function. For the case when n is a realization of a random variable N whose distribution is known and does not depend on the unknown parameter $(m_1, \ldots, m_k)'$, Amrhein (1995) has shown recently that the minimax estimator is given by

$$\delta_i^\alpha(\boldsymbol{N}, N) = \frac{1}{m} \{N_i + (m - N)(N_i + \alpha)/(N + k\alpha)\} ,$$

where N is now an ancillary statistic and α is uniquely determined by the distribution of N.

Methods of estimation of parameters of the distributions discussed in this section are unfortunately in early stages and present serious problems. Some initial results in this direction are presented in the monograph by Johnson, Kotz, and Wu (1991) and some of the references cited therein.

BIBLIOGRAPHY

Amrhein, P. (1995). Minimax estimation of proportions under random sample size, *Journal of the American Statistical Association*, **90**, 1107–1111.

Balakrishnan, N., and Ma, Y. (1996). Empirical Bayes rules for selecting the most and least probable multivariate hypergeometric event, *Statistics & Probability Letters*, **27**, 181–188.

Boland, P. J., and Proschan, F. (1987). Schur convexity of the maximum likelihood function for the multivariate hypergeometric and multinomial distributions, *Statistics & Probability Letters*, **5**, 317–322.

Charalambides, C. A. (1981). On a restricted occupancy model and its applications, *Biometrical Journal*, **23**, 601–610.

Chesson, J. (1976). A non-central multivariate hypergeometric distribution arising from biased sampling with application to selective predation, *Journal of Applied Probability*, **13**, 795–797.

Childs, A., and Balakrishnan, N. (1996). Some approximations to the multivariate hypergeometric distribution with applications to hypothesis testing, *Submitted for publication*.

Consul, P. C., and Shenton, L. R. (1972). On the multivariate generalization of the family of discrete Lagrangian distributions, *Multivariate Statistical Analysis Symposium*, Halifax, Nova Scotia.

de Montmort, P. R. (1708). *Essay d'Analyse sur les Jeux de Hazard*, Paris: Quillan (published anonymously). Second augmented edition, Quillan, Paris (1913) (published anonymously). Reprinted by Chelsea, New York (1980).

Dwass, M. (1979). A generalized binomial distribution, *The American Statistician*, **33**, 86–87.

Gerstenkorn, T. (1977). Multivariate doubly truncated discrete distributions, *Acta Universitatis Lodziensis*, 1–115.

Good, I. J. (1960). Generalizations to several variables of Lagrange expressions, with application to stochastic processes, *Proceedings of the Cambridge Philosophical Society*, **56**, 367–380.

Hald, A. (1990). *A History of Probability and Statistics and Their Applications Before 1750*, New York: John Wiley & Sons.

Hodges, J. L., Jr., and Lehman, E. L. (1950). Some problems in minimax point estimation, *Annals of Mathematical Statistics*, **21**, 182–197.

Ishii, G., and Hayakawa, R. (1960). On the compound binomial distribution, *Annals of the Institute of Statistical Mathematics*, **12**, 69–80. Errata, **12**, 208.

Jain, G. C. (1974). On power series distributions associated with Lagrange expansion, *Biometrische Zeitschrift*, **17**, 85–97.

Jain, G. C., and Singh, N. (1975). On bivariate power series associated with Lagrange expansion, *Journal of the American Statistical Association*, **70**, 951–954.

Janardan, K. G., and Patil, G. P. (1972). A unified approach for a class of multivariate hypergeometric models, *Sankhyā, Series A*, **34**, 1–14.

Johnson, N. L., and Kotz, S. (1985). Some multivariate distributions arising in faulty sampling inspection, in *Multivariate Analysis—VI* (Ed. P. R. Krishnaiah), pp. 311–325, Amsterdam, Netherlands: Elsevier.

Johnson, N. L., Kotz, S., and Wu, X. Z. (1991). *Inspection Errors for Attributes in Quality Control*, London, England: Chapman and Hall.

Lyons, N. I. (1980). Closed expressions for noncentral hypergeometric probabilities, *Communications in Statistics—Simulation and Computation*, **9**, 313–314.

Macdonald, D. G., Power, M., and Fuller, J. D. (1994). A new discovery process applicable to forecasting hydrocarbon discoveries, *Resource and Energy Economics*, **16**, 147–166.

Manly, B. F. J. (1974). A model for certain types of selection experiments, *Biometrics*, **30**, 281–294.

Mosimann, J. E. (1962). On the compound multinomial distribution, the multivariate β-distribution and correlations among proportions, *Biometrika*, **49**, 65–82.

Mosimann, J. E. (1963). On the compound negative multinomial distribution and correlations among inversely sampled pollen counts, *Biometrika*, **50**, 47–54.

Panaretos, J. (1983a). An elementary characterization of the multinomial and the multivariate hypergeometric distribution, in *Stability Problems for Stochastic Models* (Eds. V. V. Kalashnikov and V. M. Zolotarev), Lecture Notes in Mathematics, No. 982, pp. 156–164, Berlin: Springer-Verlag.

Panaretos, J. (1983b). On some bivariate discrete distributions with multivariate components, *Publicationes Mathematicae, Institutum Mathematicum Universitatis Debreceniensis, Hungaria*, **30**, 177–184.

Rao, B. R., and Janardan, K. G. (1981). On the moments of multivariate discrete distributions using finite difference operators, *Technical Report No. 81-03*, Mathematical Systems Program, Sangamore State University, Springfield, IL.

Röhmel, J., Streitberg, B., and Tismer, C. (1996). A permutation approach to configural frequency analysis (CFA) and the iterated hypergeometric distribution, *Computational Statistics*, 355–378.

Särndal, C. E. (1964). A unified derivation of some non-parametric distributions, *Journal of the American Statistical Association*, **59**, 1042–1053.

Shenton, L. R., and Consul, P. C. (1973). On bivariate Lagrange and Borel–Tanner distributions and their use in queueing theory, *Sankhyā, Series A*, **35**, 229–236.

Sibuya, M., and Shimizu, R. (1981). Classification of the generalized hypergeometric family of distributions, *Keio Science and Technology Reports*, **34**, 1–39.

Steyn, H.S. (1951). On discrete multivariate probability functions, *Proceedings Koninklijke Nederlandse Akademie van Wetenschappen, Series A*, **54**, 23–30.

Steyn, H.S. (1955). On discrete multivariate probability functions of hypergeometric type, *Proceedings Koninklijke Nederlandse Akademie van Wetenschappen, Series A*, **58**, 588–595.

Steyn, P. W. (1990). Bayes tests of hypotheses in a multivariate hypergeometric case with special reference to contingency tables, Ph.D. thesis, University of the Orange Free State, South Africa (see also *South African Statistical Journal*, **25**, 79–80, for abstract).

Trybula, S. (1958). Some problems of simultaneous minimax estimation, *Annals of Mathematical Statistics*, **29**, 245–253.

Trybula, S. (1991). Systematic estimation of parameters of multinomial and multivariate hypergeometric distribution, *Statistics*, **22**, 59–67.

Wallenius, K. T. (1963). Biased Sampling: The Noncentral Hypergeometric Distribution, Ph.D. thesis, Stanford University, Stanford, CA.

Wilczynski, M. (1985). Minimax estimation for the multinomial and multivariate hypergeometric distributions, *Sankhyā, Series A*, **47**, 128–132.

Xekalaki, E. (1977). Bivariate and multivariate extensions of the generalized Waring distribution, Ph.D. thesis, University of Bradford, England.

Xekalaki, E. (1984a). The bivariate generalized Waring distribution and its application to accident theory, *Journal of the Royal Statistical Society, Series A*, **147**, 488–498.

Xekalaki, E. (1984b). Models leading to the bivariate generalized Waring distribution, *Utilitas Mathematica*, **25**, 263–290.

CHAPTER 40

Multivariate Pólya–Eggenberger Distributions

1 GENESIS

Many discrete multivariate distributions can be generated from fairly simple urn models. Johnson and Kotz (1976) surveyed the literature on this topic at that time; also see Johnson and Kotz (1977) for a comprehensive exposition of the results up till the mid-1970s. Several of these distributions bear the name of *Pólya or Pólya-Eggenberger distributions* since they are motivated, to a great extent, by the well-known Eggenberger–Pólya (1923, 1928) urn schemes. These univariate distributions have attracted, during the past 70 years or so, substantial attention of researchers both in theoretical and applied fields, as exemplified by the discussion on this topic by Johnson, Kotz, and Kemp (1992). Reference may also be made here to the dictionaries by Patil, Joshi, and Rao (1968) and Patil *et al.* (1984).

In a typical scheme of this kind, a ball is removed "at random" from an urn containing r red balls and b black balls. The color of the removed ball is noted and then the ball is returned to the urn along with c additional balls of the same color. [If $c = 0$, the scheme becomes *sampling with replacement*; if $c = -1$, the scheme becomes *sampling without replacement*.] This operation is repeated, using the newly constituted urn, until n such operations (often called "trials") have been completed. The distribution of the number of red balls (X, say) observed at the end of n trials is an example of a Pólya–Eggenberger distribution.

It is, of course, not too difficult to devise a multitude of variants of the Eggenberger–Pólya schemes by postulating, changing, or revising various patterns of colors of the balls in the urn and different sampling procedures (direct or inverse).

2 DISTRIBUTION

A multivariate version of the above described scheme allows for $a_1, a_2, \ldots, a_k$ initial balls of k different colors, $C_1, C_2, \ldots, C_k$. In this general case, Steyn (1951) showed that the joint distribution of the numbers $X_1, X_2, \ldots, X_k$ of balls of colors

$C_1, C_2, \ldots, C_k$, respectively, at the end of n trials has the joint probability mass function

$$P(x_1, x_2, \ldots, x_k) = \Pr\left[\bigcap_{i=1}^{k}(X_i = x_i)\right]$$
$$= \binom{n}{\boldsymbol{x}}\left\{\prod_{i=1}^{k} a_i^{[x_i,c]}\right\}\Big/ a^{[n,c]}, \qquad (40.1)$$

where $\sum_{i=1}^{k} x_i = n$, $\sum_{i=1}^{k} a_i = a$,

$$a^{[x,c]} = a(a+c)\cdots(a+\overline{x-1}\,c) \quad \text{with } a^{[0,c]} = 1\,, \qquad (40.2)$$

and

$$\binom{n}{\boldsymbol{x}} = \binom{n}{x_1, x_2, \ldots, x_k} = \frac{n!}{x_1!x_2!\ldots x_k!} \qquad (40.3)$$

is the multinomial coefficient.

As in the case of multinomial distributions (see Chapter 35), the distribution in (40.1) is really a $(k-1)$-dimensional multivariate distribution. An alternative formulation, leading to a truly k-dimensional multivariate distribution, is obtained by introducing an additional color (C_0) initially represented by a_0 balls, but still considering only the variables $X_1, X_2, \ldots, X_k$ representing the numbers of balls of colors $C_1, C_2, \ldots, C_k$, respectively, at the end of n trials. Then, in place of (40.1), we have

$$P(x_1, x_2, \ldots, x_k) = \Pr\left[\bigcap_{i=1}^{k}(X_i = x_i)\right]$$
$$= \binom{n}{x_1, \ldots, x_k, n - \sum_{i=1}^{k} x_i} a_0^{[n-\Sigma_{i=1}^{k} x_i, c]}\left\{\prod_{i=1}^{k} a_i^{[x_i,c]}\right\}\Big/ a^{[n,c]}, \qquad (40.4)$$

where $\sum_{i=0}^{k} a_i = a$, $0 \le \sum_{i=1}^{k} x_i \le n$ and $x_i \ge 0$ $(i = 1, 2, \ldots, k)$.

Gerstenkorn (1977a–c), based on Dyczka (1973), represented the above probability mass function in the form

$$P(x_1, x_2, \ldots, x_k)$$
$$= \frac{n!}{1\left(1+\frac{c}{a}\right)\left(1+\frac{2c}{a}\right)\cdots\left(1+\frac{\overline{n-1}\,c}{a}\right)} \prod_{i=0}^{k}\left\{p_i^{[x_i,c/a]}\Big/ x_i!\right\}$$
$$= \frac{n!}{1^{[n,c/a]}} \prod_{i=0}^{k}\left\{p_i^{[x_i,c/a]}\Big/ x_i!\right\} \qquad (40.5)$$

with $p_i = a_i/a$ (= initial proportion of balls of color C_i) and $x_0 = n - \sum_{i=1}^n x_i$. To ensure that the number of balls of any one color remains non-negative (when $c < 0$), we need to impose the condition

$$- nc \le \min(p_0, p_1, \ldots, p_k) . \tag{40.6}$$

Note that $\sum_{i=0}^k p_i = 1$.

Denoting $\boldsymbol{x}$ for $(x_0, x_1, \ldots, x_k)'$, $\boldsymbol{a}$ for $(a_0, a_1, \ldots, a_k)'$, a for $\sum_{i=0}^k a_i$, and $MP(\boldsymbol{x}, \boldsymbol{a}; c; n)$ for the multivariate Pólya–Eggenberger probability mass function in (40.4), Janardan and Patil (1970, 1972) presented several equivalent forms of this mass function:

a. $$MP(\boldsymbol{x}, \boldsymbol{r}, n) = \binom{n}{\boldsymbol{x}} \frac{\prod_{i=0}^k \{r_i(r_i+1)\cdots(r_i+x_i-1)\}}{r(r+1)\cdots(r+n-1)} , \tag{40.7}$$

where $r_i = \frac{a_i}{c}$ for $c \neq 0$, and $r = \sum_{i=0}^k r_i = \frac{a}{c}$;

b. $$MP(\boldsymbol{x}, \boldsymbol{a}; c; n) = \binom{n}{\boldsymbol{x}} \frac{a_0^{[n-x_1,c]} \prod_{i=0}^k a_i^{[x_i,c]}}{a^{[n,c]}} ; \tag{40.8}$$

c. $$MP(\boldsymbol{x}, \boldsymbol{r}, n) = \binom{n}{\boldsymbol{x}} \frac{\prod_{i=0}^k r_i^{[x_i]}}{r^{[n]}} ; \tag{40.9}$$

d. $$MP(\boldsymbol{x}, \boldsymbol{p}; \nu; n) = \binom{n}{\boldsymbol{x}} \frac{\prod_{i=0}^k \{p_i(p_i+\nu)\cdots(p_i+\overline{x_i-1}\,\nu)\}}{1(1+\nu)\cdots(1+\overline{n-1}\,\nu)} , \tag{40.10}$$

where $p_i = \frac{a_i}{a} = \frac{r_i}{r}$ and $\nu = \frac{c}{a} = \frac{1}{r}$;

e. $$MP(\boldsymbol{x}, \boldsymbol{p}; \nu; n) = \binom{n}{\boldsymbol{x}} \frac{\prod_{i=0}^k p_i^{[x_i,\nu]}}{1^{[n,\nu]}} , \tag{40.11}$$

where p_i can be interpreted as the probability of drawing a ball of color C_i in the first draw, and the parameter ν as the measure of the degree of the "contagion";

f. $$MP(\boldsymbol{x}, \boldsymbol{r}, n) = \binom{n}{\boldsymbol{x}} \frac{\Gamma(r)}{\Gamma(r+n)} \prod_{i=0}^k \left\{ \frac{\Gamma(r_i+x_i)}{\Gamma(r_i)} \right\} ; \tag{40.12}$$

and

g. $$MP(\boldsymbol{x}, \boldsymbol{r}, n) = \frac{\prod_{i=0}^k \binom{-r_i}{x_i}}{\binom{-r}{n}} . \tag{40.13}$$

Using the basic identity

$$\sum_{x_i=0}^n \frac{a_0^{[n-x,c]} \prod_{i=1}^k a_i^{[x_i,c]}}{(n-x)! \prod_{i=1}^k x_i!} = \frac{a^{[n,c]}}{n!} , \quad x = \sum x_i \le n , \tag{40.14}$$

the following distributional properties can be established:

1. $(X_1, X_2, \ldots, X_i)' \rightsquigarrow MP(a, a_1, a_2, \ldots, a_i; c; n)$ for $1 \le i \le k$;
2. $X_1 + X_2 + \cdots + X_k \rightsquigarrow MP\left(a, \sum_{i=1}^k a_i; c; n\right) \equiv P\left(a, \sum_{i=1}^k a_i; c; n\right)$, where P denotes the univariate Pólya–Eggenberger distribution;
3. $(X_1, X_2, \ldots, X_i)' \mid \left(\sum_{j=1}^i X_j = m\right) \rightsquigarrow SMP(a_1, a_2, \ldots, a_i; c; m)$, where SMP denotes the singular multivariate Pólya–Eggenberger distribution as given in (40.1);
4. $(X_1, X_2, \ldots, X_i)' \mid (X_{i+1}, \ldots, X_k)' \rightsquigarrow MP(a^*, a_1, \ldots, a_i; c; n^*)$, where $a^* = a - \sum_{j=i+1}^k a_j$ and $n^* = n - \sum_{j=i+1}^k n_j$;
5. $(Z_1, Z_2, \ldots, Z_i)' \rightsquigarrow MP(a, b_1, b_2, \ldots, b_i; c; n)$, where $Z_j = \sum_j X_\ell$, $b_j = \sum_j a_\ell$ with $\sum_j$ denoting summation taken over $\ell \in H_j$ for $j = 1, 2, \ldots, i$, such that $\bigcup_{j=1}^i H_j = \{1, 2, \ldots, k\}$, $H_j \cap H_{j'} = \emptyset$ for $j \neq j' = 1, 2, \ldots, i$; and
6. $(X_1, X_2, \ldots, X_k)' \mid (Z_1, Z_2, \ldots, Z_i)' \rightsquigarrow i$ independent singular multivariate Pólya–Eggenberger distributions of the type $SMP(\{b_{pq}\}, c; z_q)$ for $q = 1, 2, \ldots, i$, where $b_{pq} = a_p$ for $p \in H_q$, $q = 1, 2, \ldots, i$.

Janardan (1974) has discussed some characterization results, while Marshall and Olkin (1990) have discussed three urn models, each of which leads to a bivariate Pólya–Eggenberger distribution.

Urn Model I [The Steyn (1951) model]: From an urn containing a red, b black, and c white balls, a ball is chosen at random, its color is noted, and the ball is returned to the urn along with s additional balls of the same color. Always starting with the newly constituted urn, this experiment is continued n times; the number x of red balls drawn and the number y of black balls drawn are noted.

Urn Model II [The Kaiser and Stefansky (1972) model]: From an urn containing a_{ij} balls labeled (i, j), $i = 0, 1$ and $j = 0, 1$, a ball is chosen at random, its label is noted, and the ball is returned to the urn along with s additional balls with the same label. Always starting with the newly constituted urn, this experiment is continued n times; the number x of balls with the first label 1 and the number y of balls with second label 1 are recorded. (Note that this model reduces to Model I when $a_{10} = a$, $a_{01} = b$, $a_{00} = c$, and $a_{11} = 0$.)

Urn Model III [A generalization of the Hald (1960) and Bosch (1963) model]: Consider the classical Eggenberger–Pólya urn scheme defined at the beginning of this chapter. Let $Z_i = 1$ if a red ball is drawn on the ith trial and $Z_i = 0$ otherwise. The numbers $X = \sum_{i=1}^{n_1} Z_i$ and $Y = \sum_{i=n_1-k+1}^{n_1-k+n_2} Z_i$ are recorded.

For Model I, the joint probability mass function (with $s = 1$) is

$$P(x, y) = \binom{n}{x, y, n-x-y} \frac{\Gamma(a+x)\Gamma(b+y)\Gamma(c+n-x-y)}{\Gamma(a)\Gamma(b)\Gamma(c)} \times \frac{\Gamma(a+b+c)}{\Gamma(a+b+c+n)}$$

$$= \binom{n}{x, y, n-x-y} \frac{B(a+x, b+y, c+n-x-y)}{B(a, b, c)},$$

$$x, y = 0, 1, \ldots, \tag{40.15}$$

where $B(t_1, \ldots, t_k) = \{\prod_{i=1}^k \Gamma(t_i)\}/\Gamma(\sum_{i=1}^k t_i)$. Alternatively, $P(x, y)$ is a Dirichlet mixture of multinomial distributions with

$$P(x, y) = \int \binom{n}{x, y, n-x-y} \theta_1^x \theta_2^y (1-\theta_1-\theta_2)^{n-x-y}$$

$$\times \frac{1}{B(a, b, c)} \theta_1^{a-1} \theta_2^{b-1} (1-\theta_1-\theta_2)^{c-1} \, d\theta_1 \, d\theta_2 \tag{40.16}$$

with integration over the region $0 \le \theta_1 \le 1$, $0 \le \theta_2 \le 1$, and $0 \le \theta_1 + \theta_2 \le 1$.

Similarly, for Model II, the joint probability mass function (with $s = 1$) is

$$P(x, y) = \sum_\alpha \binom{n}{\alpha, x-\alpha, y-\alpha, n-x-y+\alpha}$$

$$\times \frac{B(a_{11}+\alpha, a_{10}+x-\alpha, a_{01}+y-\alpha, a_{00}+n-x-y+\alpha)}{B(a_{11}, a_{10}, a_{01}, a_{00})}$$

$$= \int P_1(x, y \mid \theta_{11}, \theta_{10}, \theta_{01}) P_2(\theta_{11}, \theta_{10}, \theta_{01} \mid a_{11}, a_{10}, a_{01}, a_{00})$$

$$d\theta_{11} \, d\theta_{10} \, d\theta_{01}, \tag{40.17}$$

where integration extends over the region $\theta_{11}, \theta_{10}, \theta_{01} \ge 0$, $\theta_{00} = 1-\theta_{10}-\theta_{01}-\theta_{11} \ge 0$, P_1 is the bivariate binomial probability mass function given by

$$P_1(x, y \mid \theta_{11}, \theta_{10}, \theta_{01})$$

$$= \sum_\alpha \binom{n}{\alpha, x-\alpha, y-\alpha, n-x-y+\alpha} \theta_{11}^\alpha \theta_{10}^{x-\alpha} \theta_{01}^{y-\alpha} \theta_{00}^{n-x-y+\alpha},$$

and P_2 is the Dirichlet density given by

$$P_2(\theta_{11}, \theta_{10}, \theta_{01} \mid a_{11}, a_{10}, a_{01}, a_{00}) = \frac{\theta_{11}^{a_{11}-1} \theta_{10}^{a_{10}-1} \theta_{01}^{a_{01}-1} \theta_{00}^{a_{00}-1}}{B(a_{11}, a_{10}, a_{01}, a_{00})},$$

$$\theta_{11}, \theta_{10}, \theta_{01}, \theta_{00} \ge 0.$$

Finally, for Model III, the joint probability mass function (with $s = 1$) is

$$P(x, y) = \sum \binom{k}{\ell} \binom{n_1-k}{x-\ell} \binom{n_2-k}{y-\ell}$$

$$\times \frac{B(a+x+y-\ell, b+n_1+n_2-x-y-k+\ell)}{B(a, b)}$$

$$= \int_0^1 P_1(x, y \mid \theta) \, \frac{\theta^{a-1}(1-\theta)^{b-1}}{B(a, b)} \, d\theta \, , \tag{40.18}$$

where

$$P_1(x, y \mid \theta) = \sum_{\ell} \binom{k}{\ell} \theta^{\ell}(1-\theta)^{k-\ell} \binom{n_1 - k}{x - \ell} \theta^{x-\ell}(1-\theta)^{n_1-k-x+\ell}$$
$$\times \binom{n_2 - k}{y - \ell} \theta^{y-\ell}(1-\theta)^{n_2-k-y+\ell}$$

is a bivariate binomial distribution of $(U+W, V+W)'$, with $U \rightsquigarrow$ Binomial(n_1-k, θ), $V \rightsquigarrow$ Binomial$(n_2 - k, \theta)$, and $W \rightsquigarrow$ Binomial(k, θ) (U, V and W are mutually independent). [Here, $P(x, y)$ is a beta-mixture of bivariate binomial distributions different from those arising with Model II.]

3 MOMENTS AND MODE

The $\boldsymbol{r} = (r_1, r_2, \ldots, r_k)'$th (descending) factorial moment of the distribution (40.5) is [Dyczka (1973)]

$$\mu'_{(\boldsymbol{r})}(\boldsymbol{X}) = \mu'_{(r_1, r_2, \ldots, r_k)}(X_1, X_2, \ldots, X_k) = E\left[\prod_{i=1}^{k} X_i^{(r_i)}\right]$$
$$= \frac{n^{(r)}}{1^{[r, c/a]}} \prod_{i=1}^{k} p_i^{[r_i, c/a]} \, , \tag{40.19}$$

where $r = \sum_{i=1}^{k} r_i$.

The result in (40.19) is established by using the relationship

$$A^{[x,B]} = A^{[r,B]}(A + rB)^{[x-r,B]}, \qquad r \le x \, , \tag{40.20}$$

and, in particular,

$$1^{[n,B]} = 1^{[r,B]}(1 + rB)^{[n-r,B]}, \qquad r \le n \, , \tag{40.21}$$

and then writing [see (40.19)]

$$\mu'_{(\boldsymbol{r})}(\boldsymbol{X}) = \frac{n!}{1^{[n,c/a]}} \sum_{x_0} \cdots \sum_{x_k} \left[\frac{p_0^{[x_0,c/a]}}{x_0!} \prod_{i=1}^{k} \left\{ x_i^{(r_i)} p_i^{[x_i,c/a]} / x_i! \right\}\right]$$
$$= \frac{n^{(r)}}{1^{[r,c/a]}} \prod_{i=1}^{k} p_i^{[r_i,c/a]}$$
$$\times \left[\frac{(n-r)!}{\left(1 + \frac{rc}{a}\right)^{[n-r,c/a]}} \sum_{x_0} \cdots \sum_{x_k} \frac{p_0^{[x_0,c/a]}}{x_0!} \prod_{i=1}^{k} \frac{p_i'^{[x_i - r_i, c/a]}}{(x_i - r_i)!}\right] \tag{40.22}$$

with $p_i' = p_i + r_i c/a$ $(i = 1, 2, \ldots, k)$. The result in (40.19) follows readily from (40.22) upon noting that $1 + rc/a = p_0 + \sum_{i=1}^k p_i'$ and then observing that the term inside the brackets is 1 [being the entire sum of all the probabilities of a Pólya–Eggenberger distribution].

The raw (crude) product-moments about zero, $\mu_{\boldsymbol{r}}'(\boldsymbol{X}) = E\left[\prod_{i=1}^k X_i^{r_i}\right]$, can be obtained from general formulas relating factorial moments to raw product-moments. For example, we have

$$\mu_{\boldsymbol{r}}'(\boldsymbol{X}) = E\left[\prod_{i=1}^k X_i^{r_i}\right] = \sum_{m_1=0}^{r_1} \cdots \sum_{m_k=0}^{r_k} \left(\prod_{i=1}^k S(r_i, m_i)\right) \mu_{(\boldsymbol{r})}'(\boldsymbol{X}), \tag{40.23}$$

where $S(r_i, m_i)$ is a Stirling number of the second kind [see Chapter 34, Eq. (34.13) for its definition].

In particular, we have the results

$$E[X_i] = np_i = na_i/a, \tag{40.24}$$

$$\operatorname{var}(X_i) = \frac{na_i(a - a_i)(a + nc)}{a^2(a + c)}, \tag{40.25}$$

$$\operatorname{cov}(X_i, X_j) = -\frac{na_i a_j(a + nc)}{a^2(a + c)}, \tag{40.26}$$

and

$$\operatorname{corr}(X_i, X_j) = -\sqrt{\frac{a_i a_j}{(a - a_i)(a - a_j)}} = -\sqrt{\frac{p_i p_j}{(1 - p_i)(1 - p_j)}}. \tag{40.27}$$

The property

$$\begin{aligned} MP(\boldsymbol{x}, \boldsymbol{a}; c; n) &= \frac{n^{(r)}}{\prod_{i=1}^k x_i^{(r_i)}} \binom{n - r}{\boldsymbol{x} - \boldsymbol{r}} \frac{a_0^{[n-x,c]} \prod_{i=1}^k a_i^{[r_i,c]} \prod_{i=1}^k (a_i + r_i c)^{[x_i - r_i, c]}}{a^{[r,c]}(a + rc)^{[n-r,c]}} \\ &= \frac{n^{(r)}}{\prod_{i=1}^k x_i^{(r_i)}} \frac{\prod_{i=1}^k a_i^{[r_i,c]}}{a^{[r,c]}} MP(\boldsymbol{x} - \boldsymbol{r}, \boldsymbol{a} + \boldsymbol{r}c; c; n - r) \end{aligned} \tag{40.28}$$

facilitates calculation of descending factorial moments. Using (40.28), we get

$$\begin{aligned} \mu_{(\boldsymbol{r})}'(\boldsymbol{X}) &\doteq E\left[\prod_{i=1}^k X_i^{(r_i)}\right] \\ &= \sum_{x_1, \ldots, x_k} \left\{\prod_{i=1}^k x_i^{(r_i)}\right\} MP(\boldsymbol{x}, \boldsymbol{a}; c; n) \end{aligned}$$

$$= \frac{n^{(r)} \prod_{i=1}^{k} a_i^{[r_i,c]}}{a^{[r,c]}} \sum_{x_1,\ldots,x_k} MP(\boldsymbol{x} - \boldsymbol{r}, \boldsymbol{a} + \boldsymbol{r}c; c; n - r)$$

$$= \frac{n^{(r)} \prod_{i=1}^{k} a_i^{[r_i,c]}}{a^{[r,c]}} . \tag{40.29}$$

Equation (40.29) can also be written as

$$\mu'_{(\boldsymbol{r})}(\boldsymbol{X}) = \frac{n^{(r)} \prod_{i=1}^{k} \left(\frac{a_i}{c} + r_i - 1\right)^{(r_i)}}{\left(\frac{a}{c} + r - 1\right)^{(r)}} \tag{40.30}$$

and

$$\mu'_{(\boldsymbol{r})}(\boldsymbol{X}) = \frac{n^{(r)} \prod_{i=1}^{k} p_i^{[r_i,\nu]}}{1^{[r,\nu]}} \qquad \text{[cf. (40.22)]}, \tag{40.31}$$

where, as before, $p_i = \frac{a_i}{a}$ and $\nu = \frac{c}{a}$.

The regression of X_k on the other variables $X_1, \ldots, X_{k-1}$ is linear, and is given by

$$E[X_k \mid X_1, \ldots, X_{k-1}]$$
$$= np_k - \frac{p_k}{p_0 + p_k} \{(x_1 - np_1) + \cdots + (x_{k-1} - np_{k-1})\}, \tag{40.32}$$

which clearly shows that the regression function depends only on the sum of the conditioning variables and not on their individual values. Next, we have the conditional variance as

$$\mathrm{var}(X_k \mid X_1, \ldots, X_{k-1})$$
$$= (n - x_1 - \cdots - x_{k-1}) \frac{p_0 p_k}{(p_0 + p_k)^2} \frac{1 + (n - x_1 - \cdots - x_{k-1})\nu}{1 + \nu}, \tag{40.33}$$

which shows that the regression is heteroscedastic.

Following Finucan (1964), Janardan and Patil (1970, 1972) introduced the following definitions:

(i) $\boldsymbol{y}$ is an $[\alpha, \beta]$ *neighbor of* $\boldsymbol{x}$ if $y_\alpha = x_\alpha + 1$ and $y_\beta = x_\beta - 1$, with $y_i = x_i$ for all $i \neq \alpha, \beta$;

(ii) $\boldsymbol{x}$ is a *strict local mode* if $P(x_1, x_2, \ldots, x_k) > P(y_1, y_2, \ldots, y_k)$ for all $\boldsymbol{y}$ which are $[\alpha, \beta]$ neighbors of $\boldsymbol{x}$ (for all possible pairs of values of α and β); and

(iii) a *joint mode* is one with at least one equiprobable neighbor.

Conditions for a strict local mode in this case are

$$\frac{x_i + 1}{(a_i/c) - 1} > \frac{x_j}{(a_j/c) - 1} \qquad \text{for all } i, j . \tag{40.34}$$

Conditions for a joint mode are

$$\frac{x_i + 1}{(a_i/c) - 1} \geq \frac{x_j}{(a_j/c) - 1} \qquad \text{for all } i, j \tag{40.35}$$

with

$$\frac{x_\beta + 1}{(a_\beta/c - 1)} > \frac{x_\alpha}{(a_\alpha/c) - 1} \quad \text{if} \quad \frac{x_\alpha + 1}{(a_\alpha/c) - 1} > \frac{x_\beta}{(a_\beta/c) - 1}\,. \tag{40.36}$$

[For Finucan's (1964) conditions for the multinomial case, see Chapter 35.]

Modes may be found by obtaining a constant κ (say) such that x_i = integer part of $\kappa(a_i - c)$ with $\sum_{i=1}^{k} x_i = n$. If κ is unique, then there exists a set of joint modes. If κ is not unique, then there is a unique $\boldsymbol{x}$ such that $\sum_{i=1}^{k} x_i = n$, and this is the unique mode.

4 GENERALIZATIONS

From the genesis of distribution (40.4), it follows that the quantities $\{a_i\}$ are nonnegative integers. The requirement that they be integers can be relaxed, leading to generalizations of the multivariate Pólya–Eggenberger distributions.

Sen and Jain (1994) proposed a generalized multivariate Pólya–Eggenberger model by considering, for given integers $\mu_1, \mu_2, \ldots, \mu_k$ and $n > 0$,

$$\begin{aligned} P_X = \Pr\Bigg[& {}_jX_{n+\Sigma(\mu_i+1)x_i} = x_j \ \ (j = 1, 2, \ldots, k), \\ & {}_0X_i < n + \Sigma\mu_j\ {}_jX_i \ \ (i = 1, 2, \ldots, n + \Sigma(\mu_i + 1)x_i - 1), \\ & {}_0X_{n+\Sigma(\mu_i+1)x_i} = n + \Sigma\mu_i x_i \Bigg], \end{aligned} \tag{40.37}$$

where

$${}_jX_i = \text{number of balls of } j\text{th color drawn among the first } i \text{ balls drawn } (j = 1, 2, \ldots, k;\ i = 1, 2, \ldots),$$

$${}_0X_i = \text{number of white (color } C_0\text{) balls drawn among the first } i \text{ balls drawn,}$$

and Σ is used to denote $\sum_1^k$.

Their derivation involves a combinatorial method of counting multidimensional lattice paths [see, for example, Mohanty (1979)] representing sequences of $n + \Sigma(\mu_i + 1)x_i$ drawings satisfying the condition that there are x_i balls of color i $(i = 1, 2, \ldots, k)$. Equation (40.37) represents a true probability distribution for some selected values of $\boldsymbol{\mu}$ $(\boldsymbol{\mu} \geq -\mathbf{1})$.

Sen and Jain's (1994) generalized multivariate Pólya–Eggenberger model in (40.37) generates many multivariate discrete distributions for different choices of the parameters n, μ_i, a_i, and c. These include multinomial, negative multinomial, and multivariate Pólya–Eggenberger distributions. In fact, Table 1 of Sen and Jain (1994) presents 19 particular cases.

5 ESTIMATION

Janardan and Patil (1972) have considered moment estimators of the parameters of the multivariate Pólya–Eggenberger distribution under the assumptions $c > 0$, n is a known positive integer, and a is unknown. If a is known, the problem becomes trivial, and in this case the estimators are given by

$$\hat{a}_i = \frac{a}{n}\,\bar{x}_i, \qquad i = 1, 2, \ldots, k\,. \tag{40.38}$$

The $\hat{a}_i$'s are also the maximum likelihood estimators of the a_i's.

Estimators of the parameters $a_0, a_1, \ldots, a_k$ by the method of moments on the basis of a random sample of size m with x_{ij}, $i = 1, 2, \ldots, k$, $j = 1, 2, \ldots, m$, taken on $(X_1, \ldots, X_k)' \rightsquigarrow MP(a, a_1, \ldots, a_k; c; n)$, follow from Mosimann's (1963) moment equations:

$$\bar{x}_i = \frac{n\,\hat{r}_i}{\Sigma r_i}, \qquad i = 1, 2, \ldots, k,$$

$$\hat{r}_0 + \hat{r}_1 + \cdots + \hat{r}_k = \frac{n - \hat{R}}{\hat{R} - 1}, \tag{40.39}$$

where, as before, $r_i = a_i/c$, which can be written as

$$\hat{a}_0 + \hat{a}_1 + \cdots + \left(\frac{\bar{x}_i - n}{\bar{x}_i}\right)\hat{a}_i + \cdots + \hat{a}_k = 0\,,$$

$$\hat{a}_0 + \hat{a}_1 + \cdots + \hat{a}_k = \frac{c(n - \hat{R})}{\hat{R} - 1}, \tag{40.40}$$

where $\hat{R}$ is an estimator of

$$R = \frac{\sum_{i=0}^{k} a_i + nc}{\sum_{i=0}^{k} a_i + c}\,. \tag{40.41}$$

The moment estimators, obtained from (40.40), are

$$\hat{a}_i = \bar{x}_i \left[\frac{c(n - \hat{R})}{n(\hat{R} - 1)}\right] \tag{40.42}$$

and

$$\hat{a}_0 = \left(n - \sum_{i=1}^{k} \bar{x}_i\right)\left[\frac{c(n - \hat{R})}{n(\hat{R} - 1)}\right]. \tag{40.43}$$

There are several ways of getting an estimate of R in (40.41). Mosimann (1963) suggested an estimate of R to be based on either a ratio of a number of variances and covariances, or as

$$\hat{R} = \left(\frac{|\boldsymbol{W}|}{|\boldsymbol{W}^*|} \right)^{1/k}, \tag{40.44}$$

where $\boldsymbol{W}$ and $\boldsymbol{W}^*$ are consistent estimators of $\boldsymbol{\Sigma}$ and $\boldsymbol{\Sigma}^*$, or an estimator based on one variable only. Here, $\boldsymbol{\Sigma}$ denotes the variance–covariance matrix of the multivariate Pólya–Eggenberger distribution [see Eqs. (40.25) and (40.26)] and

$$\boldsymbol{\Sigma} = \left(\frac{a + nc}{a + c} \right) \boldsymbol{\Sigma}^* = R\boldsymbol{\Sigma}^* . \tag{40.45}$$

Janardan and Patil (1972) proposed a consistent estimator of R based on the variances of the sum of the components. Since ΣX_i is distributed as univariate Pólya $(a, \Sigma a_i; c; n)$, its variance equals $nRP(1 - P)$, where $P = \Sigma a_i/a$. Thus,

$$\hat{R} = \frac{nS^2}{\bar{x}(n - \bar{x})}, \tag{40.46}$$

where S^2 is the sample variance and $\bar{x}$ is the mean of the total number of observations.

Burr (1960) has discussed the use of this distribution in deriving the sampling distribution of Kendall's score S for a pair of tied rankings. Freeman and Halton (1951) have used the distribution for an exact analysis of contingency tables, goodness-of-fit and other related inferential problems.

6 SPECIAL CASES

Bose–Einstein Uniform Distribution

This distribution is obtained from the multivariate Pólya–Eggenberger distribution by taking $p_i = 1/(k + 1)$ for $i = 0, 1, \ldots, k$, and $\nu = 1/(k + 1)$, with the probability mass function given by

$$MP(\boldsymbol{x}; n) = \frac{1}{\binom{n+k}{k}}, \qquad x_i = 0, 1, \ldots, n, \ i = 0, 1, \ldots, k, \tag{40.47}$$

with $\sum_{i=0}^{k} x_i = n$. This has been used by Särndal (1964) as a prior distribution in the derivation of certain distributions through Bayes approach.

Multinomial Distribution

This distribution is obtained from the multivariate Pólya–Eggenberger distribution by taking $c = 0$. Since $p_i^{[x_i,0]} = p_i^{x_i}$, we have the probability mass function

$$MP(\boldsymbol{x};\boldsymbol{p};n) = \binom{n}{\boldsymbol{x}} \prod_{i=0}^{k} p_i^{x_i}, \; x_i = 0, 1, \ldots, n, \; i = 0, 1, \ldots, k, \tag{40.48}$$

so that $\sum_{i=0}^{k} x_i = n$. This is also evident from Eggenberger–Pólya's sampling scheme when $c = 0$, in which case the ball drawn is simply replaced but no more additional balls are added or removed.

Multivariate Hypergeometric Distribution

This distribution is obtained from the multivariate Pólya–Eggenberger distribution by taking $c = -1$, or by taking $\nu = -1/a$ and $p_i = a_i/a$, with the probability mass function

$$MP(\boldsymbol{x};\boldsymbol{a};n) = \frac{\prod_{i=0}^{k} \binom{a_i}{x_i}}{\binom{a}{n}}, \quad x_i = 0, 1, \ldots, n, \quad i = 0, 1, \ldots, k\,, \tag{40.49}$$

so that $\sum_{i=0}^{k} x_i = n$.

Multivariate Negative Hypergeometric Distribution

This distribution is obtained from the multivariate Pólya–Eggenberger distribution by taking $c = 1$, or by taking $\nu = 1/a$ and $p_i = a_i/a$, with the probability mass function

$$MP(\boldsymbol{x};\boldsymbol{a};n) = \frac{\prod_{i=0}^{k} \binom{-a_i}{x_i}}{\binom{-a}{n}}\,. \tag{40.50}$$

Dirichlet Distribution

The multivariate Pólya–Eggenberger distribution tends to the Dirichlet distribution [see Chapter 40 of Johnson and Kotz (1972)] as $n \longrightarrow \infty$ such that $\frac{x_i}{n}$ $(i = 1, 2, \ldots, k)$ remains finite and $r, r_1, \ldots, r_k$ remain fixed. The probability mass function is

$$MP(\boldsymbol{x};\boldsymbol{r};n) = \frac{\Gamma(r)}{\prod_{i=0}^{k} \Gamma(r_i)} \; \frac{\prod_{i=0}^{k} \left\{ \left(\frac{x_i}{n} + \frac{1}{n}\right) \cdots \left(\frac{x_i}{n} + \frac{r_i - 1}{n}\right) \right\}}{n^k \left(1 + \frac{1}{n}\right) \cdots \left(1 + \frac{r-1}{n}\right)}\,, \tag{40.51}$$

which, when letting $\frac{x_i}{n} \longrightarrow p_i$ and $dx_i = n\,dp_i$, gives the density function as

$$\frac{\Gamma(r)}{\prod_{i=0}^{k} \Gamma(r_i)} \prod_{i=0}^{k} p_i^{r_i - 1}\,. \tag{40.52}$$

Doubly Generalized Pólya Distribution

Morgenstern (1976) considered the joint distribution of the numbers of balls of colors $C_1, C_2, \ldots, C_k$ drawn from an urn with Eggenberger–Pólya system of replacement ($c + 1$ balls of color drawn) in r *successive* series $n_1, n_2, \ldots, n_r$ drawings. He referred to the joint distribution (of kr variables) as *doubly generalized Pólya distribution*; it could perhaps be called *generalized multivariate Pólya distribution*.

7 POLYTOMIC MULTIVARIATE PÓLYA DISTRIBUTIONS

Kriz (1972) provided a far-reaching generalization of multivariate Pólya–Eggenberger distributions which is described in this section.

Let us start with the following model:

V1: Given k urns.

V2: Each urn contains the same number, N, of balls.

V3: (U, V) means *there are U balls of color V*.

The ith urn contains the following balls:

$$(G_{1_i}, A_{1_i}), \dots, (G_{j_i}, A_{j_i}), \dots, (G_{m_i}, A_{m_i})$$

for $i = 1, 2, \dots, k$. From **V2**, we have

$$\sum_{j=1}^{m} G_{j_i} = N, \qquad i = 1, 2, \dots, k\,. \tag{40.53}$$

Also, no two colors in the urns are alike, so that there are $\sum_{i=1}^{k} m_i$ different colors in all.

V4: Balls are drawn from the urns in the following way:

V4.1: At each drawing, k balls are chosen, one from each urn.

V4.2: Before the next drawing, all balls are returned to their urns, together with g balls of the same color.

V4.3: This procedure is repeated n times.

V5: Now denote by $B = (b_{j_1 j_2 \dots j_k})$, $j_i = 1, 2, \dots, m_i$, the frequencies (more precisely, the *k-dimensional frequency matrix*) with which each possible combination of (colors of) the k balls has been drawn. This includes combinations with frequency zero (in common terms, *not occurring*).

We have $\prod_{i=1}^{k} m_i$ possible combinations and the relationship

$$\sum_{j_1=1}^{m_1} \cdots \sum_{j_k=1}^{m_k} b_{j_1 j_2 \dots j_k} = n\,. \tag{40.54}$$

We now determine the probability distribution of the multidimensional random variable $\boldsymbol{X}$ taking the values $\boldsymbol{B} = (b_{j_1 j_2 \dots j_k})$. To simplify the calculations, we introduce Figure 40.1 taken from Kriz (1972).

We now note that in this procedure the results of a drawing are recorded as follows:

a. At each drawing there will be an entry, for each urn, in the column corresponding to the color of the chosen ball.

Urn	1st urn		ith urn		kth urn
Color	$A_{1_1} \cdots A_{j_1} \cdots A_{m_1}$	$\cdots$	$A_{1_i} \cdots A_{j_i} \cdots A_{m_i}$	$\cdots$	$A_{1_k} \cdots A_{j_k} \cdots A_{m_k}$
Row					
1		$\cdots$		$\cdots$	
2		$\cdots$		$\cdots$	
.		$\cdots$		$\cdots$	
.		$\cdots$		$\cdots$	
.		$\cdots$		$\cdots$	
n		$\cdots$		$\cdots$	
	$S_{1_1} \cdots S_{j_1} \cdots S_{m_1}$	$\cdots$	$S_{1_i} \cdots S_{j_i} \cdots S_{m_i}$	$\cdots$	$S_{1_k} \cdots S_{j_k} \cdots S_{m_k}$

Figure 40.1. Possible results of a drawing in Kriz's (1972) model.

b. Each row will contain k entries.

c. There are n rows.

d. Each entry is a ratio, with the numerator being the number of balls of the chosen color in the urn before the drawing and the denominator being the total number of balls in the urn at that time. The ratio is therefore the probability that a ball of the relevant color would be chosen from that urn.

e. After the final (nth) drawing, the total numbers of drawings of each color are shown. Thus, S_{j_i} is the frequency with which a ball of color A_{j_i} is chosen from urn i in the n drawings.

Hence, according to assumption **V2** we have

$$\sum_{j=1}^{m} S_{j_i} = n \qquad \text{for all } i = 1, 2, \ldots, k\,. \tag{40.55}$$

The probability for one complete series of drawings is easily obtained because in each row (i.e., for each drawing) the k ratios are multiplied together to give the probability for that drawing, and these products are again multiplied together to give the probability for the whole series. Altogether there are nk factors, being all the entries described in **d.**

In the first column of the diagram there are S_{1_1} ratios (entries). The numerators of these ratios change only when a ball of color A_{1_1} is chosen. The product of these in column 1 is then

$$\prod_{r_{1_1}=1}^{S_{1_1}} \{G_{1_1} + (r_{1_1} - 1)g\}\,. \tag{40.56}$$

Similar formulas hold for other columns. For the first urn, the product of the numerators is

$$\prod_{j_1=1}^{m_1} \prod_{r_{j_1}=1}^{S_{j_1}} \{G_{j_1} + (r_{j_1} - 1)g\}\,. \tag{40.57}$$

From **V4** and (40.55), there are n balls chosen from the first urn. At the first drawing there are N balls in the urn, increasing by g at each drawing. The product of the denominators for the first urn is

$$\prod_{r=1}^{n} \{N + (r - 1)g\}\,. \tag{40.58}$$

From Eqs. (40.57) and (40.58), we obtain the product of all ratios for the first urn as

$$\frac{\prod_{j_1=1}^{m_1} \prod_{r_{j_1}=1}^{S_{j_1}} \{G_{j_1} + (r_{j_1} - 1)g\}}{\prod_{r=1}^{n} \{N + (r - 1)g\}}\,. \tag{40.59}$$

Similar expressions hold for all $i = 1, 2, \ldots, k$ urns. The probability for the whole set of drawings is thus

$$P = \frac{1}{\left[\prod_{r=1}^{n} \{N + (r - 1)g\}\right]^k} \prod_{i=1}^{k} \prod_{j_i=1}^{m_i} \prod_{r_{j_i}=1}^{S_{j_i}} \{G_{j_i} + (r_{j_i} - 1)g\}\,. \tag{40.60}$$

From **V5**, the n drawings can be arranged in $\prod_{i=1}^{k} m_i$ groups of $(b_{j_1 j_2 \ldots j_k})$ drawings with identical numbers of color combinations. The number of such possible combinations forms a multidimensional frequency matrix

$$\frac{n!}{\prod_{j_1=1}^{m_1} \prod_{j_2=1}^{m_2} \cdots \prod_{j_k=1}^{m_k} b_{j_1 j_2 \cdots j_k}!}\,. \tag{40.61}$$

Before evaluating the probability that $\boldsymbol{X} = \boldsymbol{B}$ we note that $\boldsymbol{X}$ and $\boldsymbol{B}$ are each k-dimensional matrices. Given the frequencies for each possible combination (i.e., $b_{j_1 j_2 \ldots j_k}$, $j_i = 1, 2, \ldots, m_i$), one can construct a k-dimensional diagram. The diagram in Figure 40.1 is represented on a two-dimensional sheet of paper and consequently the variables (urns) appear side-by-side, and only the "natural" events (i.e., those with $b_{j_i} \geq 1$) appear in this diagram. As we will now establish, this is permissible, as on the one hand all of the combinations not appearing in the diagram have frequency zero with the result that the diagram contains all the information, and on the other hand (since $0! = 1$) these contributions do not appear numerically in either (40.55) or (40.56).

We obtain from (40.60) and (40.61) the probability mass function for the random variable $\boldsymbol{X}$ as

$$\Pr[\boldsymbol{X} = \boldsymbol{B}] = \frac{n! \prod_{i=1}^{k} \prod_{j_i=1}^{m_i} \prod_{r_{j_i}=1}^{S_{j_i}} \{G_{j_i} + (r_{j_i} - 1)g\}}{\left\{\prod_{j_1=1}^{m_1} \prod_{j_2=1}^{m_2} \cdots \prod_{j_k=1}^{m_k} b_{j_1 j_2 \cdots j_k}!\right\} \left[\prod_{r=1}^{n} \{N + (r - 1)g\}\right]^k}\,, \tag{40.62}$$

where, as before,

- k: number of urns,
- N: number of balls in each urn before the first drawing,
- G_{j_i}: number of balls with color A_{j_i} in urn i before the first drawing,
- g: number of balls of the chosen color added to the respective urns,
- m_i: number of balls of different colors in urn i,
- n: number of drawings,
- S_{j_i}: frequency with which a ball of color A_{j_i} is drawn from the ith urn in n drawings.

We will call the distribution defined by (40.62) the *Polytomic Multivariate Pólya distribution* or the *PMP distribution*.

We now describe some special cases of the PMP distribution.

1. We first note that the Pólya distribution is a special case of the PMP distribution when $k = 1$ and $m = 2$. From (40.62) we have

$$\frac{n!}{\prod_{j=1}^{2} b_j!} \frac{\prod_{j=1}^{2} \prod_{r_j=1}^{S_j} \{G_j + (r_j - 1)g\}}{\prod_{r=1}^{n} \{N + (r-1)g\}}.$$

Since $m = 2$, there are only two possible combinations. If we take $b_1 = u$, then because $k = 1$, $b_2 = n - u$. Hence, S_j can only take values u and $n - u$, and so we can write

$$\Pr[X = u] = \binom{n}{u} \frac{\prod_{r=1}^{u} \{G + (r-1)g\} \prod_{r=1}^{n-u} \{N - G + (r-1)g\}}{\prod_{r=1}^{n} \{N + (r-1)g\}}, \tag{40.63}$$

which is just the Pólya distribution.

Since the Pólya distribution is itself a generalization of other distributions, we have

2. The binomial distribution is a special case of the PMP distribution when $k = 1$, $m = 2$, and $g = 0$.

3. The hypergeometric distribution is a special case of the PMP distribution when $k = 1, m = 2$, and $g = -1$.

4. Also, the multinomial distribution is a special case of the PMP distribution with $k = 1$ and $g = 0$. From (40.62) we have

$$\frac{n! \prod_{j=1}^{m} \prod_{r_j=1}^{S_j} G_j}{\prod_{j=1}^{m} b_j! \prod_{r=1}^{n} N}$$

and since $b_j = S_j$ for all $j = 1, 2, \ldots, n$, we also have $\prod_{r=1}^{n} N = N^n$ and because $n = \sum_{j=1}^{m} S_j$, also $N^n = \prod_{j=1}^{m} N^{S^j}$; we have

$$\frac{n!\prod_{j=1}^{m} G_j^{S_j}}{\prod_{j=1}^{m} S_j!\prod_{j=1}^{m} N^{S_j}}.$$

Defining $P_j = G_j/S_j$, we finally obtain

$$\Pr[\boldsymbol{X} = \boldsymbol{S}] = \frac{n!}{\prod_{j=1}^{m} S_j!}\prod_{j=1}^{m} P_j^{S_j}, \tag{40.64}$$

which is nothing but the multinomial distribution.

5. We now put $g = -1$ and $n = N$, meaning that drawings continue until all urns are empty. These kinds of models were used by R. A. Fisher in the construction of Fisher's exact probability tests—just, however, for the case $k = 2$ and $m = 2$, for which the distribution is equivalent to a hypergeometric distribution. This result—that for $k = 2, m = 2, n = N$, and $g = -1$, where the PMP distribution becomes a hypergeometric distribution—can be derived, requiring only slight mathematical proof, from the already noted fact that (see **3**) the hypergeometric distribution is obtained by putting $k = 1$, $m = 2$, and $g = -1$. It possibly contributes to overall understanding of the PMP distribution if one notes how a single distribution can be obtained in different ways from the polytomic diagram, as the following short example indicates.

The hypergeometric distribution (as a special case of the PMP distribution with $k = 1$, $m = 2$, and $g = -1$) applies to a model with a single urn ($k = 1$) containing N balls from which n are chosen (*without replacement*). This sampling process, however, can itself be represented in terms of a dichotomous variable with the two possible properties *appearing in the sample* and *not appearing* in the sample. When one speaks of $k = 1$ variables (urns) with the additional assumption that exactly n balls out of N are chosen, one is also speaking of $k = 2$ variables, with the additional condition (see **V4.1**) that all balls are chosen, so naturally one is led to the same conclusion.

For $N = n$ and $g = -1$, thus the PMP distribution gives the probability distribution of the $b_{j_1 j_2 \ldots j_k}$'s, conditional on the S_j's, and one can therefore interpret them as multivariate polytomic generalizations of Fisher's test. In this case, to be sure, we have to remember that for $k > 2$ and/or $m > 2$ the random variables are multivariate and so the construction of a test of significance—for example, the definition of "extreme cases"—encounters difficulties. We will not, however, consider this issue any further.

Finally, we collect our results in Table 40.1, which shows the values of m, k, and g and exhibits the PMP as a logical sequence.

In Table 40.1, only those cases of the PMP distribution corresponding to already established distributions are shown; however, Kriz (1972) is of the opinion that further special cases of the PMP distribution can be employed in models of interest in social sciences.

Table 40.1. Special Cases of the PMP Distribution

Distribution	m	k	g
Binomial	2	1	0
Hypergeometric	2	1	-1
Multinomial	Arbitrary	1	0
Fisher	2	2	-1 $(n = N)$
Burr (1960)	$m_1 = 3,\ m_2 = 4$	2	-1 $(n = N)$
Freeman and Halton (1951)	Arbitrary	Arbitrary	-1 $(n = N)$
Pólya	2	1	Arbitrary
PMP	Arbitrary	Arbitrary	Arbitrary

8 TRUNCATED MULTIVARIATE PÓLYA–EGGENBERGER DISTRIBUTIONS

Gerstenkorn (1977a–c, 1978) restricted the range of values of one of the random variables (X_i) in the basic multivariate Pólya–Eggenberger model so that $m + 1 \le X_i \le n$, where m is an integer. In view of the condition $\sum_{i=0}^{k} X_i = n$, the restriction on X_i readily yields

$$0 \le X_j \le n - (m+1) \quad \text{for } j = 0, 1, 2, \dots, i-1, i+1, \dots, k\,.$$

The probability function of the *truncated* multivariate Pólya–Eggenberger distribution, when one of the variables X_i is truncated below by m, is

$$P^T(x_1, x_2, \dots, x_k) = \frac{P(x_1, x_2, \dots, x_k)}{1 - F_i(m)}\,, \tag{40.65}$$

where $P(x_1, x_2, \dots, x_k)$ is the probability mass function of the multivariate Pólya–Eggenberger distribution and

$$F_i(c) = \sum_{x_i=0}^{m} \binom{n}{x_i} \frac{p_i^{[x_i, -a]}}{1^{[n,-a]}} (1 - p_i)^{[n-x_i, -a]}\,. \tag{40.66}$$

Note that

$$\binom{n}{x_i} \frac{p_i^{[x_i, -a]}}{1^{[n,-a]}} (1 - p_i)^{[n-x_i, -a]} \tag{40.67}$$

is the marginal distribution of the *untruncated* X_i. Similarly, by restricting the values of ℓ random variables $X_1, X_2, \dots, X_\ell$ $(1 \le \ell \le k)$ by the condition

$$m_i + 1 \le X_i \le n \ \text{ for } i = 1, 2, \dots, \ell\,,$$

where $\sum_{i=1}^{\ell} m_i < n$, we have the probability mass function of the corresponding truncated multivariate Pólya–Eggenberger distribution as

$$P^T(x_1, x_2, \ldots, x_k) = \frac{P(x_1, x_2, \ldots, x_k)}{1 - F_{1,2,\ldots,\ell}(m_1, m_2, \ldots, m_\ell)} . \tag{40.68}$$

In the above,

$$F_{1,2,\ldots,\ell}(m_1, m_2, \ldots, m_\ell) = \sum_{x_1, x_2, \ldots, x_\ell = 0}^{m_1, m_2, \ldots, m_\ell} P_{1,2,\ldots,\ell}(x_1, x_2, \ldots, x_\ell) , \tag{40.69}$$

where

$$P_{1,2,\ldots,\ell}(x_1, x_2, \ldots, x_\ell) = \frac{n^{(x_1+x_2+\cdots+x_\ell)}}{1^{[n,-a]}} \left\{ \prod_{i=1}^{\ell} \frac{p_i^{[x_i,-a]}}{x_i!} \right\} \times \left(1 - \sum_{i=1}^{\ell} p_i\right)^{[n-\sum_{i=1}^{\ell} x_i, -a]} \tag{40.70}$$

is the probability mass function of the ℓ-dimensional marginal distribution of the multivariate Pólya–Eggenberger distribution. See also Dyczka (1973), who has described the corresponding urn models.

It should be noted that although these results are natural, their derivation is far from being simple and requires careful consideration of the regions of variation of both restricted and originally unrestricted variables. Observe that

$$P^T(x_1, x_2, \ldots, x_k) \geq 0 \qquad \text{and} \qquad \sum_{R_{k+1,n}} P^T(x_1, x_2, \ldots, x_k) = 1 ,$$

where $R_{k+1,n}$ denotes the class of all $(k+1)$-tuples $(x_0, x_1, \ldots, x_k)$ of non-negative integers satisfying

$$m + 1 \leq x_i \leq n \quad \text{for } i = 0, 1, 2, \ldots, \ell,$$
$$0 \leq x_i \leq n - (m+1) \quad \text{for } i = \ell + 1, \ldots, k,$$

and

$$\sum_{i=0}^{k} x_i = n.$$

With

$$\mu'_{(\boldsymbol{r})} = n^{(r)} \left\{ \prod_{i=1}^{k} p_i^{[r_i,-a]} \right\} \Big/ 1^{[r,-a]}, \qquad r = \sum_{i=1}^{k} r_i , \tag{40.71}$$

being the rth descending factorial moment of the multivariate Pólya–Eggenberger distribution, the corresponding expression for the truncated distribution is

$$\mu'_{(r)T} = \frac{\mu'_{(r)}}{1 - F_{1,2,\dots,\ell}(m_1, m_2, \dots, m_\ell)} (1 - H), \tag{40.72}$$

where

$$H = \frac{1}{(1 + ra)^{[n-r,-a]}} \sum_{i_1, i_2, \dots, i_\ell = 0}^{m_1 - r_1, \dots, m_\ell - r_\ell} \frac{(n - r)^{(i_1 + \cdots + i_\ell)}}{i_1! i_2! \cdots i_\ell!} \left\{ \prod_{j=1}^{\ell} (p_j + r_j a)^{[i_j, -a]} \right\}$$
$$\times \left(1 - \sum_{j=1}^{\ell} p_j + a \sum_{j=\ell+1}^{k} r_j \right)^{[n-r-\Sigma_{j=1}^{\ell} i_j, -a]}. \tag{40.73}$$

Of course, from these results the corresponding expressions for the truncated multivariate binomial distribution (corresponding to $a = 0$) and for the truncated multivariate hypergeometric distribution can be easily derived.

9 MULTIVARIATE INVERSE PÓLYA–EGGENBERGER DISTRIBUTIONS

In the sampling situation described earlier in Section 2, suppose that there are initially $a_0, a_1, \dots, a_k$ balls of colors $C_0, C_1, \dots, C_k$ in the urn, respectively, and that the sampling continues until a specified number m_0 of balls of color C_0 have been chosen. The joint distribution of the numbers $X_1, X_2, \dots, X_k$ of balls of colors $C_1, \dots, C_k$, respectively, drawn when this requirement is achieved can be derived in the following manner.

When $X_i = x_i$ (for $i = 1, 2, \dots, k$) and $x_\bullet = \sum_{i=1}^{k} x_i$, there must have been a total of $(m_0 + x_\bullet)$ balls drawn, with a C_0 color ball being chosen at the final drawing; the first $(m_0 + x_\bullet - 1)$ drawings must include $m_0 - 1, x_1, x_2, \dots, x_k$ balls of colors $C_0, C_1, \dots, C_k$, respectively. Since there are $a_0 + (m_0 - 1)c, a_1 + x_1 c, \dots, a_k + x_k c$ balls of colors $C_0, C_1, \dots, C_k$, respectively, in the urn when the final drawing takes place, the joint probability mass function is [Janardan and Patil (1970)]

$$P(x_1, x_2, \dots, x_k) = \Pr\left[\bigcap_{i=1}^{k} (X_i = x_i) \right]$$
$$= \left[\binom{m_0 - 1 + x_\bullet}{m_0 - 1, x_1, \dots, x_k} \left\{ \prod_{i=1}^{k} a_i^{[x_i, c]} \right\} a_0^{[m_0 - 1, c]} \Big/ a^{[m_0 - 1 + x_\bullet, c]} \right]$$
$$\times \frac{a_0 + (m_0 - 1)c}{a + (m_0 - 1 + x_\bullet)c}, \tag{40.74}$$

where $a = \sum_{i=0}^{k} a_i$ and $-cx_i \le a_i$ $(i = 1, 2, \ldots, k)$. Of course, we also have $-c(m_0 - 1) < a_0$. The distribution (40.74) is the *multivariate inverse Pólya–Eggenberger distribution*. The special cases $c = -1$ and $c = 0$ correspond to the multivariate inverse hypergeometric distribution and the negative multinomial distribution, respectively.

The $\boldsymbol{r}$th (descending) factorial moment of distribution (40.74) is [Gerstenkorn (1977c)]

$$\mu'_{(\boldsymbol{r})}(\boldsymbol{X}) = \frac{m_0^{(r)}}{(a - a_0)^{[r]}} \prod_{i=1}^{k} a_i^{[r_i, c]} \tag{40.75}$$

with $r = \sum_{i=1}^{k} r_i$. From (40.75), we obtain in particular that

$$E[X_i] = \frac{m_0}{a - a_0}\, a_i\,, \tag{40.76}$$

$$\begin{aligned} \operatorname{var}(X_i) &= \frac{m_0(m_0 - 1)}{(a - a_0)(a - a_0 + 1)}\, a_i(a_i + c) \\ &\quad + \frac{m_0}{a - a_0}\, a_i - \left(\frac{m_0}{a - a_0}\, a_i\right)^2 \\ &= \frac{m_0 a_i}{(a - a_0)^2(a - a_0 + 1)} \{(a - a_0 + m_0)(a - a_0 - a_i) \\ &\quad + (c - 1)(m_0 - 1)(a - a_0)\}\,, \end{aligned} \tag{40.77}$$

$$E[X_i X_j] = \frac{m_0(m_0 - 1)}{(a - a_0)(a - a_0 + 1)}\, a_i a_j\,, \tag{40.78}$$

whence

$$\operatorname{cov}(X_i, X_j) = -\frac{m_0 a_i a_j}{(a - a_0)^2(a - a_0 + 1)} (a - a_0 + m_0) \tag{40.79}$$

and

$$\operatorname{corr}(X_i, X_j) = -\sqrt{\frac{a_i a_j}{\left\{(a - a_0 - a_i) + \frac{(c-1)(m_0-1)(a-a_0)}{a-a_0+m_0}\right\}\left\{(a - a_0 - a_j) + \frac{(c-1)(m_0-1)(a-a_0)}{a-a_0+m_0}\right\}}}\,. \tag{40.80}$$

Gerstenkorn (1977c) also obtained expressions for the incomplete moments from which the moments of the truncated multivariate inverse Pólya–Eggenberger distributions can be derived.

Sibuya (1980) derived the *multivariate digamma distribution* with probability mass function

$$P(\boldsymbol{x}) = \frac{1}{\psi(\alpha + \gamma) - \psi(\gamma)} \frac{(x-1)!}{(\alpha + \gamma)^{[x]}} \prod_{i=1}^{k} \left\{ \frac{\alpha_i^{[x_i]}}{x_i!} \right\}, \tag{40.81}$$

$$\alpha_i, \gamma > 0, \; x = \sum_{i=1}^{k} x_i > 0, \; x_i = 0, 1, \ldots,$$

as the limit of a truncated [by omission of $(0, 0, \ldots, 0)$] multivariate Pólya–Eggenberger distribution

$$P(\boldsymbol{x}) = \frac{\gamma^{[\alpha]}\gamma^{[\beta]}}{\gamma^{[\alpha+\beta]}} \cdot \frac{\beta^{[x]}}{(\alpha + \beta + \gamma)^{[x]}} \prod_{i=1}^{k} \left\{ \frac{\alpha_i^{[x_i]}}{x_i!} \right\}, \tag{40.82}$$

$$x = \sum_{i=1}^{k} x_i > 0, \; x_i = 0, 1, \ldots, \alpha_i > 0,$$

as $\beta \to 0$. $X = \sum_{i=1}^{k} X_i$ in (40.81) has the univariate digamma distribution; see Chapter 11, p. 436. The conditional joint distribution of $X_1, X_2, \ldots, X_k$, given $X = \sum_{i=1}^{k} X_i$, is the singular multivariate hypergeometric distribution with probability mass function

$$P(\boldsymbol{x}) = \frac{\prod_{i=1}^{k} \binom{-\alpha_i}{x_i}}{\binom{-\alpha}{x}} = \binom{x}{\boldsymbol{x}} \left\{ \prod_{i=1}^{k} \alpha_i^{[x_i]} \right\} \Big/ \alpha^{[x]}. \tag{40.83}$$

Sibuya (1980) has shown that the multivariate digamma distribution is unimodal with mode at $(0, \ldots, 0, 1, 0, \ldots, 0)$, where 1 is the value of x_i with i defined as the least i for which $\alpha_i = \max(\alpha_1, \ldots, \alpha_m)$. Properties analogous to those of univariate digamma distributions include:

1. The distribution (40.81) can be obtained by compounding the multivariate logarithmic series distribution with probability mass function

$$P(\boldsymbol{x}) = -\frac{(x-1)!}{\log(1-\theta)} \prod_{i=1}^{k} \left\{ \frac{\theta_i^{x_i}}{x_i!} \right\}, \; x = \sum_{i=1}^{k} x_i > 0, \; x_i = 0, 1, \ldots,$$

$$\theta_i > 0, \; \theta = \sum_{i=1}^{k} \theta_i < 1, \tag{40.84}$$

with an "end-accented" Dirichlet distribution for $\alpha_1, \ldots, \alpha_k$ with probability density function

$$\frac{1}{\{\psi(\alpha+\gamma)-\psi(\gamma)\}B(\alpha,\gamma)}\,\frac{\Gamma(\alpha)}{\prod_{i=1}^{k}\Gamma(\alpha_i)}\left\{-\log\left(1-\sum_{i=1}^{k}t_i\right)\right\}$$
$$\times\left(1-\sum_{i=1}^{k}t_i\right)^{\gamma-1}\prod_{i=1}^{k}t_i^{\alpha_i-1}. \tag{40.85}$$

2. If $\alpha_i \to \infty$ and $\gamma \to \infty$ with $\frac{\alpha_i}{\alpha_i+\gamma} \to \theta_i$ $(i = 1, 2, \ldots, k)$, then the multivariate digamma distribution in (40.81) tends to the multivariate logarithmic series distribution in (40.84).
3. If $\alpha_i \to 0$ $(i = 1, 2, \ldots, k)$ with $\frac{\alpha_i}{\alpha} \to \tau_i$ and γ constant, the limit (a *multivariate trigamma*) distribution is degenerate.

Tripsiannis and Philippou (1994) describe *multivariate inverse Pólya distributions of order s* (see Chapter 10, Section 6). These arise from the requirement that the sampling continues until m_0 *nonoverlapping* runs of length s, composed entirely of balls of color C_0, are achieved. Under these conditions, Tripsiannis and Philippou (1994) show that the joint probability mass function of $X_1, X_2, \ldots, X_k$ is

$$\Pr\left[\bigcap_{i=1}^{k}(X_i = x_i)\right] = \sum_{\sum_{j=1}^{s} j\, y_{ij}=x_i}\cdots\sum \binom{y+m_0-1}{m_0-1, y_{11}, \ldots, y_{ks}}$$
$$\times\left\{\prod_{i=1}^{k} a_i^{[y_i,c]}\right\}\frac{a_0^{[y-x+m_0 s,c]}}{a^{[y+m_0 s,c]}}, \tag{40.86}$$

where

$$y_i = \sum_{j=1}^{s} y_{ij}, \quad y = \sum_{i=1}^{k} y_i, \quad \text{and} \quad x = \sum_{i=1}^{k} x_i.$$

If $s = 1$, the distribution becomes an ordinary multivariate inverse Pólya–Eggenberger distribution; if $m_0 = 1$, the distribution becomes a *generalized Waring distribution of order s* [Panaretos and Xekalaki (1986)]; if $c = -1$, we have a *multivariate inverse hypergeometric distribution of order s*. The univariate distribution $(k = 1)$ is the *β-compound negative multinomial distribution* of Mosimann (1963). For some properties of distributions of order s, one may also refer to Hirano (1986).

Philippou and Tripsiannis (1991) had earlier derived distributions of the form (40.86) from the following model. An urn contains $s(a_0 + a_1 + \cdots + a_k)$ balls of which there are sa_0 balls of color C_0 and a_i balls of color C_{ij} for each $j = 1, 2, \ldots, s$ and $i = 1, 2, \ldots, k$. A ball is drawn at random from the urn and replaced. If it is of color C_{ij} for any $(i = 1, 2, \ldots, k)$ $j = 1, 2, \ldots, s$, then c balls of each of colors $C_{i1}, C_{i2}, \ldots, C_{is}$ are added to the urn. This procedure is repeated until m_0 balls of color C_0 have appeared. The joint distribution of $X_1, X_2, \ldots, X_k$, where X_i is the total number of occasions when a ball of one of colors $C_{i1}, C_{i2}, \ldots, C_{is}$ has been drawn, is then of the form (40.86).

An alternative form

$$P(x_1, x_2, \ldots, x_k) = \sum_{\sum_{j=1}^{s} j y_{ij} = x_i} \cdots \sum \frac{\Gamma(m_0 + y)}{\Gamma(m_0) \prod_{i=1}^{k} \prod_{j=1}^{s} \Gamma(y_{ij} + 1)}$$

$$\times \frac{\Gamma(\alpha)\Gamma(\alpha_0 + y - x + m_0 s)}{\Gamma(\alpha + y + m_0 s)\Gamma(\alpha_0)} \prod_{i=1}^{k} \frac{\Gamma(\alpha_i + y_i)}{\Gamma(\alpha_i)} \tag{40.87}$$

is equivalent to (40.86) with $\alpha_i = a_i/c$ and $\alpha = a/c$. As shown by Philippou, Antzoulakos, and Tripsiannis (1988), the formula in (40.87) can be reached starting from a Dirichlet $(\alpha_0, \alpha_1, \ldots, \alpha_k)$ compound of multivariate negative binomial distributions of order s. Moments of this distribution may be derived directly; for example,

$$E[X_i] = m_0 \alpha_i \sum_{j=0}^{m_0} j(\alpha - 1)^{m_0 - j} / (\alpha_0 - 1)^{m_0 - j + 1}, \quad \alpha_0 > m_0 . \tag{40.88}$$

As α, α_i, and $m_0 \to \infty$ with $m_0\alpha_i/\alpha \to \lambda_i$, $\alpha_i/\alpha \to 0$, and $m_0/\alpha \to 0$, the distribution (40.87) tends to a multivariate Poisson with parameters $\lambda_1, \lambda_2, \ldots, \lambda_k$ (see Chapter 37). If $\alpha_i/\alpha \to q_i$ $(0 < q_i < 1, \sum_{i=1}^{k} q_i < 1)$ as $\alpha_i \to \infty$ and $\alpha_0 \to k$, the limit distribution is a multivariate negative binomial distribution (see Chapter 36) with probability mass function

$$P(x_1, x_2, \ldots, x_k) = \left(1 - \sum_{i=1}^{k} q_i\right)^{x - y + m_0 s} \prod_{i=1}^{k} q_i^{y_i} . \tag{40.89}$$

Distributions discussed by Philippou and Tripsiannis (1991) are called *multivariate Pólya distributions of order s*.

Tripsiannis (1993) has also discussed an even more elaborate modification of the distribution, with the numbers of balls of colors $C_{i1}, C_{i2}, \ldots, C_{is}$ added when the chosen ball has any one of these colors depending on the second subscript; that is, c_{ij} balls of colors C_{ij} $(j = 1, 2, \ldots, s)$ are added. These distributions discussed by Tripsiannis (1993), called *modified multivariate Pólya distributions of order s*, have potential applications in several fields (e.g., contagious diseases and growth population) whenever there are multiple groups of items and one is interested in the distribution of the total number of items. The modified multivariate Pólya distributions of order s (Type II) introduced by Tripsiannis (1993), denoted by $\text{MoMP}_{s,\text{II}}(n; d; c_0, c_{11}, \ldots, c_{ks})$, include the following distributions as special cases:

(i) the Type II modified multivariate hypergeometric distribution of order s [Tripsiannis (1993)] when $d = -ks$;

(ii) the Type II modified multivariate negative hypergeometric distribution of order s [Tripsiannis (1993)] when $d = ks$;

(iii) the modified multivariate Pólya distribution [Janardan and Patil (1974); also see Section 9] when $s = 1$;

(iv) the multiparameter Pólya distribution of order s [Tripsiannis (1993)] when $k = 1$;

(v) the Type II Pólya distribution of order s [Philippou, Tripsiannis, and Antzoulakos (1989)] when $k = s = 1$; and

(vi) the Pólya distribution when $k = 1$ and $c_i = c/s$.

Tripsiannis (1993) also introduced the *modified multivariate inverse Pólya distributions of order s* (Type II), denoted by $\text{MoMIP}_{s,\text{II}}(r; d; c_0, c_{11}, \ldots, c_{ks})$, which include the following distributions as special cases:

(i) the Type II modified multivariate inverse hypergeometric distribution of order s [Tripsiannis (1993)] when $d = -ks$;

(ii) the modified multivariate inverse Pólya distribution [Janardan and Patil (1974); see also Section 9] when $s = 1$;

(iii) the multiparameter inverse Pólya distribution of order s [Tripsiannis (1993)] when $k = 1$;

(iv) the inverse Pólya distribution when $k = s = 1$; and

(v) the Type II inverse Pólya distribution of order s [Philippou, Tripsiannis, and Antzoulakos (1989)] when $k = 1$ and $c_i = c/s$.

Tripsiannis and Philippou (1994) have similarly defined *Type I multivariate inverse Pólya distributions of order s* and discussed their properties.

For more details on multivariate distributions of order s, one may refer to Chapter 42.

10 MULTIVARIATE MODIFIED PÓLYA DISTRIBUTIONS

10.1 Genesis

Suppose an urn contains $a_1, a_2, \ldots, a_k$ ($\sum_{i=1}^k a_i = a$) balls of k different colors as well as a_0 white balls. We draw a ball: if it is colored, it is replaced with additional $c_1, c_2, \ldots, c_k$ balls of the k-colors; but if it is white, it is replaced with additional $d = \sum_{i=1}^k c_i$ white balls, where $\frac{c_i}{a_i} = \frac{d}{a} = \frac{1}{K}$, a constant. This is repeated n times. If X_i denotes the number of balls of ith color and X_0 denotes the number of white balls such that $\sum_{i=0}^k X_i = n$, then the random variable $\boldsymbol{X} = (X_0, X_1, \ldots, X_k)'$ has the joint probability mass function

$$MMP(\boldsymbol{x}; d, n, \boldsymbol{a}) = \binom{n}{\boldsymbol{x}} \frac{a_0^{[x_0,d]}\, a^{[x,d]}}{(a_0 + a)^{[n,d]}} \prod_{i=1}^{k} \left(\frac{a_i}{a}\right)^{x_i}, \tag{40.90}$$

where $x_i = 0, 1, \ldots, n$ for $i = 1, 2, \ldots, k$ so that $x = \sum_{i=1}^k x_i$. This result was derived by Sibuya, Yoshimura, and Shimizu (1964).

Janardan and Patil (1974) pointed out an alternate genesis of the multivariate modified Pólya distribution as follows. Let $\boldsymbol{X} = (X_1, X_2, \ldots, X_k)'$ and $X = \sum_{i=1}^k X_i$.

If the conditional distribution of $\boldsymbol{X}$, given X, is singular multinomial distribution with parameters x and $\boldsymbol{\phi} = (\phi_1 \phi_2 \dots \phi_k)'$ and the distribution of X is univariate Pólya with parameters n, ν, and p_0, then the distribution of $(\boldsymbol{X}, X)'$ is multivariate modified Pólya as in (40.90). Note that

$$P_{\boldsymbol{X},X}(\boldsymbol{x}, x) = P_{\boldsymbol{X}|X}(\boldsymbol{x}|x)\, P_X(x)$$

$$= \frac{x!}{\prod_{i=1}^k x_i!} \left\{ \prod_{i=1}^k \phi_i^{x_i} \right\} \binom{-q_0/\nu}{x} \binom{-p_0/\nu}{n-x} \Big/ \binom{-1/\nu}{n}, \tag{40.91}$$

where $q_0 = 1 - p_0$. Equation (40.91) can be easily shown to be equivalent to the probability mass function (40.90) by taking $\phi_i = \frac{p_i}{1-p_0}$, $p_i = \frac{a_i}{a_0+a}$ and $\nu = \frac{d}{a_0+a}$.

10.2 Properties

The probability generating function of the distribution (40.90) is

$$G(\boldsymbol{t}) = \frac{(-p_0/\nu)^{(n)}}{(-1/\nu)^{(n)}}\, {}_2F_1\left(-n; \frac{q_0}{\nu}, -n+1; z\right), \tag{40.92}$$

where $z = \sum_{i=1}^k \phi_i t_i$ with ϕ_i, p_0, and ν as defined above, and ${}_2F_1(\cdot)$ is as defined in (1.104). Thus, the probability generating function is the same as the probability generating function of univariate Pólya distribution (with parameters n, ν, and p_0) evaluated at z.

The factorial moment generating function of the distribution (40.90) is

$$\frac{(-p_0/\nu)^{(n)}}{(-1/\nu)^{(n)}}\, {}_2F_1(-n; \frac{q_0}{\nu}, \frac{p_0}{\nu} - n + 1; v), \tag{40.93}$$

where $v = 1 + \sum_{i=1}^k \phi_i t_i$. Thus, the $\boldsymbol{r}$th descending factorial moment is

$$n^{(r)}\, a^{[r,d]} \prod_{i=1}^k \left(\frac{a_i}{a}\right)^{r_i}, \qquad r = \sum_{i=1}^k r_i. \tag{40.94}$$

If a random variable $\boldsymbol{X}$ has the multivariate modified Pólya distribution (40.90), then:

1. The sum of the components of $\boldsymbol{X}$ follows univariate Pólya distribution with parameters n, d, and a.
2. The conditional distribution of $\boldsymbol{X}$, given $\sum_{i=1}^k X_i = x$, is a singular multinomial with parameters x and $(\phi_1, \dots, \phi_k)$, where $\phi_i = \frac{p_i}{1-p_0}$ and $p_i = \frac{a_i}{a_0+a}$.
3. The random variable $(Z_1, \dots, Z_\ell)'$ follows multivariate modified Pólya with parameters n, d, and $\boldsymbol{b}$, where $Z_j = \Sigma_j X_i$, $b_j = \Sigma_j a_i$ with Σ_j denoting the summation taken over $i \in G_j$ for $j = 1, 2, \dots, \ell$ such that $\bigcup_{j=1}^{\ell} G_j = \{1, 2, \dots, k\}$ and $G_j \cap G_{j'} = \emptyset$ (empty set) for $j \neq j' = 1, 2, \dots, \ell$.

4. The marginal distribution of $\boldsymbol{X}_1 = (X_1, \ldots, X_\ell)'$, $1 \le \ell \le k$, is

$$\binom{n}{\boldsymbol{x}_1} \frac{a_0^{[n-y,d]}\, a^{[y,d]}}{(a_0+a)^{[n,d]}} \prod_{i=1}^{\ell} \left(\frac{a_i}{a}\right)^{x_i}$$

$$\times\, {}_2F_1\left(-n+y;\ \frac{a+yd}{d},\ -\frac{a_0}{d}+n-y-1;\ \sum_{i=\ell+1}^{k} \frac{a_i}{a}\right), \tag{40.95}$$

where $y = \sum_{i=1}^{\ell} x_i$. Further, the conditional distribution of $\boldsymbol{X}_2 = (X_{\ell+1}, \ldots, X_k)'$, given $\boldsymbol{X}_1$, is

$$\binom{n}{\boldsymbol{x}_2} \frac{a_0^{[n-x,d]}\, a^{[x,d]}}{a_0^{[n-y,d]}\, a^{[y,d]}\, {}_2F_1(\cdots)} \prod_{i=\ell+1}^{k} \left(\frac{a_i}{a}\right)^{x_i}, \tag{40.96}$$

where $x = \sum_{i=\ell+1}^{k} x_i$.

11 MULTIVARIATE MODIFIED INVERSE PÓLYA DISTRIBUTIONS

11.1 Genesis

In the urn sampling scheme described in the last section, suppose that we draw balls until we observe exactly w white balls following the same procedure. Let X_i denote the number of balls of ith color drawn in the process of getting w white balls. Then the joint probability mass function of the random variable $\boldsymbol{X} = (X_1, \ldots, X_k)'$ is

$$MMIP(\boldsymbol{x}; d, w, \boldsymbol{a}) = \binom{w+x-1}{\boldsymbol{x}} \frac{a_0^{[w,d]}\, a^{[x,d]}}{(a_0+a)^{[w+x,d]}} \prod_{i=1}^{k} \left(\frac{a_i}{a}\right)^{x_i}, \tag{40.97}$$

where $x_i = 0, 1, 2, \ldots$ for $i = 1, 2, \ldots, k$, $x = \sum_{i=1}^{k} x_i$, and $a = \sum_{i=1}^{k} a_i$.

The case $k = 1$ corresponds to the univariate inverse Pólya distribution (see Chapter 6, Section 2.4). A derivation of the multivariate modified inverse Pólya distribution in (40.97) is due to Sibuya, Yoshimura, and Shimizu (1964).

Janardan and Patil (1974) pointed out an alternate genesis of the multivariate modified inverse Pólya distribution as follows. Let $\boldsymbol{X} = (X_1, X_2, \ldots, X_k)'$ and $X = \sum_{i=1}^{k} X_i$. If the conditional distribution of $\boldsymbol{X}$, given X, is the singular multinomial distribution and the distribution of X is a univariate inverse Pólya distribution with parameters w, ν, and p_0, then the distribution of $(\boldsymbol{X}, X)'$ is multivariate modified inverse Pólya as in (40.97). In fact,

$$\begin{aligned} P_{\boldsymbol{X},X}(\boldsymbol{x}, x) &= P_{\boldsymbol{X}|X}(\boldsymbol{x}|x)\, P_X(x) \\ &= \frac{x!}{\prod_{i=1}^{k} x_i!} \left\{\prod_{i=1}^{k} \phi_i^{x_i}\right\} \frac{w}{w+x} \frac{\binom{-p_0/\nu}{w}\binom{-q_0/\nu}{x}}{\binom{-1/\nu}{w+x}}, \end{aligned} \tag{40.98}$$

where $q_0 = 1 - p_0$. The probability mass function (40.98) is equivalent to the probability mass function (40.97) with $\phi_i = \frac{p_i}{1-p_0}$, $p_i = \frac{a_i}{a_0+a}$, and $\nu = \frac{d}{a_0+a}$.

11.2 Properties

The probability generating function of the multivariate modified inverse Pólya distribution is

$$G(\boldsymbol{t}) = \frac{(-p_0/\nu)^{(w)}}{(-1/\nu)^{(w)}}\, {}_2F_1\left(w; \frac{q_0}{\nu}, \frac{1}{\nu} + w; z\right), \tag{40.99}$$

where $z = \sum_{i=1}^{k} \phi_i t_i$. Thus, the probability generating function is the same as the probability generating function of univariate inverse Pólya distribution (with parameters w, ν and p_0) evaluated at z.

The factorial moment generating function is

$$\frac{(-p_0/\nu)^{(w)}}{(-1/\nu)^{(w)}}\, {}_2F_1\left(w; \frac{q_0}{\nu}, \frac{1}{\nu} + w; v\right), \tag{40.100}$$

where, as in the last section, $v = 1 + \sum_{i=1}^{k} \phi_i t_i$. Thus, the $\boldsymbol{r}$th descending factorial moment is

$$\frac{w^{[r]}\, a^{[r,d]}}{(a_0 - rd)^{[r,d]}} \prod_{i=1}^{k} \left(\frac{a_i}{a}\right)^{r_i}, \tag{40.101}$$

where $r = \sum_{i=1}^{k} r_i$.

If a random variable $\boldsymbol{X}$ has the multivariate modified inverse Pólya distribution (40.98), then:

1. The sum of the components of $\boldsymbol{X}$ follows univariate inverse Pólya distribution with parameters w, d, and a.
2. The conditional distribution of $\boldsymbol{X}$, given $\sum_{i=1}^{k} X_i = x$, is singular multinomial with parameters x and $(\phi_1, \ldots, \phi_k)$, where $\phi_i = \frac{p_i}{1-p_0}$ and $p_i = \frac{a_i}{a_0+a}$.
3. The random variable $(Z_1, \ldots, Z_\ell)'$ follows multivariate modified inverse Pólya with parameters w, d, and $\boldsymbol{b}$, where $Z_j = \Sigma_j X_i$ and $b_j = \Sigma_j a_i$ with Σ_j denoting the summation taken over $i \in G_j$ for $j = 1, 2, \ldots, \ell$ such that $\bigcup_{j=1}^{\ell} G_j = \{1, 2, \ldots, k\}$ and $G_j \cap G_{j'} = \emptyset$ for $j \neq j' = 1, 2, \ldots, \ell$.
4. The marginal distribution of $\boldsymbol{X}_1 = (X_1, \ldots, X_\ell)'$, $1 \leq \ell \leq k$, is

$$\binom{w + y - 1}{x_1, \ldots, x_\ell} \frac{a_0^{[w,d]}\, a^{[y,d]}}{(a_0 + a)^{[w+y,d]}} \prod_{i=1}^{\ell} \left(\frac{a_i}{a}\right)^{x_i}$$
$$\times\, {}_2F_1\left(w + y; \frac{a + yd}{d}, \frac{a_0 + a}{d} + w + y; \sum_{i=\ell+1}^{k} \frac{a_i}{a}\right), \tag{40.102}$$

where $y = \sum_{i=1}^{\ell} x_i$. Further, the conditional distribution of $\boldsymbol{X}_2 = (X_{\ell+1}, \ldots, X_k)'$, given $\boldsymbol{X}_1$, is

$$\binom{w+y+z-1}{x_{\ell+1},\ldots,x_k} \frac{(a+yd)^{[z,d]}}{(a_0+a+\overline{w+y}\,d)^{[z,d]}\,{}_2F_1(\cdots)} \prod_{i=\ell+1}^{k} \left(\frac{a_i}{a}\right)^{x_i}, \tag{40.103}$$

where $y = \sum_{i=1}^{\ell} x_i$ and $z = \sum_{i=\ell+1}^{k} x_i$.

The multivariate Pólya–Eggenberger distributions have also found a number of interesting applications. Wagner and Taudes (1991), for example, used it in a model incorporating both marketing and consumer-specific variables in the analysis of new product purchase. Incidentally, they observed that the use of the traditional negative binomial-Dirichlet model (see Chapter 36) of brand choice and purchase incidence in this context has a drawback in the sense that it postulates stationary market conditions; see also Wagner and Taudes (1986) on this topic.

Athreya and Ney (1972) described a generalized Pólya urn model. Durham and Yu (1990) studied some special cases in detail. They considered the situation described in Section 2, when $c = 1$ but with probability p_i of adding a ball when color C_i is chosen. (In Section 2, of course, $p_i = 1$ for all i.) They showed that if $p_i > p_j$, and N_{ij} is the number of times color C_j has been chosen before the n_ith choice of C_i, then $N_{ij}/n_i^{p_j/p_i}$ has a proper limiting distribution as $n_i \to \infty$.

Li, Durham, and Flournoy (1996) showed that if there is a unique maximum, p_h, among $p_1, \ldots, p_k$, then the proportion P_i of balls of color C_i in the urn tends to 1 for P_h and 0 for P_i as the number of trials increases. They also showed that if there are r (≥ 2) equal maximum values $p_{h_1}, \ldots, p_{h_r}$ among $p_1, \ldots, p_k$, then the limiting distribution of P_{h_g} is beta$\left(a_{h_g}, \sum_{j\neq 1,\ldots,r} a_{h_j}\right)$. Their analysis exploits a precise analogy with Yule processes.

One can envision many extensions of this model. For example, the number of balls added need not be restricted to 1, and there could be probabilities $\{p_{iw}\}$ for adding w balls when the chosen color is C_i. (Of course, $\sum_w p_{iw} = p_i$.) Athreya and Ney (1972) developed limiting theory in which balls of colors other than that of the chosen ball may be added to the urn. Wei (1979) and Wei and Durham (1978) described applications of this model in clinical trials.

The multivariate Pólya–Eggenberger distributions discussed here are clearly very rich models for theoretical as well as applied work, and it is our hope that this chapter will prompt much more work and also hasten coordination of isolated results in this fruitful area of research.

BIBLIOGRAPHY

Athreya, K. B., and Ney, P. E. (1972). *Branching Processes*, New York: Springer-Verlag.

Bosch, A. J. (1963). The Pólya distribution, *Statistica Neerlandica*, **17**, 201–203.

Burr, E. J. (1960). The distribution of Kendall's score S for pair of tied rankings, *Biometrika*, **47**, 151–172.

Durham, S. D., and Yu, K. F. (1990). Randomized play-the-leader rules for sequential sampling from two populations, *Probability in the Engineering and Informational Sciences*, **4**, 355–367.

Dyczka, W. (1973). On the multidimensional Pólya distribution, *Annales Societatis Mathematicae Polonae, Seria I: Commentationes Mathematicae*, **17**, 43–63.

Eggenberger, F., and Pólya, G. (1923). Über die Statistik verketteter Vorgänge, *Zeitschrift für Angewandte Mathematik und Mechanik*, **1**, 279–289.

Eggenberger, F., and Pólya, G. (1928). Sur l'interpretation de certaines courbes de fréquence, *Comptes Rendus de l'Académie des Sciences, Paris*, **187**, 870–872.

Finucan, H. M. (1964). The mode of a multinomial distribution, *Biometrika*, **51**, 513–517.

Freeman, G. H., and Halton, J. H. (1951). Note on an exact treatment of contingency, goodness of fit and other problems of significance, *Biometrika*, **38**, 141–148.

Gerstenkorn, T. (1977a). The multidimensional truncated Pólya distribution, *Annales Societatis Mathematicae Polonae, Seria I: Commentationes Mathematicae*, **19**, 189–210.

Gerstenkorn, T. (1977b). Multivariate doubly truncated discrete distributions, *Acta Universitatis Lodziensis*, 1–115.

Gerstenkorn, T. (1977c). The multivariate truncated Pólya distribution, contributed paper at the *Fourth International Symposium on Multivariate Analysis.*

Gerstenkorn, T. (1978). Wielowymiarowe ucięte rozklady dyskretne, *Roczniki Polskiego Towarzystwa, Matematycznego, Seria III: Matematyka Stosowania*, **12**, 77–90 (in Polish).

Hald, A. (1960). The compound hypergeometric distribution and a system of single sampling inspection plans based on prior distributions and costs, *Technometrics*, **2**, 275–340.

Hirano, K. (1986). Some properties of the distributions of order k, in *Fibonacci Numbers and Their Properties* (Ed. A. N. Philippou), pp. 43–53, Dordrecht, Netherlands: Reidel.

Janardan, K. G. (1974). Characterizations of certain discrete distributions, in *Statistical Distributions in Scientific Work*, Vol. 3 (Eds. G. P. Patil, S. Kotz, and J. K. Ord), pp. 359–364, Dordrecht, Netherlands: Reidel.

Janardan, K. G., and Patil, G. P. (1970). On the multivariate Pólya distributions: A model of contagion for data with multiple counts, in *Random Counts in Physical Science II, Geo-Science and Business* (Ed. G. P. Patil), pp. 143–162, University Park, PA: Pennsylvania State University Press.

Janardan, K. G., and Patil, G. P. (1972). The multivariate inverse Pólya distribution: A model of contagion for data with multiple counts in inverse sampling, *Studi di Probabilita, Statistica e Ricerca Operativa in Onore di G. Pompiei Oderisi, Guggie*, pp. 1–15.

Janardan, K. G., and Patil, G. P. (1974). On multivariate modified Pólya and inverse Pólya distributions and their properties, *Annals of the Institute of Statistical Mathematics*, **26**, 271–276.

Johnson, N. L., and Kotz, S. (1972). *Continuous Multivariate Distributions*, New York: John Wiley & Sons.

Johnson, N. L., and Kotz, S. (1976). Two variants of Pólya urn models, *The American Statistician*, **30**, 186–188.

Johnson, N. L., and Kotz, S. (1977). *Urn Models and Their Application: An Approach to Modern Discrete Probability Theory*, New York: John Wiley & Sons.

Johnson, N. L., Kotz, S., and Kemp, A. W. (1992). *Univariate Discrete Distributions*, second edition, New York: John Wiley & Sons.

Kaiser, H. F., and Stefansky, W. (1972). A Pólya distribution for teaching, *The American Statistician*, **26**, 40–43.

Kriz, J. (1972). Die PMP-Verteilung, *Statistische Hefte*, **15**, 211–218.

Li, W., Durham, S. D., and Flournoy, N. (1996). A randomized Pólya urn design, *Proceedings of the Biometric Section of the American Statistical Association* (to appear).

Marshall, A. W., and Olkin, I. (1990). Bivariate distributions generated from Pólya–Eggenberger urn models, *Journal of Multivariate Analysis*, **35**, 48–65.

Mohanty, S. G. (1979). *Lattice Path Counting and Applications*, New York: Academic Press.

Morgenstern, D. (1976). Eine zweifache Verallgemeinerung der Pólya–Verteilung und ihre verschiedenen Deutungen, *Metrika*, **23**, 117–122.

Mosimann, J. E. (1963). On the compound negative multinomial distribution and correlations among inversely sampled pollen counts, *Biometrika*, **50**, 47–54.

Panaretos, J., and Xekalaki, E. (1986). On some distributions arising from certain generalized sampling schemes, *Communications in Statistics—Theory and Methods*, **15**, 873–891.

Patil, G. P., Boswell, M. T., Joshi, S. W., and Ratnaparkhi, M. V. (1984). *Dictionary and Classified Bibliography of Statistical Distributions in Scientific Work—1: Discrete Models*, Fairland, MD: International Co-operative Publishing House.

Patil, G. P., Joshi, S. W., and Rao, C. R. (1968). *A Dictionary and Bibliography of Discrete Distributions*, New York: Hafner, for International Statistical Institute.

Philippou, A. N., Antzoulakos, D. L., and Tripsiannis, G. A. (1988). Multivariate distributions of order *k*, *Statistics & Probability Letters*, **7**, 207–216.

Philippou, A. N., and Tripsiannis, G. A. (1991). Multivariate Pólya and inverse Pólya distributions of order *k*, *Biometrical Journal*, **33**, 225–236.

Philippou, A. N., Tripsiannis, G. A., and Antzoulakos, D.L. (1989). New Pólya and inverse Pólya distributions of order *k*, *Communications in Statistics—Theory and Methods*, **18**, 2125–2137.

Särndal, C. E. (1964). A unified derivation of some non-parametric distributions, *Journal of the American Statistical Association*, **59**, 1042–1053.

Sen, K., and Jain, R. (1994). A generalized multivariate Pólya–Eggenberger model generating various distributions, unpublished manuscript, Department of Statistics, University of Delhi, India.

Sibuya, M. (1980). Multivariate digamma distribution, *Annals of the Institute of Statistical Mathematics*, **32**, 25–36.

Sibuya, M., Yoshimura, I., and Shimizu, R. (1964). Negative multinomial distributions, *Annals of the Institute of Statistical Mathematics*, **16**, 409–426.

Steyn, H. S. (1951). On discrete multivariate probability functions, *Proceedings, Koninklijke Nederlandse Akademie Van Wetenschappen, Series A*, **54**, 23–30.

Tripsiannis, G. A. (1993). Modified multivariate Pólya and inverse Pólya distributions of order *k*, *Journal of the Indian Society for Statistics and Operations Research*, **14**, 1–11.

Tripsiannis, G. A., and Philippou, A. N. (1994). A new multivariate inverse Pólya distribution of order *k*, *Technical Report No. 10*, Department of Mathematics, University of Patras, Greece.

Wagner, U., and Taudes, A. (1986). A multivariate Pólya model of brand choice and purchase incidence, *Marketing Science*, **5**, 219–244.

Wagner, U., and Taudes, A. (1991). Microdynamics of new product purchase: A model incorporating both marketing and consumer-specific variables, *International Journal of Research in Marketing*, **8**, 223–249.

Wei, L. J. (1979). The generalized Pólya urn for sequential medical trials, *Annals of Statistics*, **7**, 291–296.

Wei, L. J., and Durham, S. D. (1978). The randomized play-the-winner rule in medical trials, *Journal of the American Statistical Association*, **73**, 840–843.

CHAPTER 41

Multivariate Ewens Distribution[1]

1 GENESIS AND HISTORY

The *Multivariate Ewens Distribution* (MED), called in genetics the Ewens Sampling Formula (ESF), describes a specific probability for the partition of the positive integer n into parts. It was discovered by Ewens (1972) as providing the probability of the partition of a sample of n selectively equivalent genes into a number of different gene types (alleles), either exactly in some models of genetic evolution or as a limiting distribution (as the population size becomes indefinitely large) in others. It was discovered independently by Antoniak (1974) in the context of Bayesian statistics.

The impetus for the derivation of the formula came from the non-Darwinian theory of evolution. It is claimed, under this theory, that the quite extensive genetical variation observed in natural populations is, on the whole, not due to natural selection, but arises rather as a result of purely stochastic changes in gene frequency in finite populations. The MED describes the partition distribution of a sample of n genes into allelic types when there are no selective differences between types, and thus provides the null hypothesis distribution for the non-Darwinian theory.

The distribution contains one parameter, usually denoted by θ, which in the genetic context is related to (a) the rate of mutation of the genes to new allelic types, (b) the population size, and (c) the details of the evolutionary model, being extremely robust with respect to these details. For the case $\theta = 1$ the distribution is quite old, going back in effect to Cauchy, since it then describes the partition into cycles of the numbers $(1, 2, \dots, n)$ under a random permutation, each possible permutation being given probability $(n!)^{-1}$. As noted below, the distribution arises for a much wider variety of combinatorial objects besides permutations.

[1] We thank Professors S. Tavaré and W. J. Ewens for providing us an original write-up of this chapter, and we thank J. W. Pitman for comments on early drafts. The distribution described in this chapter, which originated from applications in genetics and also independently in Bayesian statistical methodology, serves as a striking example of adaptability and universality of statistical methodology for scientific explorations in various seemingly unrelated fields.

2 DISTRIBUTION, MOMENTS, AND STRUCTURAL PROPERTIES

The MED is most easily described in terms of sequential sampling of animals from an infinite collection of distinguishable species [Fisher, Corbet, and Williams (1943), McCloskey (1965), and Engen (1978)]. We use this example throughout, except where specific genetic or other properties are discussed. Suppose that the species have (random) frequencies $P = (P_1, P_2, \ldots)$ satisfying

$$0 < P_i < 1, \quad i = 1, 2, \ldots, \quad \sum_{i=1}^{\infty} P_i = 1 . \tag{41.1}$$

Let $\eta_1, \eta_2, \ldots$ denote the species of the first, second, ... animal sampled. Conditional on P, the η_i are independent and identically distributed, with $\Pr[\eta_1 = k \mid P] = P_k$, $k = 1, 2, \ldots$. The sequence $I_1, I_2, \ldots$ of distinct values observed in $\eta_1, \eta_2, \ldots$ induces a random permutation $P^\# = (P_{I_1}, P_{I_2}, \ldots)$ of P. The vector $P^\#$ is known as the *size-biased permutation* of P.

Consider the sample of n individuals determined by $\eta_1, \ldots, \eta_n$, and write $A_1(n)$ for the number of animals of first species to appear, $A_2(n)$ for the number of animals of the second species to appear, and so on. The number of distinct species to appear in the sample is denoted by K_n. Another way to describe the sample is to record the counts $C_j(n)$, the number of species represented by j animals in the sample. The vector $C(n) = (C_1(n), \ldots, C_n(n))$ satisfies $\sum_{j=1}^n jC_j(n) = n$ and $K_n = \sum_{j=1}^n C_j(n)$.

In what follows, we consider the case where P satisfies

$$P_1 = W_1, P_r = (1 - W_1)(1 - W_2) \cdots (1 - W_{r-1})W_r, \quad r = 2, 3, \ldots, \tag{41.2}$$

where, for some $0 < \theta < \infty$,

$$W_1, W_2, \ldots \text{ are i.i.d. with density } \theta(1 - x)^{\theta - 1}, \quad 0 < x < 1. \tag{41.3}$$

The MED gives the distribution of the vector $C(n)$ as

$$\Pr[C(n) = a(n)] = \frac{n!}{\theta^{[n]}} \prod_j \frac{(\theta/j)^{a_j}}{a_j!}, \tag{41.4}$$

where, as earlier, $\theta^{[n]} = \theta(\theta + 1) \cdots (\theta + n - 1)$ and $a(n) = (a_1, a_2, \ldots, a_n)$ is a vector of non-negative integers satisfying $a_1 + 2a_2 + \cdots + na_n = n$.

The distribution of K_n is [Ewens (1972)]

$$\Pr[K_n = k] = \bar{s}(n, k)\theta^k / \theta^{[n]}. \tag{41.5}$$

Here $\bar{s}(n, k)$ is the coefficient of θ^k in $\theta^{[n]}$—that is, a Stirling number of the third kind (see Chapter 34). The distribution of the vector $A(n) = (A_1(n), A_2(n), \ldots)$ is determined by [Donnelly and Tavaré (1986)]

$$\Pr[K_n = k, A_i(n) = n_i, i = 1, 2, \ldots, k] \\ = \frac{\theta^k (n-1)!}{\theta^{[n]} n_k (n_k + n_{k-1}) \cdots (n_k + n_{k-1} + \cdots + n_2)}, \tag{41.6}$$

for $n_1 + \cdots + n_k = n$.

The conditional distribution of $C(n)$, given $K_n = k$, is

$$\Pr[C(n) = a(n) \mid K_n = k] = \frac{n!}{\bar{s}(n,k) \prod_j j^{a_j} a_j!} . \tag{41.7}$$

An alternative expression for this probability is as follows [due to Ewens (1972)]. Label the K_n species observed in an arbitrary way (independently of the sampling mechanism), and denote the number of animals of species i by N_i, $i = 1, 2, \ldots, K_n$. Then

$$\Pr[N_i = n_i,\ i = 1, \ldots, K_n \mid K_n = k] = \frac{n!}{k! \bar{s}(n,k) n_1 \cdots n_k} . \tag{41.8}$$

This conditional distribution is used in the statistical testing of the non-Darwinian theory (see Section 6.1 on page 239).

2.1 Moments

The joint factorial moments of $C(n)$, of arbitrary order, are

$$\mathrm{E}\left[\prod_{j=1}^{n} (C_j(n))^{(r_j)}\right] = \frac{n!}{m!} \frac{\theta^{[m]}}{\theta^{[n]}} \prod_{j=1}^{n} \left(\frac{\theta}{j}\right)^{r_j} \tag{41.9}$$

when $m = n - \sum j r_j \geq 0$ and are 0 when $m < 0$ [Watterson (1974)]; here $x^{(r)} = x(x-1)\cdots(x-r+1)$ for $r = 0, 1, 2, \ldots$.

The number of singleton species is of particular interest. The distribution of this number is

$$\Pr[C_1(n) = a] = \frac{\theta^a}{a!} \left[\sum_{j=0}^{n-a} (-1)^j \frac{\theta^j}{j!} \frac{(n+1-a-j)^{[a+j]}}{(n+\theta-a-j)^{[a+j]}}\right], \tag{41.10}$$

so that the mean and the variance of the number of singleton species are, respectively,

$$\frac{n\theta}{n+\theta-1}, \quad \frac{n(n-1)(n-2+2\theta)\theta}{(n+\theta-2)(n+\theta-1)^2} . \tag{41.11}$$

It follows from the structure of the urn model in the next section that

$$K_n = \xi_1 + \xi_2 + \cdots + \xi_n, \tag{41.12}$$

where $\xi_1, \ldots, \xi_n$ are independent Bernoulli random variables with

$$\Pr[\xi_i = 1] = 1 - \Pr[\xi_i = 0] = \frac{\theta}{\theta + i - 1}. \tag{41.13}$$

From this [for example, Cauchy (1905)],

$$\mathrm{E}[K_n] = \sum_{i=0}^{n-1} \frac{\theta}{\theta + i}, \qquad \mathrm{var}(K_n) = \sum_{i=1}^{n-1} \frac{\theta i}{(\theta + i)^2}. \tag{41.14}$$

2.2 Urn Models

Now we consider the properties of (41.4) and (41.6) for two consecutive sample sizes, n and $n + 1$. We denote the history of the sample of size n by $\mathcal{H}_n = (A(1), A(2), \ldots, A(n))$ and ask: Given $\mathcal{H}_n$, what is the conditional probability that the next animal will be of a new species? This probability is found from (41.4) as

$$\Pr[(n + 1)\text{th animal of a new species} \mid \mathcal{H}_n] = \frac{\theta}{n + \theta}. \tag{41.15}$$

The representation (41.12) follows immediately from this. If a given species has been observed m times ($m > 0$) in the sample of n, the conditional probability that the $(n + 1)$th animal will be of this species is

$$\begin{aligned} &\Pr[(n + 1)\text{th animal of a particular species seen} \\ &\qquad m \text{ times in the sample} \mid \mathcal{H}_n] = \frac{m}{n + \theta}. \end{aligned} \tag{41.16}$$

The probabilities (41.15) and (41.16) may be used to generate the process $A(n)$, $n = 1, 2 \ldots$ by a sequential urn scheme, starting from $A(1) = 1$. This model is a special case of an urn scheme of Blackwell and MacQueen (1973) that arises in the context of sampling from a Dirichlet process (see Section 6.2). Hoppe (1984, 1987) exploited a similar urn model in genetics.

2.3 Species Deletion (Noninterference)

Let μ_n denote the distribution of the partition vector $C(n)$ when sampling from the species model in (41.1). We say the sample has the species deletion property if, when an animal is taken at random from the sample, and it is observed that in all there are r animals of this species in the sample, then the partition distribution of the remaining $n - r$ animals is μ_{n-r}. Kingman (1978a,b) shows that the species deletion property holds for the MED [when μ_n is given by (41.4)].

2.4 Characterizations

The urn probabilities (41.15) and (41.16) and the species deletion property may be used to *characterize* the MED in the context of sampling from the model (41.1).

1. If the species deletion property in Section 2.3 holds, then the vector $C(n)$ has distribution μ_n given by the ESF [Kingman (1978a,b)].
2. *The law of succession.* Suppose that the sample history $\mathcal{H}_n$ is given. If the conditional probability that the next animal be of a new species depends only on n, then this probability must be of the form $\theta/(\theta + n)$ for some non-negative constant θ [Donnelly (1986)]. If, further, the conditional probability that this animal be of a specific species seen m times in the sample depends only on m [the sufficientness principle of Johnson (1932)], then the species partition probability is given by the MED [Zabell (1996)].

There is a theory of exchangeable random partitions that describes sampling from models slightly more general than (41.1); see Kingman (1978a), Aldous (1985), and Zabell (1992).

3 ESTIMATION

Equation (41.4) shows that the MED is a member of the exponential family of distributions; see, for example, Chapter 34. The complete sufficient statistic for θ is K_n. The maximum likelihood estimator $\hat{\theta}$ is, from (41.5), given implicitly as the solution of the equation $\sum_{i=0}^{n-1} \hat{\theta}/(\hat{\theta} + i) = K_n$. This estimator is biased, but the bias decreases as n increases. For large n, the variance of $\hat{\theta}$ is $\theta/(\sum_{i=1}^{n-1} i/(\theta + i)^2)$ [Ewens (1972)].

The only functions of θ admitting unbiased estimation are linear combinations of expressions of the form

$$[(i + \theta)(j + \theta) \cdots (m + \theta)]^{-1}, \tag{41.17}$$

where $i, j, \ldots, m$ are integers with $1 \le i < j < \cdots < m \le n - 1$.

The "law of succession" probability (41.15) thus does not admit unbiased estimation. However, bounds to unbiased estimation are readily provided by using the inequalities

$$\frac{(n-1)p_n}{n} < \frac{\theta}{n + \theta} < p_n \tag{41.18}$$

and the MVU estimate $\bar{s}(n - 1, k - 1)/\bar{s}(n, k)$ of p_n.

In genetics one frequently wishes to estimate the homozygosity probability, which in the species context is the probability $(1 + \theta)^{-1}$ that two animals taken at random are of the same species. Given $C(n) = a(n)$, it is natural to estimate this probability by $\sum a_i i(i - 1)/n(n - 1)$, an estimator occurring often in the genetics literature. The sufficiency of K_n for θ shows, however, that this estimator uses precisely the uninformative part of the data and that, given $K_n = k$, the MVU estimator is $T(n, k)/\bar{s}(n, k)$, where $T(n, k)$ is the coefficient of θ^k in $\theta(\theta + 2)(\theta + 3) \cdots (\theta + n - 1)$.

4 RELATIONS WITH OTHER DISTRIBUTIONS

The MED can be derived from other classical distributions [Watterson (1974)]. The first of these is the logarithmic (see, for example, Chapter 8). Suppose k is fixed and we observe k i.i.d. random variables $N_1, \ldots, N_k$ having the logarithmic distribution $\Pr[N_1 = j] \propto x^j/j$, $j = 1, 2, \ldots$, for $0 < x < 1$. Given that $\sum N_i = n$, the distribution of $(N_1, \ldots, N_k)'$ is (41.8). For a second representation, suppose that $Z_1, Z_2, \ldots$ are independent Poisson random variables with $\mathrm{E}[Z_j] = \theta/j$. Then

$$(C_1, \ldots, C_n)' \stackrel{d}{=} \left(Z_1, \ldots, Z_n \,\middle|\, \sum_{j=1}^{n} jZ_j = n \right)', \qquad (41.19)$$

where $\stackrel{d}{=}$ denotes equality in distribution.

Another representation, called the *Feller Coupling*, is useful for deriving asymptotic results for the MED [Arratia, Barbour, and Tavaré (1992)]. Let ξ_i, $i \geq 1$, be independent Bernoulli random variables with distribution (41.13), and let $C_j(n)$ be the number of spacings of length j between the 1s in the sequence $\xi_1 \xi_2 \cdots \xi_n 1$. Then the distribution of the vector $C(n)$ is the MED. Further, if Z_j is the number of spacings of length j in the infinite sequence $\xi_1 \xi_2 \cdots$, then the Z_j are independent Poisson random variables with mean $\mathrm{E}[Z_j] = \theta/j$.

4.1 The GEM Distribution

The distribution of the vector $P = (P_1, P_2, \ldots)$ determined by (41.2) and (41.3) is known as the GEM distribution (*Generalized Engen–McCloskey distribution*). It is named after McCloskey (1965) and Engen (1978), who introduced it in the context of ecology, and Griffiths (1980), who first noted its genetic importance.

The GEM distribution is a residual allocation model (RAM) [Halmos (1944), Patil and Taillie (1977)]—that is, a model of the form (41.2) where $W_1, W_2, \ldots$ are independent. It is the only RAM P with identically distributed residual fractions for which the size-biased permutation $P^{\#}$ has the same distribution as P [McCloskey (1965), Engen (1975)]. For the analog of the noninterference property in Section 2.3 for the GEM, see McCloskey (1965) and Hoppe (1986). For further discussion of size-biasing, see Donnelly and Joyce (1989), Perman, Pitman, and Yor (1992), and Chapters 3 (p. 146) and 43 (Section 5).

The decreasing order statistics $(P_{(1)}, P_{(2)}, \ldots)$ of P have the *Poisson–Dirichlet distribution* with parameter θ [Kingman (1975)]. The GEM is the size-biased permutation of the Poisson–Dirichlet [Patil and Taillie (1977)]. For further details about the Poisson-Dirichlet distribution, see Watterson (1976), Ignatov (1982), Tavaré (1987), Griffiths (1988), Kingman (1993), and Perman (1993).

4.2 The Pitman Sampling Formula

The MED is a particular case of the *Pitman Sampling Formula* [Pitman (1992, 1995)], which gives the probability of a species partition $C(n) = a(n)$ of n animals as

$$\Pr[C(n) = a(n), K_n = k]$$
$$= \frac{n!}{(\theta + 1)^{[n-1]}} [(\theta + \alpha)(\theta + 2\alpha) \cdots (\theta + (k-1)\alpha)]$$
$$\times \prod_{j=1}^{n} \left(\frac{(1-\alpha)^{[j-1]}}{j!} \right)^{a_j} \frac{1}{a_j!}. \tag{41.20}$$

Since we are considering only the infinitely many species case, we have the restrictions $0 \leq \alpha < 1$, $\theta > -\alpha$. [The other parameter range for which (41.20) defines a proper distribution is $\alpha = -\kappa$, $\theta = m\kappa$ for some positive integer m. This corresponds to sampling from a population with m species.] The MED is then the particular case of the Pitman Sampling Formula when $\alpha = 0$.

The Pitman distribution has several important properties, of which we note here one. Suppose in the RAM model (41.2) we no longer assume that $W_1, W_2, \ldots$ are identically distributed. Then the most general distribution of W_i for which the distribution of $(P_1, P_2, P_3, \ldots)$ is invariant under size-biased sampling [Pitman (1996)] is that for which W_i has probability density proportional to $w^{-\alpha}(1-w)^{\theta+i\alpha-1}$. This model for (41.2) yields the sampling distribution (41.20). The analogue of the Poisson–Dirichlet distribution in the two-parameter setting appears in Pitman and Yor (1995).

5 APPROXIMATIONS

It follows from (41.10) and the method of moments that random variables $C(n)$ with the MED (41.4) satisfy, for each fixed b,

$$(C_1(n), \ldots, C_b(n))' \Rightarrow (Z_1, \ldots, Z_b)', \tag{41.21}$$

as $n \to \infty$, $\Rightarrow$ denoting convergence in distribution. For $\theta = 1$ see Goncharov (1944), and for arbitrary θ see Arratia, Barbour, and Tavaré (1992). The Feller Coupling may be used to show that the total variation distance between $(C_1(n), \ldots, C_b(n))'$ and $(Z_1, \ldots, Z_b)'$ is at most $c(\theta)b/n$, where $c(\theta)$ is an explicit constant depending on θ alone. For $\theta \neq 1$, the rate is sharp.

The approximation in (41.21) covers the case of species represented a small number of times. A functional central limit theorem is available for the number of species represented at most n^t times, for $0 < t \leq 1$ [Hansen (1990)]. In particular, the number K_n of species in the sample has asymptotically a normal distribution with mean and variance $\theta \log n$.

It follows directly from the strong law of large numbers that the proportions $A(n)/n$ converge almost surely as $n \to \infty$ to $P^{\#}$, which has the GEM distribution with parameter θ. The decreasing order statistics of $A(n)/n$ converge almost surely to the Poisson–Dirichlet distribution with parameter θ [Kingman (1975)].

6 APPLICATIONS

6.1 Genetics

The original aim in devising (41.4) was to obtain a testing procedure for the non-Darwinian theory, since (41.4) provides the null hypothesis distribution for this theory. The parameter θ depends, in this context, on an unknown mutation parameter, an unknown population size, and unknown details about the evolutionary model. However, the conditional distribution (41.9) does not depend on θ and hence may be used as an objective basis for a test of the non-Darwinian theory. Watterson (1978) shows that a suitable test statistic is $\sum a_i i^2/n^2$ and provides various examples of the application of this at different gene loci. Anderson [see Ewens (1979), Appendix C] provides charts allowing rapid testing.

The MED was derived directly by a genealogical argument by Karlin and McGregor (1972). The Poisson–Dirichlet distribution arises as the stationary distribution of the ranked allele frequencies in the infinitely-many-alleles model [Watterson (1976)]. Equation (41.6) provides the distribution of alleles frequencies when the alleles are ordered by decreasing age [Donnelly and Tavaré (1986)], and this provides significant evolutionary information. See also Kelly (1979, Chapter 7). Correspondingly, the GEM distribution is the stationary distribution of the infinitely-many-alleles model when the types are ordered by age [Griffiths (1980)]. The MED may also be derived directly as a consequence of mutation in the coalescent [Kingman (1980, 1982a–c)]. See also Hoppe (1987) and Ewens (1990).

6.2 Bayesian Statistics

Dirichlet processes on a set S [Ferguson (1973)] are often used as priors over spaces of probability distributions on S. Suppose that the measure α of the process is nonatomic, and assume $\theta = \alpha(S) < \infty$. Let $P = (P_1, P_2, \ldots)$ have the GEM distribution with parameter θ and let $X_1, X_2, \ldots$ be i.i.d. random elements of S with distribution $\alpha(\cdot)/\theta$, independent of P. Sethuraman and Tiwari (1981) represent the Dirichlet process as atoms of height P_i at locations X_i, $i = 1, 2, \ldots$. A similar representation arises as the stationary distribution of the infinitely-many-alleles measure-valued diffusion in population genetics [Ethier and Kurtz (1994)]. Thus the Bayesian setting is essentially the same as sampling animals from a GEM population where the labels (determined by the X_i) of the animals are recorded as well. Antoniak (1974) showed that the MED gives the distribution of the partition induced by a sample from a Dirichlet process. See Ferguson, Phadia, and Tiwari (1992) and Sethuraman (1994) for recent developments.

6.3 Permutations

A permutation of the integers $1, 2, \ldots, n$ may be decomposed into an ordered product of cycles by beginning the first cycle with the integer 1, the second with the smallest integer not in the first cycle, and so on. For any $\theta > 0$, a random permutation, decomposed in this way, may be generated by Dubins and Pitman's *Chinese restaurant process* [cf. Aldous (1985)]: Integer 1 begins the first cycle. With probability $\theta/(\theta+1)$ integer 2 starts the second cycle, and with probability $1/(\theta+1)$ it joins the first cycle, to the right of 1. Once the first $r-1$ integers have been placed in cycles, integer r starts a new cycle with probability $\theta/(\theta+r-1)$, or is placed in an existing cycle, to the right of a random chosen one of $1, 2, \dots, r-1$. After n steps of this process, the probability of obtaining a particular permutation π with k cycles is $\theta^k/\theta^{[n]}$. Since the number of n-permutations having a_i cycles of size i is $n!/\prod j^{a_j}a_j!$, it follows that the joint distribution of the numbers C_j of cycles of size j, $j = 1, 2, \dots, n$, is given by the MED (41.4).

The case $\theta = 1$ corresponds to random permutations, which have been widely studied in the literature. Shepp and Lloyd (1966) show that the proportions in the largest, second largest, ... cycle lengths, asymptotically as $n \to \infty$, have a limiting Poisson–Dirichlet distribution. Erdős and Turán (1967) showed that the logarithm of the order (the least common multiple of its cycle lengths) of such a random permutation has asymptotically a normal distribution with mean $\log^2 n/2$ and variance $\log^3 n/3$. See also Vershik and Shmidt (1977) for a connection with the GEM distribution. The functional central limit theorem for the medium-sized cycles is given by DeLaurentis and Pittel (1985). When $\theta = 1$, random permutations are intimately connected to the theory of records [Ignatov (1981) and Goldie (1989)].

For arbitrary θ, Eq. (41.6) describes the joint distribution of the ordered cycle lengths. It follows that asymptotically the proportions in these cycles have the GEM distribution. Other approximations follow directly from Section 5. For the Erdős–Turán law for arbitrary θ, see Barbour and Tavaré (1994).

6.4 Ecology

In ecology, a long-standing problem concerned the species allocation of animals when species do not interact, in the sense that removal of one species does not affect the relative abundances of other species. Several attempts in the ecological literature, notably the "broken stick" model of MacArthur (1957), attempted to resolve this question. The noninterference property of the MED shows that this distribution provides the required partition, and (41.4) has been applied in various ecological contexts [Caswell (1976), Lambshead (1986), and Lambshead and Platt (1985)] where non-interference can be assumed.

The description of species diversity, through conditioned or unconditioned logarithmic distributions, has a long history in ecology [Fisher (1943), Fisher, Corbet, and Williams (1943), McCloskey (1965), Engen (1975), and Chapter 8 of Johnson, Kotz, and Kemp (1992)]. For a summary, see Watterson (1974).

6.5 Physics

The urn representation (41.15) and (41.16) is related to an urn representation of three classical partition formulae in physics [Bose–Einstein, Fermi–Dirac, and Maxwell–Boltzmann; for details see Johnson and Kotz (1977)] where a ball represents a "particle" and an urn represents a "cell," or energy level. Constantini (1987) considers the case where balls are placed sequentially into a collection of m urns so that, if among the first n balls there are n_j in urn j, the probability that ball $n+1$ is placed in this urn is

$$\frac{n_j + \delta}{n + m\delta} \tag{41.22}$$

for some constant δ. The Maxwell–Boltzmann, Bose–Einstein and Fermi–Dirac statistics follow when $\delta \to \infty$, $\delta = 1$, $\delta = -1$ respectively, while (41.15) and (41.16) show that the MED follows when $\delta \to 0$, $m \to \infty$ with $m\delta = \theta$. See also Keener, Rothman, and Starr (1987).

None of the physics partition formulae satisfy the noninterference property. Direct application of the MED in physics, in cases where the noninterference property is required, are given by Sibuya, Kawai, and Shida (1990), Mekjian (1991), Mekjian and Lee (1991), and Higgs (1995).

6.6 The Spread of News and Rumors

Bartholomew (1973) describes a simple model of the spread of news (or a rumor) throughout a population of n individuals. It is supposed that there is a source (e.g., a radio station) broadcasting the news and that each person in the population first hears the news either from the source or from some other individual. A person not knowing the news hears it from the source at rate α, as well as from a person who has heard the news at rate β. The analogy with (41.15) and (41.16) is apparent, and Bartholomew shows that, when all persons in the population have heard the news, the probability that k heard it directly from the source is given by (41.5), with $\theta = \alpha/\beta$.

This model is a Yule process with immigration [see Karlin and Taylor (1975)] and much more can be said. Individuals can be grouped into components, each consisting of exactly one person who first heard the news from the source, together with those individuals who first heard the news through some chain of individuals deriving from this person. Joyce and Tavaré's (1987) analysis applies directly to show among other things that the joint distribution of the component sizes is given by the MED.

6.7 The Law of Succession

The law of succession problem is perhaps the most classical in all of probability theory [see, for example, Zabell (1989) for a lucid historical account of this rule]. In the sampling of species context, we ask, given a sample of n animals, for the probability that animal $n+1$ is of a previously unobserved species and also for the probability that this animal is of a species seen m (> 0) times in the sample.

Clearly further assumptions are necessary to obtain concrete answers. For simplicity, we continue in the setting of (41.1) and we assume the sufficientness postulate. If we assume also that the probability that animal $n+1$ is of a new species depends only on n and the number k of species seen in the sample, then [Pitman (1995) and Zabell (1996)] the species partition in the sample must be given by Pitman Sampling Formula (41.20). This implies that the probability that animal $n+1$ is of a previously unobserved species is $(k\alpha + \theta)/(n + \theta)$, and that it is of a particular species seen m times in the sample is $(m - \alpha)/(n + \theta)$, where $0 \leq \alpha < 1$, $\theta > -\alpha$. This remarkable result represents the most significant recent advance in the theory of the law of succession. If we further require the probability that animal $n + 1$ be of a new species depends only on n, then $\alpha = 0$ and the species probability structure of the sample reduces to the MED.

6.8 Prime Numbers

Let N be an integer drawn at random from the set $1, 2, \ldots, n$, and write $N = p_1 p_2 p_3 \cdots$, where $p_1 \geq p_2 \geq p_3 \cdots$ are the prime factors of N. Writing $L_i = \log p_i / \log N$, $i \geq 1$, Billingsley (1972) showed that $(L_1, L_2, \ldots)$ has asymptotically as $n \to \infty$ the Poisson–Dirichlet distribution with parameter $\theta = 1$. One of the earliest investigations along these lines is Dickman (1930); see also Vershik (1986) for more recent results. Donnelly and Grimmett (1993) provide an elementary proof using size-biasing and the GEM distribution.

6.9 Random Mappings

The partition probability (41.4) appears also in the field of random mappings. Suppose random mapping of $(1, 2, \ldots, N)$ to $(1, 2, \ldots, N)$ is made, each mapping having probability N^{-N}. Any mapping defines a number of *components*, where i and j are in the same component if some functional iterate of i is identical to some functional iterate of j. In the limit $N \to \infty$, the normalized component sizes have a Poisson–Dirichlet distribution with $\theta = 1/2$ [Aldous (1985)], and the images of the components in the set $\{1, 2, \ldots, n\}$, for any fixed n, have the distribution (41.4), again with $\theta = 1/2$ [Kingman (1977)].

6.10 Combinatorial Structures

The joint distribution of the component counting process of many decomposable combinatorial structures satisfies the relation (41.19) for appropriate independent random variables Z_i [Arratia and Tavaré (1994)]. Examples include random mappings (discussed in the last section), factorization of polynomials over a finite field, and forests of labeled trees. When $i\,\mathrm{E}[Z_i] \to \theta$, $i \Pr[Z_i = 1] \to \theta$ for some $\theta \in (0, \infty)$ as $i \to \infty$, the counts of large components are close, in total variation distance, to the corresponding counts for the MED with parameter θ [Arratia, Barbour, and Tavaré (1995)]. Polynomial factorization satisfies $\theta = 1$. Poisson–Dirichlet approximations for a related class of combinatorial models are given by Hansen (1994).

BIBLIOGRAPHY

Aldous, D. J. (1985). Exchangeability and related topics, in *École d'été de probabilités de Saint-Flour XIII–1983* (Ed. P. L. Hennequin), Lecture Notes in Mathematics, Vol. 1117, pp. 2–198, Berlin: Springer-Verlag.

Antoniak, C. E. (1974). Mixtures of Dirichlet processes with applications to Bayesian nonparametric problems, *Annals of Statistics*, **2**, 1152–1174.

Arratia, R. A., and Tavaré, S. (1994). Independent process approximations for random combinatorial structures, *Advances in Mathematics*, **104**, 90–154.

Arratia, R. A., Barbour, A. D., and Tavaré, S. (1992). Poisson process approximations for the Ewens sampling formula, *Annals of Applied Probability*, **2**, 519–535.

Arratia, R. A., Barbour, A. D., and Tavaré, S. (1995). Logarithmic combinatorial structures, preprint.

Barbour, A. D., and Tavaré, S. (1994). A rate for the Erdős–Turán law, *Combinatorics, Probability and Computing*, **3**, 167–176.

Bartholomew, D. J. (1973). *Stochastic Models for Social Processes*, second edition, London: John Wiley & Sons.

Billingsley, P. (1972). On the distribution of large prime divisors, *Periodica Mathematica Hungarica*, **2**, 283–289.

Blackwell, D., and MacQueen, J. B. (1973). Ferguson distributions via Pólya urn schemes, *Annals of Statistics*, **1**, 353–355.

Caswell, H. (1976). Community structure: A neutral model analysis, *Ecological Monographs*, **46**, 327–353.

Cauchy, A. (1905). *Oeuvres Complètes. II Série, Tom 1,* Paris: Gautier-Villars.

Constantini, D. (1987). Symmetry and distinguishability of classical particles, *Physics Letters A*, **123**, 433–436.

DeLaurentis, J. M., and Pittel, B. (1985). Random permutations and Brownian motion, *Pacific Journal of Mathematics*, **119**, 287–301.

Dickman, K. (1930). On the frequency of numbers containing prime factors of a certain relative magnitude, *Arkiv för Matematik, Astronomi och Fysik*, **22**, 1–14.

Donnelly, P. (1986). Partition structures, Pólya urns, the Ewens sampling formula, and the ages of alleles, *Theoretical Population Biology*, **30**, 271–288.

Donnelly, P., and Grimmett, G. (1993). On the asymptotic distribution of large prime factors, *Journal of the London Mathematical Society*, **47**, 395–404.

Donnelly, P., and Joyce, P. (1989). Continuity and weak convergence of ranked and size-biased permutations on an infinite simplex, *Stochastic Processes and Their Applications*, **31**, 89–103.

Donnelly, P., and Tavaré, S. (1986). The ages of alleles and a coalescent, *Advances in Applied Probability*, **18**, 1–19.

Engen, S. (1975). A note on the geometric series as a species frequency model, *Biometrika*, **62**, 697–699.

Engen, S. (1978). *Stochastic Abundance Models with Emphasis on Biological Communities and Species Diversity*, London: Chapman and Hall.

Erdős, P., and Turán, P. (1967). On some problems of a statistical group theory III, *Acta Mathematica Academiae Scientiarum Hungaricae*, **18**, 309–320.

Ethier, S. N., and Kurtz, T. G. (1994). Convergence to Fleming-Viot processes in the weak atomic topology, *Stochastic Processes and Their Applications*, **54**, 1–27.

Ewens, W. J. (1972). The sampling theory of selectively neutral alleles, *Theoretical Population Biology*, **3**, 87–112.

Ewens, W. J. (1979). *Mathematical Population Genetics*, Berlin: Springer-Verlag.

Ewens W. J. (1990). Population genetics theory—the past and the future, in *Mathematical and Statistical Developments of Evolutionary Theory* (Ed. S. Lessard), pp. 177–227, Amsterdam: Kluwer.

Ferguson, T. S. (1973). A Bayesian analysis of some nonparametric problems, *Annals of Statistics*, **1**, 209–230.

Ferguson, T. S., Phadia, E. G., and Tiwari, R. C. (1992). Bayesian nonparametric inference, in *Current Issues in Statistical Inference: Essays in Honor of D. Basu* (Eds. M. Ghosh and P. K. Patnak), IMS Lecture Notes-Monograph Series, **17**, 127–150.

Fisher, R. A. (1943). A theoretical distribution for the apparent abundance of different species, *Journal of Animal Ecology*, **12**, 54–57.

Fisher, R. A., Corbet, A. S., and Williams, C. B. (1943). The relation between the number of species and the number of individuals in a random sample from an animal population, *Journal of Animal Ecology*, **12**, 42–58.

Goldie, C. M. (1989). Records, permutations and greatest convex minorants, *Mathematical Proceedings of the Cambridge Philosophical Society*, **106**, 169–177.

Goncharov, V. L. (1944). Some facts from combinatorics, *Izvestia Akad. Nauk. SSSR, Ser. Mat.*, **8**, 3–48. See also: On the field of combinatory analysis, *Translations of the American Mathematical Society*, **19**, 1–46.

Griffiths, R. C. (1980). Unpublished notes.

Griffiths, R. C. (1988). On the distribution of points in a Poisson-Dirichlet process, *Journal of Applied Probability*, **25**, 336–345.

Halmos, P. R. (1944). Random alms, *Annals of Mathematical Statistics*, **15**, 182–189.

Hansen, J. C. (1990). A functional central limit theorem for the Ewens Sampling Formula, *Journal of Applied Probability*, **27**, 28–43.

Hansen, J. C. (1994). Order statistics for decomposable combinatorial structures, *Random Structures and Algorithms*, **5**, 517–533.

Higgs, P. G. (1995). Frequency distributions in population genetics parallel those in statistical physics, *Physical Review E*, **51**, 95–101.

Hoppe, F. M. (1984). Pólya–like urns and the Ewens sampling formula, *Journal of Mathematical Biology*, **20**, 91–99.

Hoppe, F. M. (1986). Size-biased filtering of Poisson–Dirichlet samples with an application to partition structures in genetics, *Journal of Applied Probability*, **23**, 1008–1012.

Hoppe, F. M. (1987). The sampling theory of neutral alleles and an urn model in population genetics, *Journal of Mathematical Biology*, **25**, 123–159.

Ignatov, Z. (1981). Point processes generated by order statistics and their applications, in *Point Processes and Queueing Problems* (Eds. P. Bartfái and J. Tomkó), pp. 109–116, Amersterdam: North-Holland.

Ignatov, T. (1982). On a constant arising in the asymptotic theory of symmetric groups, and on Poisson–Dirichlet measures, *Theory of Probability and its Applications*, **27**, 136–147.

Johnson, N. L., and Kotz, S. (1977). *Urn Models and Their Application: An Approach to Modern Discrete Probability Theory*, New York: John Wiley & Sons.

Johnson, N. L., Kotz, S., and Kemp, A. W. (1992). *Univariate Discrete Distributions*, second edition, New York: John Wiley & Sons.

Johnson, W. E. (1932). *Logic, Part III: The Logical Foundations of Science*, Cambridge: Cambridge University Press.

Joyce, P., and Tavaré, S. (1987). Cycles, permutations and the structure of the Yule process with immigration, *Stochastic Processes and Their Applications*, **25**, 309–314.

Karlin, S., and McGregor, J. (1972). Addendum to a paper of W. Ewens, *Theoretical Population Biology*, **3**, 113–116.

Karlin, S., and Taylor, H. M. (1975). *A First Course in Stochastic Processes*, second edition, New York: Academic Press.

Keener, R., Rothman, E., and Starr, N. (1987). Distributions of partitions, *Annals of Statistics*, **15**, 1466–1481.

Kelly, F. P. (1979). *Reversibility and Stochastic Networks*, New York: John Wiley & Sons.

Kingman, J. F. C. (1975). Random discrete distributions, *Journal of the Royal Statistical Society, Series B*, **37**, 1–22.

Kingman, J. F. C. (1977). The population structure associated with the Ewens sampling formula, *Theoretical Population Biology*, **11**, 274–283.

Kingman, J. F. C. (1978a). Random partitions in population genetics, *Proceedings of the Royal Society London, Series A*, **361**, 1–20.

Kingman, J. F. C. (1978b). The representation of partition structures, *Journal of the London Mathematical Society*, **18**, 374–380.

Kingman, J. F. C. (1980). *Mathematics of Genetic Diversity*, Philadelphia: SIAM.

Kingman, J. F. C. (1982a). On the genealogy of large populations, *Journal of Applied Probability*, **19**, 27–43.

Kingman, J. F. C. (1982b). The coalescent, *Stochastic Processes and Their Applications*, **13**, 235–248.

Kingman, J. F. C. (1982c). Exchangeability and the evolution of large populations, in *Exchangeability in Probability and Statistics* (Eds. G. Koch and F. Spizzichino), pp. 97–112, Amsterdam: North-Holland.

Kingman, J. F. C. (1993). *Poisson Processes*, Oxford: Clarendon Press.

Lambshead, P. J. D. (1986). Sub-catastrophic sewage and industrial waste contamination as revealed by marine nematode faunal analysis, *Marine Ecology Progress Series*, **29**, 247–260.

Lambshead, P. J. D., and Platt, H. M. (1985). Structural patterns of marine benthic assemblages and their relationship with empirical statistical models, in *Nineteenth European Marine Biology Symposium*, pp. 371–380, Cambridge: Cambridge University Press.

MacArthur, R. H. (1957). On the relative abundance of bird species, *Proceedings of the National Academy of Sciences, USA*, **43**, 293–295.

McCloskey, J. W. (1965). A model for the distribution of individuals by species in an environment, unpublished Ph.D. thesis, Michigan State University.

Mekjian, A. Z. (1991). Cluster distributions in physics and genetic diversity, *Physical Review A*, **44**, 8361–8374.

Mekjian, A. Z., and Lee, S. J. (1991). Models of fragmentation and partitioning phenomena based on the symmetric group S_n and combinatorial analysis, *Physical Review A*, **44**, 6294–6311.

Patil, G. P., and Taillie, C. (1977). Diversity as a concept and its implications for random communities, *Bulletin of the International Statistical Institute*, **47**, 497–515.

Perman, M. (1993). Order statistics for jumps of normalized subordinators, *Stochastic Processes and Their Applications*, **46**, 267–281.

Perman, M., Pitman, J., and Yor, M. (1992). Size-biased sampling of Poisson point processes and excursions, *Probability Theory and Related Fields*, **92**, 21–39.

Pitman, J. (1992). The two-parameter generalization of Ewens' random partition structure, *Technical Report No. 345*, Department of Statistics, University of California, Berkeley.

Pitman, J. (1995). Exchangeable and partially exchangeable random partitions, *Probability Theory and Related Fields*, **12**, 145–158.

Pitman, J. (1996). Random discrete distributions invariant under size-biased permutation, *Journal of Applied Probability* (to appear).

Pitman, J., and Yor, M. (1995). The two-parameter Poisson–Dirichlet distribution derived from a stable subordinator, *Technical Report No. 427*, Department of Statistics, University of California, Berkeley.

Sethuraman, J. (1994). A constructive definition of Dirichlet priors, *Academica Sinica*, **4**, 639–650.

Sethuraman, J., and Tiwari, R. C. (1981). Convergence of Dirichlet measures and the interpretation of their parameter, in *Statistical Decision Theory and Related Topics—III*, Vol. 2 (Eds. S. S. Gupta and J. O. Berger), pp. 305–315, New York: Academic Press.

Shepp, L. A., and Lloyd, S. P. (1966). Ordered cycle lengths in a random permutation, *Transactions of the American Mathematical Society*, **121**, 340–357.

Sibuya, M., Kawai, T., and Shida, K. (1990). Equipartition of particles forming clusters by inelastic collisions, *Physica A*, **167**, 676–689.

Tavaré, S. (1987). The birth process with immigration and the genealogical structure of large populations, *Journal of Mathematical Biology*, **25**, 161–168.

Vershik, A. M. (1986). The asymptotic distribution of factorizations of natural numbers into prime divisors, *Soviet Math. Doklady*, **34**, 57–61.

Vershik, A. M., and Shmidt, A. A. (1977). Limit measures arising in the asymptotic theory of symmetric groups I, *Theory of Probability and its Applications*, **22**, 70–85.

Watterson, G. A. (1974). Models for the logarithmic species abundance distributions, *Theoretical Population Biology*, **6**, 217–250.

Watterson, G. A. (1976). The stationary distribution of the infinitely-many neutral alleles diffusion model, *Journal of Applied Probability*, **13**, 639–651.

Watterson, G. A. (1978). The homozygosity test of neutrality, *Genetics*, **88**, 405–417.

Zabell, S. L. (1989). The rule of succession, *Erkenntnis*, **31**, 283–321.

Zabell, S. L. (1992). Predicting the unpredictable, *Synthèse*, **90**, 205–232.

Zabell, S. L. (1996). The continuum of inductive methods revisited, *Pittsburgh–Konstanz Series in the History and Philosophy of Science* (to appear).

CHAPTER 42

Multivariate Distributions of Order s

The relatively recently discovered and investigated univariate distributions of order s have been discussed in some detail in Chapter 10, Section 6. In this chapter, we present a brief survey of multivariate generalizations of these distributions. Some of these multivariate distributions of order s have already been described in different chapters of this volume. Most of the generalizations are quite straightforward, and it is indeed remarkable that the class of distributions of order s is rather easily extended to the multivariate case. Nevertheless, there are some novel results which shed additional light on the structure and flexibility of the classical multivariate discrete distributions, and they certainly provide an avenue for further generalizations.

As explained earlier in Section 9 of Chapter 36, these distributions are widely known as *order k* distributions. But we have deliberately used the name *order s* since we have used k throughout this book to denote the dimension of a multivariate distribution. Also, as mentioned there, we feel that either *run* or *consecutive* would have been a better choice than the word *order* in the name of these distributions.

Since there has been considerable activity in this area of research aiming on both theory and applications, we have restricted ourselves to some basic distributional definitions and properties. We have deliberately refrained from discussions on approximations, inequalities, and limit theorems (but the references included should be of valuable assistance to readers interested in these topics).

1 MULTIVARIATE GEOMETRIC DISTRIBUTIONS OF ORDER s

Suppose $\{X_i\}_1^\infty$ are independent and identically distributed Bernoulli random variables with

$$\Pr[X_i = 1] = p \qquad \text{and} \qquad \Pr[X_i = 0] = 1 - p = q\,. \tag{42.1}$$

Then, as mentioned earlier in Section 9 of Chapter 36, the waiting time T until s consecutive 1's are observed has a geometric distribution of order s with probability generating function

$$G_T(t) = \frac{p^s t^s (1 - pt)}{1 - t + qp^s t^{s+1}}\,. \tag{42.2}$$

Multivariate versions of these geometric distributions of order s in (42.2) have been considered by Aki and Hirano (1994, 1995). For example, with M_0 denoting the number of 0's in the sequence $X_1, X_2, \ldots$ before s consecutive 1's are observed, they obtain the joint probability generating function of T and M_0 as

$$G_{T,M_0}(t, t_0) = \frac{p^s t^s (1 - pt)}{1 - pt - q t_0 t\{1 - (pt)^s\}} . \tag{42.3}$$

Aki and Hirano (1995) extended this distribution in (42.3) by starting with random variables X_i's which have a m-point distribution with probability mass function

$$\Pr[X_i = j] = p_j \qquad \text{for } j = 1, 2, \ldots, m . \tag{42.4}$$

By denoting T as the waiting time until s consecutive 1's are observed and M_i as the number of times i appeared in the sequence $X_1, X_2, \ldots$ before s consecutive 1's ($i = 2, 3, \ldots, m$), Aki and Hirano (1995) obtained the joint probability generating function of $T, M_2, \ldots, M_m$ as

$$\begin{aligned} &G_{T,M_2,\ldots,M_m}(t, t_2, \ldots, t_m) \\ &\quad = \frac{(p_1 t)^s (1 - p_1 t)}{1 - p_1 t - t\left(\sum_{j=2}^m p_j t_j\right)\{1 - (p_1 t)^s\}} . \end{aligned} \tag{42.5}$$

Similarly, with M_1 denoting the number of 1's until T, these authors also established that the joint probability generating function $T, M_1, M_2, \ldots, M_m$ is

$$\begin{aligned} &G_{T,M_1,M_2,\ldots,M_m}(t, t_1, t_2, \ldots, t_m) \\ &\quad = \frac{(1 - p_1 t_1 t)(p_1 t_1 t)^s}{1 - p_1 t_1 t - t\left(\sum_{j=2}^m p_j t_j\right)\{1 - (p_1 t_1 t)^s\}} ; \end{aligned} \tag{42.6}$$

see also Balakrishnan (1996a) who, in addition to deriving the joint probability generating function in (42.6), has established the following recurrence relations for the joint probability mass function and the mixed moments:

$$P(s, s, 0, 0, \ldots, 0) = p_1^s \tag{42.7}$$

and

$$\begin{aligned} &P(b, b_1, b_2, \ldots, b_m) - p_1\, P(b - 1, b_1 - 1, b_2, \ldots, b_m) \\ &\quad - \sum_{i=2}^m p_i\, P(b - 1, b_1, \ldots, b_{i-1}, b_i - 1, b_{i+1}, \ldots, b_m) \\ &\quad + p_1^s \sum_{i=2}^m p_i\, P(b - s - 1, b_1 - s, b_2, \ldots, b_{i-1}, b_i - 1, b_{i+1}, \ldots, b_m) \\ &\quad = 0 , \end{aligned} \tag{42.8}$$

where $P(b, b_1, \ldots, b_m) = \Pr[T = b, M_1 = b_1, \ldots, M_m = b_m]$; and

$$
\begin{aligned}
E[T^{\ell} M_1^{\ell_1} M_2^{\ell_2} \cdots M_m^{\ell_m}] &- p_1 E[(T+1)^{\ell}(M_1+1)^{\ell_1} M_2^{\ell_2} \cdots M_m^{\ell_m}] \\
&- \sum_{i=2}^{m} p_i \, E\left[(T+1)^{\ell} M_1^{\ell_1} \cdots M_{i-1}^{\ell_{i-1}} (M_i+1)^{\ell_i} M_{i+1}^{\ell_{i+1}} \cdots M_m^{\ell_m}\right] \\
&+ p_1^s \sum_{i=2}^{m} p_i \, E[(T+s+1)^{\ell}(M_1+s)^{\ell_1} M_2^{\ell_2} \cdots M_{i-1}^{\ell_{i-1}} (M_i+1)^{\ell_i} \\
&\qquad \times M_{i+1}^{\ell_{i+1}} \cdots M_m^{\ell_m}] \\
&= \{p_1^s s^{\ell+\ell_1} - p_1^{s+1}(s+1)^{\ell+\ell_1}\} I(\ell_2 = \cdots = \ell_m = 0), \qquad (42.9)
\end{aligned}
$$

where $I(\ell_2 = \cdots = \ell_m = 0)$ is the indicator function taking 1 when $\ell_2 = \cdots = \ell_m = 0$ and 0 otherwise.

Next, by denoting N_i $(i = 1, 2, \ldots, s)$ to be the number of overlapping runs of 1 ("success") of length i until T, and defining the variables $T, M_2, \ldots, M_m$ as before, Aki and Hirano (1995) derived a further extended multivariate distribution with joint probability generating function

$$
\begin{aligned}
G_{T,N_1,\ldots,N_s,M_2,\ldots,M_m}&(t, t_1, \ldots, t_s, u_2, \ldots, u_m) \\
&= \frac{p_1^s t^s t_1^s t_2^{s-1} \cdots t_s}{1 - t\left(\sum_{j=2}^m p_j u_j\right) \sum_{i=0}^{s-1} p_1^i t^i t_1^i t_2^{i-1} \cdots t_i}. \qquad (42.10)
\end{aligned}
$$

Balakrishnan (1996a) has also derived the probability generating function in (42.10) and established some useful recurrence relations for the joint probability mass function as well as for the mixed moments. Aki and Hirano (1995) and Balakrishnan (1996a) have also similarly discussed the multivariate distributions involving the number of runs of 1 ("success") of length at least i (for $i = 1, 2, \ldots, s$) and the number of nonoverlapping runs of 1 ("success") of length i. For example, in the case of the latter, with N_i^* $(i = 1, 2, \ldots, s)$ denoting the number of nonoverlapping runs of 1 of length i until T and the variables $T, M_2, \ldots, M_m$ defined as before, the joint probability generating function of the multivariate distribution of $(T, N_1^*, N_2^*, \ldots, N_s^*, M_2, \ldots, M_m)'$ is

$$
\begin{aligned}
G_{T,N_1^*,\ldots,N_s^*,M_2,\ldots,M_m}&(t, t_1, \ldots, t_s, u_2, \ldots, u_m) \\
&= \frac{p_1^s t^s t_1^s t_2^{[s/2]} \cdots t_{s-1}^{[s/(s-1)]} t_s}{1 - \left(t \sum_{j=2}^m p_j u_j\right) \sum_{i=0}^{s-1} p_1^i t^i t_1^i t_2^{[i/2]} \cdots t_{s-1}^{[i/(s-1)]}}, \qquad (42.11)
\end{aligned}
$$

where $[z]$ denotes the greatest integer contained in z; see, for example, Aki and Hirano (1995) and Balakrishnan (1996a). In this case, Balakrishnan (1996a) has presented the following recurrence relations for the joint probability mass function:

$$
P\left(s, s, \left[\frac{s}{2}\right], \ldots, \left[\frac{s}{s-1}\right], 1, 0, \ldots, 0\right) = p_1^s \qquad (42.12)
$$

and

$$P(a, a_1, \ldots, a_s, b_2, \ldots, b_m) - \sum_{i=0}^{s-1} \sum_{j=2}^{m} p_1^i p_j P\left(a - i - 1, a_1 - i, a_2 - \left[\frac{i}{2}\right], \ldots, a_{s-1} - \left[\frac{i}{s-1}\right], a_s, b_2, \ldots, b_{j-1}, b_j - 1, b_{j+1}, \ldots, b_m\right) = 0, \tag{42.13}$$

where

$$P(a, a_1, \ldots, a_s, b_2, \ldots, b_m) = \Pr[T = a, N_1^* = a_1, \ldots, N_s^* = a_s, M_2 = b_2, \ldots, M_m = b_m]. \tag{42.14}$$

Similarly, Balakrishnan (1996a) has also established the following recurrence relation for moments of the distribution corresponding to (42.11):

$$\begin{aligned} E[T^{\ell} N_1^{*\ell_1} \cdots N_s^{*\ell_s} M_2^{\ell_2'} \cdots M_m^{\ell_m'}] & - \sum_{i=0}^{s-1} \sum_{j=2}^{m} p_j^i p_j E\left[(T + i + 1)^{\ell} (N_1^* + i)^{\ell_1} \left(N_2^* + \left[\frac{i}{2}\right]\right)^{\ell_2} \cdots \right. \\ & \times \left(N_{s-1}^* + \left[\frac{i}{s-1}\right]\right)^{\ell_{s-1}} N_s^{*\ell_s} M_2^{\ell_2'} \cdots M_{j-1}^{\ell_{j-1}'} (M_j + 1)^{\ell_j'} \\ & \left. \times M_{j+1}^{\ell_{j+1}'} \cdots M_m^{\ell_m'} \right] \\ = p_1^s s^{\ell + \ell_1} \left[\frac{s}{2}\right]^{\ell_2} \cdots \left[\frac{s}{s-1}\right]^{\ell_{s-1}} I(\ell_2' = \cdots = \ell_m' = 0). \end{aligned} \tag{42.15}$$

The multivariate geometric distribution of order s in (42.3) and the extended multivariate geometric distribution of order s in (42.6) have been generalized recently by Balakrishnan and Johnson (1997) in the following way. Instead of starting with either Bernoulli variables in (42.1) or m-point variables in (42.4), these authors assumed $X_1, X_2, \ldots$ to be a sequence of general i.i.d. random variables with probability mass function $\Pr[X_1 = j] = p_j$ for $j = 1, 2, \ldots, \infty$. Once again, let T be the random variable denoting the waiting time at which the first consecutive s 1's are observed, and N_i denote the number of occurrences of i until T for $i = 1, 2, \ldots, \infty$. Then, the joint probability generating function of $(T, N_1, N_2, \ldots)'$ has been derived by Balakrishnan and Johnson (1997) as

$$G(t, t_1, t_2, \ldots) = \frac{(p_1 t t_1)^s (1 - p_1 t t_1)}{1 - p_1 t t_1 - \{1 - (p_1 t t_1)^s\} \sum_{i=2}^{\infty} p_i t t_i}. \tag{42.16}$$

For the case when $p_i = 0$ for $i \geq m + 1$, it is easily observed that the above joint probability generating function reduces to that of the extended multivariate geometric distribution of order s given earlier in (42.6). Also, for the special case when X_i's are geometric random variables with probability mass function $\Pr[X = x] = pq^{x-1}$ for $x = 1, 2, \ldots, \infty$, the joint probability generating function of $(T, N_1, N_2, \ldots)'$ in (42.16) becomes

$$G(t, t_1, t_2, \ldots) = \frac{(ptt_1)^s(1 - ptt_1)}{1 - ptt_1 - \{1 - (ptt_1)^s\}\sum_{i=2}^{\infty} pq^{i-1}tt_i} . \tag{42.17}$$

The distribution corresponding to the joint probability generating function in (42.17) has been termed the *iterated multivariate geometric distribution of order s* by Balakrishnan and Johnson (1997).

2 MULTIVARIATE DISTRIBUTIONS IN CORRELATED BINARY VARIABLES

In the last section we have described various extensions of the multivariate geometric distribution of order s. In this section we elaborate on distributions which are generalizations of the multivariate geometric distribution of order s in a different direction. Specifically, these distributions are derived by assuming some dependence among the variables X_i's.

2.1 Markov Dependence

First, let us consider the simple Markovian dependence among the binary variables. In this case, we may take $\Pr[X_1 = 1] = p_0$, $\Pr[X_i = 1 \mid X_{i-1} = 1] = p_1$, and $\Pr[X_i = 1 \mid X_{i-1} = 0] = p_2$ for $i = 2, 3, \ldots$. Also, let $q_j = 1 - p_j$ for $j = 0, 1, 2$. Several papers have appeared generalizing the distributional results presented in the last section to this case of Markov dependent binary variables; see, for example, Feder (1974), Rajarshi (1974), Gerber and Li (1981), Schwager (1983), Benvenuto (1984), Lambiris and Papastavridis (1985), Papastavridis and Lambiris (1987), Chryssaphinou and Papastavridis (1990), Hirano and Aki (1993), Aki and Hirano (1993), Balasubramanian, Viveros, and Balakrishnan (1993), Viveros and Balakrishnan (1993), Wang and Liang (1993), Godbole and Papastavridis (1994), Chryssaphinou, Papastavridis, and Tsapelas (1994), Fu and Koutras (1994a, b), Viveros, Balasubramanian, and Balakrishnan (1994), Mohanty (1994), Wang and Ji (1995), Uchida and Aki (1995), Balakrishnan, Balasubramanian, and Viveros (1995), and Koutras and Alexandrou (1995). Due to the vastness of this literature, we shall present here a few key results in this direction and refer the readers to these papers for more detailed discussions.

To this end, let $S_0 = 1$ or 0 depending on whether $X_1 = 1$ or 0, $F_0 = 1 - S_0$, $S_1(F_1)$ is the number of times $X_i = 1(0)$ when $X_{i-1} = 1$, and $S_2(F_2)$ is the number of times $X_i = 1(0)$ when $X_{i-1} = 0$. Now, suppose X_i's are observed until either s consecutive 1's or s' consecutive 0's appear, whichever appears first. Then, the joint probability generating function of $(S_0, F_0, S_1, F_1, S_2, F_2)'$ has been shown to be

$$G_{S_0,F_0,S_1,F_1,S_2,F_2}(t_0, u_0, t_1, u_1, t_2, u_2) = (A + B)/Q , \tag{42.18}$$

where

$$A = (1 - p_1t_1)(p_1t_1)^{s-1}\left[p_0t_0(1 - q_2u_2) + q_0u_0p_2t_2\left\{1 - (q_2u_2)^{s'-1}\right\}\right], \quad (42.19)$$

$$B = (1 - q_2u_2)(q_2u_2)^{s'-1}\left[q_0u_0(1 - p_1t_1) + p_0t_0q_1u_1\left\{1 - (p_1t_1)^{s-1}\right\}\right], \quad (42.20)$$

and

$$Q = (1 - p_1t_1)(1 - q_2u_2) - p_2t_2q_1u_1\left\{1 - (p_1t_1)^{s-1}\right\}\left\{1 - (q_2u_2)^{s'-1}\right\}. \quad (42.21)$$

From the joint probability generating function in (42.18), probability generating functions corresponding to some other variables such as $X = S_0 + S_1 + S_2$ (number of 1's), $Y = F_0 + F_1 + F_2$ (number of 0's), and $T = X + Y$ (the waiting time) can all be derived easily. For example, the probability generating function of the waiting time T is

$$G_T(t) = (C + D)/R\,, \quad (42.22)$$

where

$$C = (1 - p_1t)p_1^{s-1}t^s\left[p_0(1 - q_2t) + q_0p_2t\left\{1 - (q_2t)^{s'-1}\right\}\right], \quad (42.23)$$

$$D = (1 - q_2t)q_2^{s'-1}t^{s}\left[q_0(1 - q_1t) + p_0q_1t\left\{1 - (p_1t)^{s-1}\right\}\right], \quad (42.24)$$

and

$$R = (1 - p_1t)(1 - q_2t) - q_1p_2t^2\left\{1 - (p_1t)^{s-1}\right\}\left\{1 - (q_2t)^{s'-1}\right\}. \quad (42.25)$$

Suppose X_i's are observed until either s consecutive 1's or s' 0's in total appear, whichever appears first. In this case, the joint probability generating function of $(S_0, F_0, S_1, F_1, S_2, F_2)'$ is

$$G_{S_0,F_0,S_1,F_1,S_2,F_2}(t_0, u_0, t_1, u_1, t_2, u_2)$$
$$= p_0t_0(p_1t_1)^{s-1} + AB^{s'-1} + p_2t_2(p_1t_1)^{s-1}A(1 - B^{s'-1})/(1 - B)\,, \quad (42.26)$$

where, in this case,

$$A = q_0u_0 + p_0t_0q_1u_1\{1 - (p_1t_1)^{s-1}\}/(1 - p_1t_1) \quad (42.27)$$

and

$$B = q_2u_2 + p_2t_2q_1u_1\{1 - (p_1t_1)^{s-1}\}/(1 - p_1t_1)\,. \quad (42.28)$$

As before, probability generating functions of X, Y, and T can be derived easily from the joint probability generating function in (42.26). For example, the probability generating function of the waiting time T in this case is

$$G_T(t) = t^{s'}CD^{s'-1} + p_1^{s-1}t^s[p_0 + p_2tC\{1 - (tD)^{s'-1}\}/(1 - tD)]\,, \quad (42.29)$$

where

$$C = q_0 + p_0 q_1 t\{1 - (p_1 t)^{s-1}\}/(1 - p_1 t) \tag{42.30}$$

and

$$D = q_2 + q_1 p_2 t\{1 - (p_1 t)^{s-1}\}/(1 - p_1 t)\,. \tag{42.31}$$

Balasubramanian, Viveros, and Balakrishnan (1993) and Uchida and Aki (1995) have discussed these distributions under some other stopping rules as well.

Such distributions derived from Markov dependent binary variables have found a wide variety of applications including consecutive systems in reliability theory, start-up demonstration testing, acceptance sampling plans in quality control, learning models, and analysis of scores in volleyball games. These distributional results have also been generalized recently to higher-order Markov-dependent binary variables; see, for example, Aki, Balakrishnan, and Mohanty (1997) and Balakrishnan, Mohanty, and Aki (1997).

2.2 Binary Sequences of Order *s*

A sequence $\{X_i\}_0^\infty$ of $\{0, 1\}$-valued random variables is said to be a *binary sequence of order s* if there exist a positive integer s and s real numbers $0 < p_1, p_2, \ldots, p_s < 1$ such that $X_0 \equiv 0$ and

$$\Pr[X_n = 1 \mid X_0 = x_0,\ X_1 = x_1, \ldots,\ X_{n-1} = x_{n-1}] = p_j\,, \tag{42.32}$$

where $j = r - \left[\frac{r-1}{s}\right] s$, and r is the smallest positive integer such that $x_{n-r} = 0$. This sequence was defined by Aki (1985) and discussed further by Hirano and Aki (1987), Aki and Hirano (1988, 1994), Aki (1992), and Balakrishnan (1996b). As pointed out by Aki and Hirano (1995), this sequence is closely related to (a) the *cluster sampling scheme* discussed by Philippou (1988) and Xekalaki and Panaretos (1989) and (b) the $(s-1)$-*step Markov dependence model* studied by Fu (1986) in the context of consecutive systems in reliability theory.

Now let T denote the waiting time (stage) until the first s consecutive 1's are observed in $X_1, X_2, \ldots,$ let N be the number of occurrences of 0, and let N_i $(i = 1, 2, \ldots, s)$ be the number of overlapping runs of 1 of length i until T. Then, Aki and Hirano (1995) derived the joint probability generating function of $(T, N, N_1, \ldots, N_s)'$ as

$$G_{T,N,N_1,\ldots,N_s}(t, u, u_1, \ldots, u_s) = \frac{p_1 p_2 \cdots p_s t^s u_1^s u_2^{s-1} \cdots u_s}{1 - q_1 t u - \sum_{i=1}^{s-1} p_1 p_2 \cdots p_i q_{i+1} t^{i+1} u u_1^i u_2^{i-1} \cdots u_i}\,, \tag{42.33}$$

where $q_i = 1 - p_i$. Recurrence relations for the joint probability mass function and for the mixed moments have been derived by Balakrishnan (1996b). He has also presented explicit expressions for the means, variances, and covariances of the variables; for example,

$$E[T] = \frac{sp_1p_2\cdots p_s}{1 - q_1 - \sum_{i=1}^{s-1} p_1p_2\cdots p_iq_{i+1}} + \frac{p_1p_2\cdots p_s\left\{q_1 + \sum_{i=1}^{s-1}(i+1)p_1p_2\cdots p_iq_{i+1}\right\}}{\left\{1 - q_1 - \sum_{i=1}^{s-1} p_1p_2\cdots p_iq_{i+1}\right\}^2}, \tag{42.34}$$

$$E[N] = \frac{\left\{q_1 + \sum_{i=1}^{s-1} p_1p_2\cdots p_iq_{i+1}\right\} p_1p_2\cdots p_s}{\left\{1 - q_1 - \sum_{i=1}^{s-1} p_1p_2\cdots p_iq_{i+1}\right\}^2}, \tag{42.35}$$

and

$$E[N_j] = \frac{(s-j+1)p_1p_2\cdots p_s}{1 - q_1 - \sum_{i=1}^{s-1} p_1p_2\cdots p_iq_{i+1}} + \frac{p_1p_2\cdots p_s\sum_{i=j}^{s-1}(i-j+1)p_1p_2\cdots p_iq_{i+1}}{\left\{1 - q_1 - \sum_{i=1}^{s-1} p_1p_2\cdots p_iq_{i+1}\right\}^2}, \tag{42.36}$$

$$j = 1, 2, \ldots, s.$$

Similarly, with N_i^* $(i = 1, 2, \ldots, s)$ denoting the number of runs of 1 of length at least i until T, the joint probability generating function of $(T, N, N_1^*, N_2^*, \ldots, N_s^*)'$ is given by

$$G_{T,N,N_1^*,\ldots,N_s^*}(t, u, u_1, \ldots, u_s) = \frac{p_1p_2\cdots p_st^su_1u_2\cdots u_s}{1 - q_1tu - \sum_{i=1}^{s-1} p_1p_2\cdots p_iq_{i+1}t^{i+1}uu_1u_2\cdots u_i}; \tag{42.37}$$

see Aki and Hirano (1995) and Balakrishnan (1996b).

In a similar way, with N_i^{**} $(i = 1, 2, \ldots, s)$ denoting the number of nonoverlapping runs of 1 of length i until T, Aki and Hirano (1995) and Balakrishnan (1996b) have derived the joint probability generating function of $(T, N, N_1^{**}, \ldots, N_s^{**})'$ as

$$G_{T,N,N_1^{**},\ldots,N_s^{**}}(t, u, u_1, \ldots, u_s) = \frac{p_1p_2\cdots p_st^su_1^su_2^{[s/2]}\cdots u_{s-1}^{[s/(s-1)]}u_s}{1 - q_1tu - \sum_{i=1}^{s-1} p_1p_2\cdots p_iq_{i+1}t^{i+1}uu_1^iu_2^{[i/2]}\cdots u_{i-1}^{[i/(i-1)]}u_i}. \tag{42.38}$$

In this case, Balakrishnan (1996b) has also established recurrence relations for the joint probability mass function and for the mixed moments.

Balakrishnan (1996b) has also extended the above results to binary sequences of order h. Clearly, the results do not change if $h > s$. However, for the case when $h < s$, the results do change. Let $\left[\frac{s-1}{h}\right] = a$ and $s - 1 = ah + b$, where $0 \le b \le h - 1$;

further, let $\left[\frac{s}{h}\right] = c$ and $s = ch + d$. Then the joint probability generating function of $(T, N, N_1, \ldots, N_s)'$ in this case is

$$\frac{(p_1 \cdots p_h)^c p_1 \cdots p_d t^s u_1^s u_2^{s-1} \cdots u_s}{1 - D}, \tag{42.39}$$

where

$$D = \sum_{i=0}^{a-1} (p_1 \cdots p_h)^i \sum_{j=0}^{h-1} (p_1 \cdots p_j q_{j+1}) t^{ih+j+1} u u_1^{ih+j} u_2^{ih+j-1} \cdots u_{ih+j}$$
$$+ (p_1 \cdots p_h)^a \sum_{j=0}^{b} p_1 \cdots p_j q_{j+1} t^{s-b+j} u u_1^{s-b+j-1} u_2^{s-b+j-2} \cdots u_{s-b+j-1}. \tag{42.40}$$

3 MULTIVARIATE NEGATIVE BINOMIAL DISTRIBUTIONS OF ORDER s

Some discussion on multivariate negative binomial distributions of order s has been presented earlier in Section 9 of Chapter 36. A random vector $\boldsymbol{N} = (N_1, \ldots, N_k)'$ is said to have a *multivariate negative binomial distribution of order s* with parameters $r, q_{11}, \ldots, q_{ks}$ $(r > 0,\ 0 < q_{ij} < 1,\ q_{11} + \cdots + q_{ks} < 1)$, denoted by $\text{MNB}_s(r; q_{11}, \ldots, q_{ks})$, if

$$\begin{aligned} P_{\boldsymbol{N}}(\boldsymbol{n}) &= \Pr[\boldsymbol{N} = \boldsymbol{n}] \\ &= p^r \sum_{\sum_j j n_{ij} = n_i} \binom{n_{11} + \cdots + n_{ks} + r - 1}{n_{11}, \ldots, n_{ks}, r-1} \prod_i \prod_j q_{ij}^{n_{ij}}, \\ &\qquad n_i = 0, 1, \ldots \ (1 \le i \le k), \end{aligned} \tag{42.41}$$

where $p = 1 - q_{11} - \cdots - q_{ks}$. In this case,

$$P(\boldsymbol{n}) = \begin{cases} 0 & \text{if some } n_i \le -1 \ (i = 1, 2, \ldots, k) \\ p^r & \text{if } n_1 = \cdots = n_k = 0 \\ \sum_i \sum_j q_{ij} P(\boldsymbol{n} - \boldsymbol{j}_i) + \frac{r-1}{n_\ell} \sum_j j\, q_{\ell j} P(\boldsymbol{n} - \boldsymbol{j}_\ell) & \\ \quad \text{if } n_i \ge 0 \text{ and some } n_\ell \ge 1 \ (1 \le i, \ell \le k), \end{cases} \tag{42.42}$$

where $\boldsymbol{j}_i$ $(1 \le i \le k,\ 1 \le j \le s)$ denotes a k-dimensional vector with a j in the ith position and zeros elsewhere. For a detailed proof of this result, one may refer to Antzoulakos and Philippou (1991). The joint probability generating function of $\boldsymbol{N}$ in this case is

$$G_{\boldsymbol{N}}(\boldsymbol{t}) = p^r \left(1 - \sum_i \sum_j q_{ij} t_i^j\right)^{-r} \qquad \text{for } |t_i| \le 1. \tag{42.43}$$

An alternative form of (42.41) is

$$P(\boldsymbol{n}) = p^r \sum_{\sum_j jn_{ij}=n_i} \frac{\Gamma(r + \sum_i \sum_j n_{ij})}{\Gamma(r) \prod_i \prod_j n_{ij}!} \prod_i \prod_j q_{ij}^{n_{ij}}, \qquad n_i = 0, 1, \ldots \ (1 \le i \le k), \tag{42.44}$$

where $p = 1 - q_{11} - \cdots - q_{ks}$. Here, r need not be an integer and hence this distribution is called a *multivariate extended negative binomial distribution of order s.*

For $s = 1$ this distribution reduces to the usual multivariate negative binomial distribution, and for $k = 1$ it reduces to the *multiparameter negative binomial distribution of order s* of Philippou (1988); see Chapter 10, Section 6. For $r = 1$, the resulting distribution was called by Philippou, Antzoulakos, and Tripsiannis (1989) a *multivariate geometric distribution of order s* with parameters $q_{11}, \ldots, q_{ks}$, and it was denoted by $\mathrm{MG}_s(q_{11}, \ldots, q_{ks})$.

For the special case when $q_{ij} = P^{j-1} Q_i$ ($0 < Q_i < 1$ for $1 \le i \le k$, $\sum_{i=1}^k Q_i < 1$ and $P = 1 - \sum_{i=1}^k Q_i$), the distribution in (42.44) reduces to

$$P(\boldsymbol{n}) = P^{rs} \sum_{\sum_j jn_{ij}=n_i} \frac{\Gamma(r + \sum_i \sum_j n_{ij})}{\Gamma(r) \prod_i \prod_j n_{ij}!} \prod_i P^{n_i} \left(\frac{Q_i}{P}\right)^{\sum_j n_{ij}}, \qquad n_i = 0, 1, \ldots, \tag{42.45}$$

which is the multivariate analogue of the (shifted) negative binomial distribution of order s of Philippou, Georghiou, and Philippou (1983) and Aki, Kuboki, and Hirano (1984); also see Chapter 10, Section 6. It is called a *Type I multivariate negative binomial distribution of order s* with parameters $r, Q_1, \ldots, Q_k$ and is denoted by $\mathrm{MNB}_{s,\mathrm{I}}(r; Q_1, \ldots, Q_k)$. Of course, for $r = 1$ the resulting distribution is called a *Type I multivariate geometric distribution of order s* with parameters $Q_1, \ldots, Q_k$ and is denoted by $\mathrm{MG}_{s,\mathrm{I}}(Q_1, \ldots, Q_k)$.

For the special case when $q_{ij} = Q_i/s$ ($0 < Q_i < 1$ for $1 \le i \le k$, $\sum_{i=1}^k Q_i < 1$ and $P = 1 - \sum_{i=1}^k Q_i$), the distribution in (42.44) reduces to

$$P(\boldsymbol{n}) = P^r \sum_{\sum_j jn_{ij}=n_i} \frac{\Gamma(r + \sum_i \sum_j n_{ij})}{\Gamma(r) \prod_i \prod_j n_{ij}!} \prod_i \left(\frac{Q_i}{s}\right)^{\sum_j n_{ij}}, \qquad n_i = 0, 1, \ldots, \tag{42.46}$$

which is the multivariate analogue of the compound Poisson (or negative binomial) distribution of order s of Philippou (1983). It is called a *Type II multivariate negative binomial distribution of order s* with parameters $r, Q_1, \ldots, Q_k$ and is denoted by $\mathrm{MNB}_{s,\mathrm{II}}(r; Q_1, \ldots, Q_k)$. Once again, for $r = 1$ the resulting distribution is called a *Type II multivariate geometric distribution of order s* with parameters $Q_1, \ldots, Q_k$ and is denoted by $\mathrm{MG}_{s,\mathrm{II}}(Q_1, \ldots, Q_k)$.

Finally, for the case when $q_{ij} = \left\{\prod_{\ell=1}^{j-1} P_\ell\right\} Q_{ij}$ with $P_0 \equiv 1$, $0 < Q_{ij} < 1$ ($1 \le i \le k$, $1 \le j \le s$), $\sum_{i=1}^k Q_{ij} < 1$ and $P_j = 1 - \sum_{i=1}^k Q_{ij}$ for $1 \le j \le s$, the distribution in (42.44) reduces to

$$P(\boldsymbol{n}) = (P_1 \cdots P_s)^r \sum_{\sum_j jn_{ij}=n_i} \frac{\Gamma(r + \sum_i \sum_j n_{ij})}{\Gamma(r) \prod_i \prod_j n_{ij}!} \times \prod_i \prod_j \left(P_1 \cdots P_{j-1} Q_{ij}\right)^{n_{ij}}, \quad n_i = 0, 1, \ldots, \tag{42.47}$$

which is the multivariate analogue of the (shifted) extended negative binomial distribution of order s of Aki (1985); also see Chapter 10, Section 6. It is called a *multivariate extended negative binomial distribution of order s* with parameters $r, Q_{11}, \ldots, Q_{ks}$ and is denoted by $\text{MENB}_s(r; Q_{11}, \ldots, Q_{ks})$. For $r = 1$, the resulting distribution is called a *multivariate extended geometric distribution of order s* with parameters $Q_{11}, \ldots, Q_{ks}$ and is denoted by $\text{MEG}_s(Q_{11}, \ldots, Q_{ks})$. This distribution, incidentally, is the multivariate analogue of the (shifted) extended geometric distribution of order s of Aki (1985); also see Chapter 10, Section 6.

Recurrence relations, similar to the one in (42.42) for the multivariate negative binomial distribution of order s, can be presented for the $\text{MNB}_{s,\text{I}}$, $\text{MNB}_{s,\text{II}}$, and MENB_s distributions. These are, respectively, given by

$$\textbf{(i)}\ P(\boldsymbol{n}) = \begin{cases} 0 & \text{if some } n_i \le -1\ (i = 1, 2, \ldots, k) \\ P^{rs} & \text{if } n_1 = \cdots = n_k = 0 \\ \sum_i \sum_j P^{j-1} Q_i P(\boldsymbol{n} - \boldsymbol{j}_i) \\ \quad + \frac{r-1}{n_\ell} \sum_j j P^{j-1} Q_\ell P(\boldsymbol{n} - \boldsymbol{j}_\ell) \\ & \text{if } n_i \ge 0 \text{ and some } n_\ell \ge 1\ (1 \le i, \ell \le k)\,; \end{cases} \tag{42.48}$$

$$\textbf{(ii)}\ P(\boldsymbol{n}) = \begin{cases} 0 & \text{if some } n_i \le -1\ (i = 1, 2, \ldots, k) \\ P^{r} & \text{if } n_1 = \cdots = n_k = 0 \\ \sum_i \sum_j \frac{Q_i}{s} P(\boldsymbol{n} - \boldsymbol{j}_i) + \frac{r-1}{n_\ell} \sum_j j \frac{Q_\ell}{s} P(\boldsymbol{n} - \boldsymbol{j}_\ell) \\ & \text{if } n_i \ge 0 \text{ and some } n_\ell \ge 1\ (1 \le i, \ell \le k)\,; \end{cases} \tag{42.49}$$

and

$$\textbf{(iii)}\ P(\boldsymbol{n}) = \begin{cases} 0 & \text{if some } n_i \le -1\ (i = 1, 2, \ldots, k) \\ (P_1 \cdots P_s)^r & \text{if } n_1 = \cdots = n_k = 0 \\ \sum_i \sum_j P_1 \cdots P_{j-1} Q_{ij} P(\boldsymbol{n} - \boldsymbol{j}_i) \\ \quad + \frac{r-1}{n_\ell} \sum_j j P_1 \cdots P_{j-1} Q_{\ell j} P(\boldsymbol{n} - \boldsymbol{j}_\ell) \\ & \text{if } n_i \ge 0 \text{ and some } n_\ell \ge 1\ (1 \le i, \ell \le k)\,. \end{cases} \tag{42.50}$$

In the above equations, $\boldsymbol{j}_i$ ($1 \le i \le k$ and $1 \le j \le s$) denotes a k-dimensional vector with a j in the ith position and zeros elsewhere.

For $r = 1$, Eq. (42.50) yields the following recurrence relation for the probability mass function of the multivariate extended geometric distribution of order s:

$$P(\boldsymbol{n}) = \begin{cases} 0 & \text{if some } n_i \le -1\ (i = 1, 2, \ldots, k) \\ P_1 \cdots P_s & \text{if } n_1 = \cdots = n_k = 0 \\ \sum_i \sum_j P_1 \cdots P_{j-1} Q_{ij} P(\boldsymbol{n} - \boldsymbol{j}_i) \\ & \text{if } n_i \ge 0\ (1 \le i \le k) \text{ and } \sum_{i=1}^k n_i > 0\,. \end{cases} \tag{42.51}$$

Furthermore, Eq. (42.48) readily yields for the case $s = 1$ the recurrence relation for the probability mass function of the ordinary multivariate negative binomial distribution (see Chapter 36) as

$$P(\boldsymbol{n}) = \begin{cases} 0 & \text{if some } n_i \le -1 \ (i = 1, 2, \ldots, k) \\ P^r & \text{if } n_1 = \cdots = n_k = 0 \\ \sum_i Q_i P(\boldsymbol{n} - \mathbf{1}_i) + \frac{r-1}{n_\ell} Q_\ell P(\boldsymbol{n} - \mathbf{1}_\ell) & \\ \quad \text{if } n_i \ge 0 \text{ and some } n_\ell \ge 1 \ (1 \le i, \ell \le k), \end{cases} \tag{42.52}$$

which is a generalization of the recurrence relation for the probability mass function of the univariate negative binomial distribution (corresponding to $k = 1$); for example, see Chapter 5, Section 5.

For the multivariate negative binomial distribution of order s in (42.41), a genesis through an urn model can be provided as follows. Consider an urn containing balls bearing the letters $C_{11}, \ldots, C_{ks}$ and C with respective proportions $q_{11}, \ldots, q_{ks}$ and p ($0 < q_{ij} < 1$ for $1 \le i \le k$ and $1 \le j \le s$, $\sum_i \sum_j q_{ij} < 1$ and $\sum_i \sum_j q_{ij} + p = 1$). Suppose balls are drawn from the urn with replacement until r balls ($r \ge 1$) bearing the letter C appear. Let N_i ($1 \le i \le k$) be the random variable denoting the sum of the second indices of the letters on the balls drawn whose first index is i. Then $\boldsymbol{N} = (N_1, N_2, \ldots, N_k)'$ has the multivariate negative binomial distribution of order s in (42.41).

For the multivariate negative binomial distribution of order s in (42.41), upon using the joint probability generating function in (42.43), it can be shown that

$$E[N_i] = \frac{r}{p} \sum_j j q_{ij}, \quad i = 1, 2, \ldots, k, \tag{42.53}$$

$$\operatorname{var}(N_i) = \frac{r}{p} \left\{ \sum_j j^2 q_{ij} + \frac{1}{p} \left(\sum_j j q_{ij} \right)^2 \right\}, \quad i = 1, 2, \ldots, k, \tag{42.54}$$

and

$$\operatorname{cov}(N_i, N_t) = \frac{r}{p^2} \left(\sum_j j q_{ij} \right) \left(\sum_j j q_{tj} \right), \quad 1 \le i \ne t \le k. \tag{42.55}$$

In addition, the following properties also hold:

(i) Let $\boldsymbol{N}_t$, $t = 1, 2, \ldots, n$, be independent $(k \times 1)$ random vectors distributed as $\text{MNB}_s(r_t; q_{11}, \ldots, q_{ks})$, respectively. Then $\boldsymbol{N}_\bullet = \boldsymbol{N}_1 + \cdots + \boldsymbol{N}_n$ is distributed as $\text{MNB}_s(r_\bullet; q_{11}, \ldots, q_{ks})$, where $r_\bullet = r_1 + \cdots + r_n$.

(ii) Let $\boldsymbol{N}_t$, $t = 1, 2, \ldots, r$, be independent $(k \times 1)$ random vectors identically distributed as $\text{MG}_s(q_{11}, \ldots, q_{ks})$. Then $\boldsymbol{N} = \boldsymbol{N}_1 + \cdots + \boldsymbol{N}_r$ is distributed as $\text{MNB}_s(r; q_{11}, \ldots, q_{ks})$.

(iii) Let $\boldsymbol{N} = (N_1, N_2, \ldots, N_k)'$ be distributed as $\text{MNB}_s(r; q_{11}, \ldots, q_{ks})$. Then $\boldsymbol{N}^* = (N_1, N_2, \ldots, N_\ell)'$, $1 \le \ell \le k$, is distributed as $\text{MNB}_s(r; q_{11}^*, \ldots, q_{\ell s}^*)$, where $q_{ab}^* = q_{ab}/(1 - \sum_{i=\ell+1}^k \sum_j q_{ij})$ for $1 \le a \le \ell$ and $1 \le b \le s$.

(iv) Let $\boldsymbol{N} = (N_1, N_2, \ldots, N_k)'$ be distributed as $\text{MNB}_s(r; q_{11}, \ldots, q_{ks})$. Then, N_t is distributed as $\text{NB}_s(r; q_1, q_2, \ldots, q_s)$, where $q_b = q_{ab}/(1 - \sum_{\substack{i=1 \\ i \neq a}}^{k} \sum_j q_{ij})$ for $1 \leq b \leq s$.

4 MULTIVARIATE POISSON DISTRIBUTIONS OF ORDER s

Multivariate Poisson distributions of order s are obtained as limiting forms of multivariate negative binomial distributions of order s. [One may refer to Philippou (1983) for some discussion on these lines for the univariate case.] Specifically, let $\boldsymbol{X}$ be a $(k \times 1)$ random vector distributed as $\text{MNB}_s(r; q_{11}, \ldots, q_{ks})$. Then, as $r \to \infty$ in such a way that $q_{ij} \to 0$ and $rq_{ij} \to \lambda_{ij}$ with $0 < \lambda_{ij} < \infty$ for $1 \leq i \leq k$ and $1 \leq j \leq s$,

$$P(x_1, x_2, \ldots, x_k) \to \sum_{\sum_j jx_{ij} = x_i} e^{-\sum_i \sum_j \lambda_{ij}} \prod_i \prod_j \left\{ \lambda_{ij}^{x_{ij}} / x_{ij}! \right\},$$

$$x_i = 0, 1, 2, \ldots, \quad 1 \leq i \leq k. \tag{42.56}$$

The probability distribution with joint probability mass function as in (42.56) is called a *multivariate Poisson distribution of order s* with parameters $\lambda_{11}, \ldots, \lambda_{ks}$ ($0 \leq \lambda_{ij} < \infty$ for $1 \leq i \leq k$ and $1 \leq j \leq s$), and is denoted by $\text{MP}_s(\lambda_{11}, \ldots, \lambda_{ks})$. For $s = 1$, this distribution reduces to the usual multivariate Poisson distribution discussed earlier in Chapter 37. For $k = 1$, it reduces to the *multiparameter Poisson distribution of order s* of Philippou (1988) and Aki (1985) (also called *extended Poisson distribution of order s*); see Chapter 10, Section 6. The distribution in (42.56) is called a *Type I multivariate Poisson distribution of order s* when $\lambda_{ij} = \lambda_i$ for $1 \leq i \leq k$, and it is denoted by $\text{MP}_{s,\text{I}}(\lambda_1, \lambda_2, \ldots, \lambda_k)$. The joint probability mass function of this distribution is

$$P(x_1, x_2, \ldots, x_k) = \sum_{\sum_j jx_{ij} = x_i} e^{-s\sum_i \lambda_i} \prod_i \lambda_i^{\sum_j x_{ij}} \Big/ \left\{ \prod_i \prod_j x_{ij}! \right\},$$

$$x_i = 0, 1, 2, \ldots, \quad 1 \leq i \leq k. \tag{42.57}$$

For the multivariate negative binomial distribution of order s in (42.41), a genesis through the multivariate Poisson distribution of order s can be provided as follows. Let $\boldsymbol{X}$ and Y be a $(k \times 1)$ random vector and a univariate random variable, respectively, such that $\boldsymbol{X}|(Y = y)$ is distributed as $\text{MP}_s(y\,\lambda_{11}, \ldots, y\,\lambda_{ks})$ and Y has a scaled gamma distribution with probability density function $p_Y(y) = \alpha^r \times y^{r-1} e^{-\alpha y}/\Gamma(r)$, $y, r, \alpha > 0$ (see Chapter 17). Then, the unconditional distribution of $\boldsymbol{X}$ is $\text{MNB}_s(r; q_{11}, \ldots, q_{ks})$, where $q_{ij} = \lambda_{ij}/(\alpha + \sum_i \sum_j \lambda_{ij})$ for $1 \leq i \leq k$ and $1 \leq j \leq s$, and $p = 1 - \sum_i \sum_j q_{ij}$.

For the multivariate Poisson distribution of order s in (42.56), the joint probability generating function is

$$G_{\boldsymbol{X}}(\boldsymbol{t}) = \exp\left\{ \sum_i \sum_j \lambda_{ij}(t_i^j - 1) \right\}, \quad |t_i| \leq 1, \ 1 \leq i \leq k, \tag{42.58}$$

using which it can be shown that

$$E[X_i] = \sum_j j\,\lambda_{ij}, \qquad 1 \le i \le k\,, \tag{42.59}$$

$$\operatorname{var}(X_i) = \sum_j j^2\,\lambda_{ij}, \qquad 1 \le i \le k\,, \tag{42.60}$$

and

$$\operatorname{cov}(X_i, X_t) = 0, \qquad 1 \le i \ne t \le k\,. \tag{42.61}$$

In addition, the following properties also hold:

(i) Let $\boldsymbol{X} = (X_1, X_2, \ldots, X_k)'$ be distributed as $\text{MP}_s(\lambda_{11}, \ldots, \lambda_{ks})$. Then $\boldsymbol{X}^* = (X_1, X_2, \ldots, X_\ell)'$, $1 \le \ell \le k$, is distributed as $\text{MP}_s(\lambda_{11}, \ldots, \lambda_{ls})$.

(ii) Let $\boldsymbol{X} = (X_1, X_2, \ldots, X_k)'$ be distributed as $\text{MP}_s(\lambda_{11}, \ldots, \lambda_{ks})$. Then X_t is distributed as $P_s(\lambda_1, \ldots, \lambda_s)$, where $\lambda_j = \lambda_{tj}$ for $1 \le j \le s$ and $1 \le t \le k$.

5 MULTIVARIATE LOGARITHMIC SERIES DISTRIBUTIONS OF ORDER s

A distribution with joint probability mass function

$$P(\boldsymbol{x}) = \alpha \sum_{\sum_j jx_{ij} = x_i} \frac{(\sum_i \sum_j x_{ij} - 1)!}{\prod_i \prod_j x_{ij}!} \prod_i \prod_j q_{ij}^{x_{ij}}\,,$$
$$x_i = 0, 1, \ldots \text{ for } 1 \le i \le k, \ \sum_i x_i > 0\,, \tag{42.62}$$

where $0 < q_{ij} < 1$ (for $1 \le i \le k$ and $1 \le j \le s$), $\sum_i \sum_j q_{ij} < 1$, $p = 1 - \sum_i \sum_j q_{ij}$ and $\alpha = -(\log p)^{-1}$, is said to be a *multivariate logarithmic series distribution of order* s. It is denoted by $\text{MLS}_s(q_{11}, \ldots, q_{ks})$. For $s = 1$, this distribution reduces to the usual multivariate logarithmic series distribution (see Chapter 38). For $k = 1$, it reduces to the *multiparameter logarithmic series distribution of order* s of Philippou (1988); see also Chapter 10, Section 6.

In the case when $q_{ij} = P_i^{j-1} Q_i$ ($Q_i = 1 - P_i$ for $1 \le i \le k$) so that $p = 1 - \sum_i (1 - P_i^s) \equiv P$ (say), the distribution in (42.62) becomes

$$P(\boldsymbol{x}) = -(\log P)^{-1} \sum_{\sum_j jx_{ij} = x_i} \frac{(\sum_i \sum_j x_{ij} - 1)!}{\prod_i \prod_j x_{ij}!} \prod_i P_i^{x_i} \left(\frac{Q_i}{P_i}\right)^{\sum_j x_{ij}},$$
$$x_i = 0, 1, \ldots \text{ for } 1 \le i \le k, \ \sum_i x_i > 0\,, \tag{42.63}$$

which is the multivariate analogue of the logarithmic series distribution of order s of Aki, Kuboki, and Hirano (1984); see also Chapter 10, Section 6. The distribution (42.63) is called a *Type I multivariate logarithmic series distribution of order s* with parameters $Q_1, \ldots, Q_k$ and is denoted by $\text{MLS}_{s,\text{I}}(Q_1, \ldots, Q_k)$.

In the special case when $q_{ij} = Q_i/s$ ($1 \le i \le k$ and $1 \le j \le s$) so that $p = 1 - \sum_{i=1}^{k} Q_i \equiv P$ (say), the distribution in (42.62) reduces to

$$P(\boldsymbol{x}) = -(\log P)^{-1} \sum_{\sum_j jx_{ij}=x_i} \frac{(\sum_i \sum_j x_{ij} - 1)!}{\prod_i \prod_j x_{ij}!} \prod_i \left(\frac{Q_i}{s}\right)^{\sum_j x_{ij}},$$

$$x_i = 0, 1, \ldots \text{ for } 1 \le i \le k, \sum_i x_i > 0, \tag{42.64}$$

which is called a *Type II multivariate logarithmic series distribution of order s* with parameters $Q_1, \ldots, Q_k$, and is denoted by $\text{MLS}_{s,\text{II}}(Q_1, \ldots, Q_k)$.

Multivariate logarithmic series distributions of order s also serve as limiting forms of the multivariate negative binomial distributions of order s. Specifically, let $\boldsymbol{X}$ be a $(k \times 1)$ random vector distributed as $\text{MNB}_{s,\text{I}}(r; Q_1, \ldots, Q_k)$. Then, as $r \to 0$ the probability mass function of $\boldsymbol{X}$ tends to $\text{MLS}_{s,\text{I}}(Q_1, \ldots, Q_k)$ in (42.63). Similarly, if $\boldsymbol{X}$ is a $(k \times 1)$ random vector distributed as $\text{MNB}_{s,\text{II}}(r; Q_1, \ldots, Q_k)$, then as $r \to 0$ the probability mass function of $\boldsymbol{X}$ tends to $\text{MLS}_{s,\text{II}}(Q_1, \ldots, Q_k)$.

Further, for the multivariate logarithmic series distribution of order s in (42.62), the joint probability generating function is

$$G_{\boldsymbol{X}}(\boldsymbol{t}) = -\alpha \log \left(1 - \sum_i \sum_j q_{ij}\, t_i^j\right), \quad |t_i| \le 1, \ 1 \le i \le k, \tag{42.65}$$

using which it can be shown that

$$E[X_i] = \frac{\alpha}{p} \sum_j j\, q_{ij}, \qquad 1 \le i \le k, \tag{42.66}$$

$$\text{var}(X_i) = \frac{\alpha}{p} \left[\sum_j j^2 q_{ij} + \left(\frac{1-\alpha}{p}\right) \left(\sum_j j\, q_{ij}\right)^2\right], \quad 1 \le i \le k, \tag{42.67}$$

and

$$\text{cov}(X_i, X_t) = -\frac{\alpha(\alpha - 1)}{p^2} \left(\sum_j j\, q_{ij}\right) \left(\sum_j j\, q_{tj}\right), \quad 1 \le i \ne t \le k. \tag{42.68}$$

It is well known that a univariate negative binomial distribution is derived if a Poisson distribution with mean $-r \log p$ is generalized by a univariate logarithmic series distribution; see, for example, Chapter 5. This result can be generalized to the multivariate case as follows. Let $\boldsymbol{X}_\ell$, $\ell = 1, 2, \ldots$, be independent $(k \times 1)$ random vectors identically distributed as $\mathrm{MLS}_s(q_{11}, \ldots, q_{ks})$, independently of a random variable N which is distributed as Poisson$(-r \log p)$. Then, the $(k \times 1)$ random vector $\sum_{\ell=1}^{N} \boldsymbol{X}_\ell$ is distributed as $\mathrm{MNB}_s(r; q_{11}, \ldots, q_{ks})$.

Additional properties of multivariate logarithmic series distributions of order s can be established through a *multivariate s-point distribution* with probability mass function [Philippou, Antzoulakos, and Tripsiannis (1990)]

$$P(\boldsymbol{x}) = \begin{cases} \nu_{ij} & \text{if } x_i = j\ (1 \le j \le s) \text{ and } x_r = 0\ (1 \le r \ne i \le k) \\ 0 & \text{elsewhere}, \end{cases} \tag{42.69}$$

which is a multivariate extension of the s-point distribution with parameters $\nu_1, \ldots, \nu_k$ introduced by Hirano (1986). This distribution is denoted by $\mathrm{MK}(\nu_{11}, \ldots, \nu_{ks})$. Evidently, if $\boldsymbol{X} = (X_1, X_2, \ldots, X_k)'$ is distributed as $\mathrm{MK}(\nu_{11}, \ldots, \nu_{ks})$, the probability generating function of $\boldsymbol{X}$ is

$$G_{\boldsymbol{X}}(\boldsymbol{t}) = \sum_i \sum_j \nu_{ij} t_i^j, \qquad |t_i| \le 1,\ 1 \le i \le k. \tag{42.70}$$

Now if $\boldsymbol{X}$ is distributed as $\mathrm{MLS}_s(q_{11}, \ldots, q_{ks})$ where $q_{ij}/\sum_i \sum_j q_{ij} \to \nu_{ij}$ as $q_{ij} \to 0$ for $1 \le i \le k$ and $1 \le j \le s$, then $\boldsymbol{X}$ weakly converges to $\mathrm{MK}(\nu_{11}, \ldots, \nu_{ks})$. This is a multivariate extension of a result due to Hirano and Aki (1987).

A *modified multivariate logarithmic series distribution of order s* with parameters $\delta, q_{11}, \ldots, q_{ks}$ $(0 < \delta < 1, 0 < q_{ij} < 1$ for $1 \le i \le k$ and $1 \le j \le s, \sum_i \sum_j q_{ij} < 1)$, denoted by $\mathrm{MoMLS}_s(\delta; q_{11}, \ldots, q_{ks})$, has joint probability mass function

$$P(\boldsymbol{x}) = \begin{cases} \delta & \text{if } x_1 = \cdots = x_k = 0 \\ (1-\delta)\alpha \displaystyle\sum_{\sum_j j x_{ij} = x_i} \frac{(\sum_i \sum_j x_{ij} - 1)!}{\prod_i \prod_j x_{ij}!} \prod_i \prod_j q_{ij}^{x_{ij}} & \\ \quad \text{if } x_i = 0, 1, \ldots \text{ for } 1 \le i \le k,\ \sum_{i=1}^k x_i > 0, \end{cases} \tag{42.71}$$

where $\alpha = -(\log p)^{-1}$ and $p = 1 - q_{11} - \cdots - q_{ks}$. The joint probability generating function of this distribution is

$$G_{\boldsymbol{X}}(\boldsymbol{t}) = \delta + (1-\delta)(\log p)^{-1} \log\left(1 - \sum_i \sum_j q_{ij} t_i^j\right),$$

$$|t_i| \le 1,\ 1 \le i \le k. \tag{42.72}$$

Philippou, Antzoulakos, and Tripsiannis (1990) proved an interesting relationship which connects the multivariate logarithmic series distributions of order s and the modified multivariate logarithmic series distributions of order s. Specifically, these authors showed that if $\boldsymbol{X} = (X_1, \ldots, X_k)'$ is distributed as $\mathrm{MLS}_s(q_{11}, \ldots, q_{ks})$, then $\boldsymbol{X} = (X_1, \ldots, X_\ell)'$ $(1 \le \ell < k)$ is distributed as $\mathrm{MoMLS}_s(\delta; Q_{11}, \ldots, Q_{\ell s})$, where

$$\delta = (\log p)^{-1} \log\left(1 - \sum_{i=\ell+1}^{k} \sum_j q_{ij}\right)$$

and

$$Q_{ab} = q_{ab} \Big/ \left(1 - \sum_{i=\ell+1}^{k} \sum_{j} q_{ij}\right), \qquad 1 \le a \le \ell,\ 1 \le b \le s\,.$$

6 MULTINOMIAL DISTRIBUTIONS OF ORDER s

A distribution with joint probability mass function

$$P(\boldsymbol{x}) = \sum_{\sum_j j x_{ij} = x_i} \binom{n}{x_{11}, \ldots, x_{ks}, n - \sum_i \sum_j x_{ij}} q^{n - \sum_i \sum_j x_{ij}} \prod_i \prod_j p_{ij}^{x_{ij}},$$
$$x_i = 0, 1, 2, \ldots, ns,\ 1 \le i \le k\,, \tag{42.73}$$

where $0 < q_{ij} < 1$ and $p + \sum_i \sum_j q_{ij} = 1$, is called a *multinomial distribution of order s* with parameters $n, p_{11}, \ldots, p_{ks}$ and is denoted by $\mathrm{M}_s(n; p_{11}, \ldots, p_{ks})$. Evidently, the joint probability generating function of this distribution is

$$G_{\boldsymbol{X}}(\boldsymbol{t}) = \left(q + \sum_i \sum_j p_{ij} t_i^j\right)^n, \qquad |t_i| \le 1,\ 1 \le i \le k\,. \tag{42.74}$$

This distribution appears in Steyn (1963) and Panaretos and Xekalaki (1986), but the *order s* terminology is due to Philippou, Antzoulakos, and Tripsiannis (1990).

One of the uses of this distribution is to provide another genesis for the multivariate negative binomial distribution of order s which is as follows. Let $\boldsymbol{X}$ and N be a random vector and a univariate random variable, respectively, such that $\boldsymbol{X} \mid (N = n)$ is distributed as $\mathrm{M}_s(n; p_{11}, \ldots, p_{ks})$, and $\Pr[N = n] = \frac{\Gamma(n+r)}{\Gamma(r)\,n!}\, \theta^r (1-\theta)^n$ for $0 < \theta < 1$, $r > 0$, and $n = 0, 1, \ldots$ Then the unconditional distribution of $\boldsymbol{X}$ is $\mathrm{MNB}_s(r; q_{11}, \ldots, q_{ks})$, where $q_{ij} = (1-\theta)p_{ij}/\{1 - q(1-\theta)\}$ for $1 \le i \le k$ and $1 \le j \le s$, and $p = 1 - \sum_i \sum_j q_{ij}$. This result is a generalization of the corresponding result for the univariate negative binomial distribution due to Sibuya, Yoshimura, and Shimizu (1964).

Antzoulakos and Philippou (1994) have presented alternative expressions (in terms of binomial coefficients) for the probability mass functions for most of the distributions discussed in this section. Finally, it needs to be mentioned once again that a number of other multivariate distributions of order s have also been discussed earlier in Chapters 35–41.

7 MULTIVARIATE PÓLYA AND INVERSE PÓLYA DISTRIBUTIONS OF ORDER s

Multivariate inverse Pólya distributions of order s arise from an Eggenberger–Pólya sampling scheme when the sampling continues until m_0 non-overlapping runs of length s, composed entirely of balls of color C_0, are obtained. Then, with X_i denoting the number of balls of color C_i ($i = 1, 2, \dots, k$), Tripsiannis and Philippou (1994) derived the joint probability mass function of $(X_1, X_2, \dots, X_k)'$ as

$$\Pr[X_1 = x_1,\ X_2 = x_2, \dots,\ X_k = x_k]$$
$$= \sum_{\sum_{j=1}^{s} j y_{ij} = x_i} \cdots \sum \binom{y + m_0 - 1}{m_0 - 1, y_{11}, \dots, y_{ks}} \left\{ \prod_{i=1}^{k} a_i^{[y_i, c]} \right\} \frac{a_0^{[y - x + m_0 s, c]}}{a^{[y + m_0 s, c]}}, \qquad (42.75)$$

where

$$y_i = \sum_{j=1}^{s} y_{ij},\ y = \sum_{i=1}^{k} y_i \quad \text{and} \quad x = \sum_{i=1}^{k} x_i\,.$$

Clearly, for the case when $s = 1$, this distribution reduces to the multivariate inverse Pólya–Eggenberger distribution seen earlier in Chapter 40. For the case when $m_0 = 1$ (which means there is only one initial ball of color C_0), the distribution becomes the *generalized Waring distribution of order s* discussed by Panaretos and Xekalaki (1986). For the case when $c = -1$ (which means the balls selected are not replaced), the distribution becomes the *multivariate inverse hypergeometric distribution of order s*.

Distributions of this form had been derived earlier by Philippou and Tripsiannis (1991) using the following sampling scheme:

> An urn contains $s(a_0 + a_1 + \cdots a_k)$ balls of which sa_0 are of color C_0 and a_i are of color C_{ij} for $i = 1, 2, \dots, k$ and $j = 1, 2, \dots, s$. A ball is drawn at random from the urn and replaced. If it is of color C_{ij} for any $j = 1, 2, \dots, s$, then c balls of each of colors $C_{i1}, \dots, C_{is}$ are added to the urn. This sampling is repeated until m_0 balls of color C_0 have been selected. With X_i denoting the total number of occasions when a ball of one of colors $C_{i1}, C_{i2}, \dots, C_{is}$ has been selected ($i = 1, 2, \dots, k$), the joint distribution of $(X_1, X_2, \dots, X_k)'$ is of the form given above.

Further, the distribution derived by Philippou, Tripsiannis, and Antzoulakos (1989), by compounding the Dirichlet $(\alpha_0, \alpha_1, \dots, \alpha_k)$ distribution with a multivariate negative binomial distribution of order s, has the joint probability generating function

$$P(x_1, x_2, \dots, x_k) = \sum_{\sum_{j=1}^{s} j y_{ij} = x_i} \cdots \sum \frac{\Gamma(m_0 + y)}{\Gamma(m_0) \prod_{i=1}^{k} \prod_{j=1}^{s} \Gamma(y_{ij} + 1)}$$
$$\times \frac{\Gamma(\alpha)\Gamma(\alpha_0 + y - x + m_0 s)}{\Gamma(\alpha + y + m_0 s)\Gamma(\alpha_0)} \prod_{i=1}^{k} \frac{\Gamma(\alpha_i + y_i)}{\Gamma(\alpha_i)}. \qquad (42.76)$$

This is equivalent to the multivariate inverse Pólya distribution of order s given above if we take $\alpha_i = a_i/c$ and $\alpha = a/c$. As α, α_i, and m_0 all tend to ∞ such that

$m_0\alpha_i/\alpha \rightarrow \lambda_i$, $\alpha_i/\alpha \rightarrow 0$ and $m_0/\alpha \rightarrow 0$, this distribution tends to a multivariate Poisson distribution with parameters $\lambda_1, \lambda_2, \ldots, \lambda_k$ discussed earlier in Chapter 37. If $\alpha_i/\alpha \rightarrow q_i$ as $\alpha_i \rightarrow \infty$ and $\alpha_0 \rightarrow k$ $(0 < q_i < 1, \sum_{i=1}^{k} q_i < 1)$, the distribution tends to a multivariate negative binomial distribution discussed earlier in Chapter 36. Also see Philippou, Tripsiannis, and Antzoulakos (1989).

Tripsiannis (1993) has also discussed a more elaborate sampling scheme in which the number of balls of colors $C_{i1}, C_{i2}, \ldots, C_{is}$ added when the chosen ball has any one of these colors also depends on the second subscript; that is, c_{ij} balls of color C_{ij} are added. These distributions discussed by Tripsiannis (1993) are called *modified multivariate Pólya distributions of order s*. He has specifically defined two forms of these distributions, called *Type I* and *Type II modified multivariate Pólya distributions of order s*, which include many known distributions as special cases.

BIBLIOGRAPHY

Aki, S. (1985). Discrete distributions of order k on a binary sequence, *Annals of the Institute of Statistical Mathematics*, **37**, 205–224.

Aki, S. (1992). Waiting time problems for a sequence of discrete random variables, *Annals of the Institute of Statistical Mathematics*, **44**, 363–378.

Aki, S., Balakrishnan, N., and Mohanty, S. G. (1997). Sooner and later waiting time problems for success and failure runs in higher order Markov dependent trials, *Annals of the Institute of Statistical Mathematics*, **49** (to appear).

Aki, S., and Hirano, K. (1988). Some characteristics of the binomial distribution of order k and related distributions, in *Statistical Theory and Data Analysis II, Proceedings of the Second Pacific Area Statistical Conference* (Ed. K. Matusita), pp. 211–222, Amsterdam: North-Holland.

Aki, S., and Hirano, K. (1993). Discrete distributions related to succession events in a two-state Markov chain, in *Statistical Sciences and Data Analysis, Proceedings of the Third Pacific Area Statistical Conference* (Eds. K. Matusita, M. L. Puri, and T. Hayakawa), pp. 467–474, Zeist: VSP International Science Publishers.

Aki, S., and Hirano, K. (1994). Distributions of numbers of failures and successes until the first consecutive k successes, *Annals of the Institute of Statistical Mathematics*, **46**, 193–202.

Aki, S., and Hirano, K. (1995). Joint distributions of numbers of success-runs and failures until the first consecutive k successes, *Annals of the Institute of Statistical Mathematics*, **47**, 225–235.

Aki, S., Kuboki, H., and Hirano, K. (1984). On discrete distributions of order k, *Annals of the Institute of Statistical Mathematics*, **36**, 431–440.

Antzoulakos, D. L., and Philippou, A. N. (1991). A note on the multivariate negative binomial distributions of order k, *Communications in Statistics—Theory and Methods*, **20**, 1389–1399.

Antzoulakos, D. L., and Philippou, A. N. (1994). Expressions in terms of binomial coefficients for some multivariate distributions of order k, in *Runs and Patterns in Probability* (Eds. A. P. Godbole and S. G. Papastavridis), pp. 1–14, The Netherlands: Kluwer Academic Publishers.

Balakrishnan, N. (1996a). On the joint distributions of numbers of success-runs and failures until the first consecutive k successes, *submitted for publication*.

Balakrishnan, N. (1996b). Joint distributions of numbers of success-runs and failures until the first consecutive k successes in a binary sequence, *Annals of the Institute of Statistical Mathematics*, **48** (to appear).

Balakrishnan, N., Balasubramanian, K., and Viveros, R. (1995). Start-up demonstration tests under correlation and corrective action, *Naval Research Logistics*, **42**, 1271–1276.

Balakrishnan, N., and Johnson, N. L. (1997). A recurrence relation for discrete multivariate distributions and some applications to multivariate waiting time problems, in *Advances in the Theory and Practice of Statistics—A Volume in Honor of Samuel Kotz* (Eds. N. L. Johnson and N. Balakrishnan), New York: John Wiley & Sons (to appear).

Balakrishnan, N., Mohanty, S. G., and Aki, S. (1997). Start-up demonstration tests under Markov dependence model with corrective actions, *Annals of the Institute of Statistical Mathematics*, **49** (to appear).

Balasubramanian, K., Viveros, R., and Balakrishnan, N. (1993). Sooner and later waiting time problems for Markovian Bernoulli trials, *Statistics & Probability Letters*, **18**, 153–161.

Benvenuto, R. J. (1984). The occurrence of sequence patterns in Ergodic Markov chains, *Stochastic Processes and Their Applications*, **17**, 369–373.

Chryssaphinou, O., and Papastavridis, S. (1990). The occurrence of sequence patterns in repeated dependent experiments, *Theory of Probability and Its Applications*, **35**, 145–152.

Chryssaphinou, O., Papastavridis, S., and Tsapelas, T. (1994). On the waiting time appearance of given patterns, in *Runs and Patterns in Probability* (Eds. A. P. Godbole and S. G. Papastavridis), pp. 231–241, The Netherlands: Kluwer Academic Publishers.

Feder, P. I. (1974). Problem solving: Markov chain method, *Industrial Engineering*, **6**, 23–25.

Fu, J. C. (1986). Reliability of consecutive-k-out-of-n:F systems with $(k-1)$–step Markov dependence, *IEEE Transactions on Reliability*, **R–35**, 602–606.

Fu, J. C., and Koutras, M. V. (1994a). Poisson approximations for 2-dimensional patterns, *Annals of the Institute of Statistical Mathematics*, **46**, 179–192.

Fu, J. C., and Koutras, M. V. (1994b). Distribution theory of runs: A Markov chain approach, *Journal of the American Statistical Association*, **89**, 1050–1058.

Gerber, H. V., and Li, S. R. (1981). The occurrence of sequence patterns in repeated experiments and hitting times in a Markov chain, *Stochastic Processes and Their Applications*, **11**, 101–108.

Godbole, A. P., and Papastavridis, S. G. (Eds.) (1994). *Runs and Patterns in Probability*, The Netherlands: Kluwer Academic Publishers.

Hirano, K. (1986). Some properties of the distributions of order k, in *Fibonacci Numbers and Their Applications* (Eds. A. N. Philippou, G. E. Bergum, and A. F. Horadam), pp. 43–53, Dordrecht: Reidel.

Hirano, K., and Aki, S. (1987). Properties of the extended distributions of order k, *Statistics & Probability Letters*, **6**, 67–69.

Hirano, K., and Aki, S. (1993). On number of occurrences of success runs of specified length in a two-state Markov chain, *Statistica Sinica*, **3**, 313–320.

Koutras, M. V., and Alexandrou, V. A. (1995). Runs, scans and urn model distributions: A unified Markov chain approach, *Annals of the Institute of Statistical Mathematics*, **47**, 743–766.

Lambiris, M., and Papastavridis, S. G. (1985). Exact reliability formulas for linear and circular consecutive-k-out-of-n:F systems, *IEEE Transactions on Reliability*, **R–34**, 124–126.

Mohanty, S. G. (1994). Success runs of length k in Markov dependent trials, *Annals of the Institute of Statistical Mathematics*, **46**, 777–796.

Panaretos, J., and Xekalaki, E. (1986). On generalized binomial and multinomial distributions and their relation to generalized Poisson distributions, *Annals of the Institute of Statistical Mathematics*, **38**, 223–231.

Papastavridis, S., and Lambiris, M. (1987). Reliability of a consecutive-k-out-of-n:F system for Markov–dependent components, *IEEE Transactions on Reliability*, **R–36**, 78–79.

Philippou, A. N. (1983). Poisson and compound Poisson distributions of order k and some of their properties, *Zapiski Nauchnykh Seminarov Leningradskogo Otdelinya Matematicheskogo Instituta im. V. A. Steklova AN SSSR*, **130**, 175–180 (in Russian, English summary).

Philippou, A. N. (1988). On multiparameter distributions of order k, *Annals of the Institute of Statistical Mathematics*, **40**, 467–475.

Philippou, A. N., Antzoulakos, D. L., and Tripsiannis, G. A. (1989). Multivariate distributions of order k, *Statistics & Probability Letters*, **7**, 207–216.

Philippou, A. N., Antzoulakos, D. L., and Tripsiannis, G. A. (1990). Multivariate distributions of order k, II, *Statistics & Probability Letters*, **10**, 29–35.

Philippou, A. N., Georghiou, C., and Philippou, G.N. (1983). A generalized geometric distribution and some of its applications, *Statistics & Probability Letters*, **1**, 171–175.

Philippou, A. N., and Tripsiannis, G. A. (1991). Multivariate Pólya and inverse Pólya distributions of order k, *Biometrical Journal*, **33**, 225–236.

Philippou, A. N., Tripsiannis, G. A., and Antzoulakos, D. L. (1989). New Pólya and inverse Pólya distributions of order k, *Communications in Statistics—Theory and Methods*, **18**, 2125–2137.

Rajarshi, M. B. (1974). Success runs in a 2–state Markov chain, *Journal of Applied Probability*, **11**, 190–192.

Schwager, S. J. (1983). Run probabilities in sequences of Markov dependent trails, *Journal of the American Statistical Association*, **78**, 168–175.

Sibuya, M., Yoshimura, I., and Shimizu, R. (1964). Negative multinomial distribution, *Annals of the Institute of Statistical Mathematics*, **16**, 409–426.

Steyn, H. S. (1963). On approximations for the discrete distributions obtained from multiple events, *Proceedings, Koninklijke Akademie Van Wetenschappen*, **66a**, 85–96.

Tripsiannis, G. A. (1993). Modified multivariate Pólya and inverse Pólya distributions of order k, *Journal of the Indian Society for Statistics and Operations Research*, **14**, 1–11.

Tripsiannis, G. A., and Philippou, A. N. (1994). A new multivariate inverse Pólya distribution of order k, *Technical Report No. 10*, Department of Mathematics, University of Patras, Greece.

Uchida, M., and Aki, S. (1995). Sooner and later waiting time problems in a two-state Markov chain, *Annals of the Institute of Statistical Mathematics*, **47**, 415–433.

Viveros, R., and Balakrishnan, N. (1993). Statistical inference from start-up demonstration test data, *Journal of Quality Technology*, **25**, 119–130.

Viveros, R., Balasubramanian, K., and Balakrishnan, N. (1994). Binomial and negative binomial analogues under correlated Bernoulli trials, *The American Statistician*, **48**, 243–247.

Wang, Y. H., and Ji, S. (1995). Limit theorems for the number of occurrences of consecutive k successes in n Markovian trials, *Journal of Applied Probability*, **32**, 727–735.

Wang, Y. H., and Liang, Z. Y. (1993). The probability of occurrences of runs of length k in n Markovian Bernoulli trials, *The Mathematical Scientist*, **18**, 105–112.

Xekalaki, E., and Panaretos, J. (1989). On some distributions arising in inverse cluster sampling, *Communications in Statistics—Theory and Methods*, **18**, 355–366.

CHAPTER 43

Miscellaneous Distributions

1 MULTIVARIATE OCCUPANCY DISTRIBUTIONS

1.1 Introduction

Occupancy problems and discrete distributions arising from them have been a fertile ground for numerous investigations of probabilists for over last 50 years stemming perhaps from the well-known classical birthday problem discussed in practically all texts on probability theory. A detailed discussion of this topic is presented in the book by Johnson and Kotz (1977), where multivariate occupancy distributions are discussed in Chapter 3, Section 3.4. Here, we provide for completeness the main results of Johnson and Kotz as reflected in their 1976 paper. We hope that this description will stimulate some additional research on this topic.

If the random variables $N_1, N_2, \ldots, N_k$ jointly have a multinomial distribution with parameters $(n; p_1, p_2, \ldots, p_k)$ $(\sum_{i=1}^{k} p_i = 1)$ (see Chapter 35 for details), the distribution of the number of N's that are equal to zero (M_0, say) is an *occupancy distribution*. In this case, a more appropriate term might be a *nonoccupancy distribution* since M_0 is after all the number of categories unoccupied or unrepresented in the sample values. The distribution of M_0 is univariate of course (see Chapter 10, Section 4), but there are some related multivariate distributions—notably the joint distributions of $M_0, M_1, \ldots, M_r$, where M_i denotes the number of N's equal to i.

1.2 Distributions and Moments

Johnson and Kotz (1976) have described some other distributions of this kind. They considered a situation wherein there are b_g categories of "class I(g)," for $g = 1, 2, \ldots, m$, each of the b_g categories having probabilities of being chosen equal to p_g, and a remainder category "class II" with probability of being chosen as p_0. Evidently,

$$p_0 + \sum_{g=1}^{m} b_g p_g = 1 . \tag{43.1}$$

Now suppose that in n independent trials there are R_g observations of class I(g) ($g = 1, 2, \ldots, m$) and R_0 observations of class II. Then

$$R_0 + \sum_{g=1}^{m} R_g = n \tag{43.2}$$

and the joint distribution of $(R_0, R_1, \ldots, R_m)'$ is clearly multinomial with parameters $(n; p_0, b_1 p_1, \ldots, b_m p_m)$. The conditional distribution of the numbers V_{ng} of classes I(g) observed $[\boldsymbol{V}_n = (V_{n1}, V_{n2}, \ldots, V_{nm})']$, given $\boldsymbol{R} = (R_1, \ldots, R_m)'$, is

$$\begin{aligned}\Pr[\boldsymbol{V}_n = \boldsymbol{v} \mid \boldsymbol{R}] &= \prod_{g=1}^{m} \Pr[V_{ng} = v_g \mid R_g] \\ &= \prod_{g=1}^{m} \binom{b_g}{v_g} (\Delta^{v_g} 0^{r_g})\, b_g^{-R_g}, \end{aligned}\tag{43.3}$$

where Δ is the forward difference operator (see Chapter 1, Section A3), and

$$\Delta^j 0^\alpha = \sum_{i=0}^{j} (-1)^{j-i} \binom{j}{i} i^\alpha = \sum_{i=0}^{j} (-1)^i \binom{j}{i} (j-i)^\alpha . \tag{43.4}$$

Note that, conditional on $\boldsymbol{R}$, the V_{ng}'s are mutually independent.

Johnson and Kotz (1976) showed that $\Pr[\boldsymbol{V}_n = \boldsymbol{v}]$ can be expressed in the four following equivalent forms:

1. $$\begin{aligned}\Pr[\boldsymbol{V}_n = \boldsymbol{v}] &= E_{\boldsymbol{R}}\left[\Pr[\boldsymbol{V}_n = \boldsymbol{v} \mid \boldsymbol{R}]\right] \\ &= \prod_{g=1}^{m} \binom{b_g}{v_g} \Delta_g^{v_g} \left(p_0 + \sum_{g=1}^{m} p_g 0_g\right)^n ,\end{aligned}\tag{43.5}$$

   where Δ_g operates on 0_g in accordance with (43.4);

2. $$\begin{aligned}\Pr[\boldsymbol{V}_n = \boldsymbol{v}] = \prod_{g=1}^{m} \Bigg[\binom{b_g}{v_g} \sum_{\boldsymbol{h}=\boldsymbol{0}}^{\boldsymbol{v}} \cdots \sum (-1)^{\sum_{g=1}^{m}(v_g - h_g)} \left\{ \prod_{g=1}^{m} \binom{v_g}{h_g} \right\} \\ \times \left(p_0 + \sum_{g=1}^{m} h_g p_g\right)^n \Bigg];\end{aligned}\tag{43.6}$$

3. $$\begin{aligned}\Pr[\boldsymbol{V}_n = \boldsymbol{v}] = \sum_{\boldsymbol{h}=\boldsymbol{0}}^{\boldsymbol{v}} \cdots \sum \Bigg[(-1)^{\sum_{g=1}^{m}(v_g - h_g)} \Bigg\{ \prod_{g=1}^{m} \binom{b_g}{b_g - v_g, h_g, v_g - h_g} \\ \times \left(p_0 + \sum_{g=1}^{m} h_g p_g\right)^n \Bigg\} \Bigg];\end{aligned}\tag{43.7}$$

   and

4. $$\Pr[\boldsymbol{V}_n = \boldsymbol{v}] = \sum_{\boldsymbol{h}=\boldsymbol{0}}^{\boldsymbol{v}} \cdots \sum (-1)^{\sum_{g=1}^{m} h_g} \left[\prod_{g=1}^{m} \binom{b_g}{b_g - v_g, h_g, v_g - h_g} \times \left\{ p_0 + \sum_{g=1}^{m} (v_g - h_g) p_g \right\}^n \right]. \tag{43.8}$$

Further, by defining

$$Z_{gt} = \begin{cases} 1 & \text{if the } t\text{th category in class I}(g) \text{ is observed} \\ 0 & \text{otherwise} \end{cases}$$

and noting that $V_{ng} = \sum_{t=1}^{b_g} Z_{gt}$, we have

$$E[V_{ng}] = b_g\, E[Z_{gt}] = b_g\{1 - (1 - p_g)^n\}, \tag{43.9}$$

$$\text{var}(V_{ng}) = b_g\{(1 - p_g)^n - (1 - 2p_g)^n\} - b_g^2\{(1 - p_g)^{2n} - (1 - 2p_g)^n\}, \tag{43.10}$$

$$\text{cov}(V_{ng}, V_{nh}) = b_g b_h \,\text{cov}(Z_{gt}, Z_{ht'}) = b_g b_h\{(1 - p_g - p_h)^n - (1 - p_g)^n (1 - p_h)^n\}, \tag{43.11}$$

and

$$\text{corr}(V_{ng}, V_{nh}) = \frac{\sqrt{b_g b_h}\{(1 - p_g - p_h)^n - (1 - p_g)^n(1 - p_h)^n\}}{\left\{ [\{(1 - p_g)^n - (1 - 2p_g)^n\} - b_g\{(1 - p_g)^{2n} - (1 - 2p_g)^n\}] \times [\{(1 - p_h)^n - (1 - 2p_h)^n\} - b_h\{(1 - p_h)^{2n} - (1 - 2p_h)^n\}] \right\}^{1/2}} \tag{43.12}$$

The regression of V_{ng} on V_{nh} is

$$E[V_{ng} \mid V_{nh} = v_{nh}] = b_g \left\{ 1 - \frac{\Delta^{v_{nh}}(1 - b_h p_h - p_g + p_h 0)^n}{\Delta^{v_{nh}}(1 - b_g p_g + p_h 0)^n} \right\}. \tag{43.13}$$

Further details are presented in Johnson and Kotz (1976).

2 MULTIVARIATE MIXTURE DISTRIBUTIONS

Multivariate mixtures of discrete distributions have been discussed in detail and in an organized manner in some recent books including Titterington, Smith, and Makov (1985). Here, we provide a brief sketch of some basic results on this topic. Of course, individual specific discrete multivariate distributions have also been discussed in Chapters 35–42.

Mosimann (1962, 1963) compounded multinomial and negative multinomial distributions with a Dirichlet distribution for the parameter $\boldsymbol{p}$ of the multinomial, thus producing multivariate Pólya and multivariate inverse Pólya distributions, respectively (see Chapter 40). The

$$\text{Multinomial}(n; \boldsymbol{p}) \bigwedge_{\boldsymbol{p}} \text{Dirichlet} \tag{43.14}$$

distribution has been used by Burtkiewicz (1971), Dyczka (1973), and Jeuland, Bass, and Wright (1980). Kunte (1977) discussed the special case when $\boldsymbol{p}$ is uniformly distributed, in relation to "characterization of Bose–Einstein statistics"; see also Chen (1979) and Levin and Reeds (1975). Taillie *et al.* (1975) presented good descriptions of these distributions in a very useful survey article on the genesis of multivariate distributions (discrete as well as continuous).

One can form a *finite mixture multivariate distribution* by mixing ℓ k-variate distributions with probabilities $\Pi_1, \ldots, \Pi_\ell$ $\left(\sum_{i=1}^{\ell} \Pi_i = 1\right)$. Specifically, let $\boldsymbol{X}_i$ ($i = 1, 2, \ldots, \ell$) be independent k-dimensional random variables with joint probability mass functions $P_{\boldsymbol{X}_i}(\boldsymbol{x})$ for $\boldsymbol{x} \in S$. Then, the random variable $\boldsymbol{X} \overset{d}{=} \boldsymbol{X}_i$ with probability Π_i ($i = 1, 2, \ldots, \ell$) has a finite mixture distribution with joint probability mass function

$$P_{\boldsymbol{X}}(\boldsymbol{x}) = \sum_{i=1}^{\ell} \Pi_i \, P_{\boldsymbol{X}_i}(\boldsymbol{x}) \qquad \text{for } \boldsymbol{x} \in S\,.$$

It is then easily seen that the joint probability generating function of $\boldsymbol{X}$ is

$$G_{\boldsymbol{X}}(\boldsymbol{t}) = \sum_{i=1}^{\ell} \Pi_i \, G_{\boldsymbol{X}_i}(\boldsymbol{t})\,.$$

Similarly, the mixed moments of $\boldsymbol{X}$ can also be derived easily. For example,

$$E[\boldsymbol{X}] = \sum_{i=1}^{k} \Pi_i \, E[\boldsymbol{X}_i]\,.$$

3 GENERALIZATIONS

For multivariate distributions, there is a variety of methods of generalization. For example, we mention a paper by Charalambides and Papageorgiou (1981) in which the authors have constructed *bivariate Poisson–binomial distributions*. [These authors have referred to the univariate distribution

$$\text{Poisson}(\theta) \bigvee \text{Binomial}(n, p) \tag{43.15}$$

as a Poisson–binomial distribution. This distribution needs to be distinguished from the Poisson–binomial distribution derived as the distribution of number of successes (under Poisson sampling scheme) in n independent trials with probability of success p varying from trial to trial, and also from the other Poisson-binomial distribution defined as

$$\text{Binomial}(n, p) \bigwedge_{n} \text{Poisson}(\theta). \;]$$

The following three types of bivariate Poisson–binomial distributions have been introduced by Charalambides and Papageorgiou (1981):

1. Poisson(θ) $\bigvee$ Bivariate binomial($n; \boldsymbol{p}$)
   The probability generating function is

$$\exp\left[\theta\{(p_{00} + p_{10}t_1 + p_{01}t_2 + p_{11}t_1t_2)^n - 1\}\right]. \tag{43.16}$$

2. Bivariate Poisson(θ) $\bigvee \begin{cases}\text{Binomial}(n_1, p_1)\\ \text{Binomial}(n_2, p_2)\end{cases}$
   The probability generating function is

$$\exp\left[\sum_{i=1}^{2} \theta_i\{(q_i + p_it_i)^{n_i} - 1\} + \theta_{12}\{(q_1 + p_1t_1)^{n_1}(q_2 + p_2t_2)^{n_2} - 1\}\right]. \tag{43.17}$$

3. $X_1 = X_1' + X$ and $X_2 = X_2' + X$, where X_i' is Poisson(θ_i) $\bigvee$ Binomial(n, p) ($i = 1, 2$) and X is Poisson(θ_{12}) $\bigvee$ Binomial(n, p), with X_1', X_2' and X being mutually independent.
   The probability generating function is

$$\exp\left[\sum_{i=1}^{2} \theta_i\{(q + pt_i)^n - 1\} + \theta_{12}\{(q + pt_1t_2)^n - 1\}\right]. \tag{43.18}$$

The above systems can all be extended to k variables in a natural way, though there is scope for considerable elaboration from hybrid forms of generalizations.

4 MULTIVARIATE GENERALIZED (RANDOM SUM) DISTRIBUTIONS

Multivariate Hermite distributions can be constructed in an analogous manner to univariate Hermite distributions [see Chapter 9, Section 4]. In the joint probability generating function

$$G_X(\boldsymbol{t}) = G_2\left(G_1(\boldsymbol{t})\right), \tag{43.19}$$

taking G_2 to be the probability generating function of a Poisson(θ) distribution so that

$$G_X(\boldsymbol{t}) = \exp[\theta\{G_1(\boldsymbol{t}) - 1\}], \tag{43.20}$$

we expand $G_1(\boldsymbol{t})$ in a Taylor series in $t_1, \ldots, t_k$ giving

$$G_X(\boldsymbol{t}) = \exp\left[\sum_{i_1=0}^{\infty} \cdots \sum_{i_k=0}^{\infty} \theta_{i_1 i_2 \ldots i_k}\left(\prod_{j=1}^{k} t_j^{i_j} - 1\right)\right]. \tag{43.21}$$

Let us now specialize by supposing $\theta_{i_1 i_2 \ldots i_k} = 0$ for $\sum_{j=1}^{k} i_j > 2$. For $m = 2$, (43.21) then leads to

$$\begin{aligned} G_{X_1,X_2}(t_1, t_2) = \exp[&\theta_{10}(t_1 - 1) + \theta_{20}(t_1^2 - 1) + \theta_{01}(t_2 - 1) \\ &+ \theta_{02}(t_2^2 - 1) + \theta_{11}(t_1 t_2 - 1)], \end{aligned} \tag{43.22}$$

which is the probability generating function of the *bivariate Hermite distribution* [Kemp and Papageorgiou (1982)]. Kemp and Loukas (1978) have pointed out that this distribution can also be generated as the joint distribution of

$$X_1 = Y_1 + Y_{11} \qquad \text{and} \qquad X_2 = Y_2 + Y_{11}, \tag{43.23}$$

where Y_1, Y_2 and Y_{11} are mutually independent random variables with $Y_1 \rightsquigarrow$ Hermite(θ_{10}, θ_{20}), $Y_2 \rightsquigarrow$ Hermite(θ_{01}, θ_{02}) and $Y_{11} \rightsquigarrow$ Poisson(θ_{11}). As a result, we also have

$$X_1 = Y_{10} + 2\,Y_{20} + Y_{11} \text{ and } X_2 = Y_{01} + 2\,Y_{02} + Y_{11}, \tag{43.24}$$

where the Y's are mutually independent random variables with $Y_{ij} \rightsquigarrow$ Poisson(θ_{ij}).

Steyn (1976) calls the multivariate Hermite distribution the *multivariate Poisson normal distribution* and gives two further geneses for it:

(i) As the limit of multivariate binomial distributions

and

(ii) As a mixture of joint distributions of k independent Poisson(θ_i) random variables ($i = 1, 2, \ldots, k$), where the θ's have a joint multinomial distribution [see Kawamura (1979)].

Milne and Westcott (1993) have discussed the problem of characterizing those functions of the general form

$$G(\boldsymbol{t}) = \exp\left\{\sum_{0 \le \boldsymbol{\ell}'\mathbf{1} \le m} \theta_{\boldsymbol{\ell}}(\boldsymbol{t}^{\boldsymbol{\ell}} - 1)\right\},$$

where $\boldsymbol{t} = (t_1, \ldots, t_k)' \in \mathbf{R}^k$, $\boldsymbol{\ell} \in N^k$ with $N = \{0, 1, 2, \ldots\}$, and $\boldsymbol{t}^{\boldsymbol{\ell}} = t_1^{\ell_1}\, t_2^{\ell_2}\, \ldots\, t_k^{\ell_k}$, which are probability generating functions. The distribution corresponding to these probability generating functions are called *generalized multivariate Hermite distributions*. These authors have discussed in particular the possibility of some of the coefficients θ_ℓ being negative.

5 MULTIVARIATE WEIGHTED DISTRIBUTIONS

A systematic study of weighted distributions started with the paper by Rao (1965), though the concept itself can be traced to the early work of Fisher (1934). Weighted distributions arise when the observations generated from a stochastic process are recorded with some weight function. Size-biased (length-biased) distributions arise when the sampling mechanism selects units with probability proportional to some measure of the unit size (length).

Let

$$\boldsymbol{X}^w = \begin{pmatrix} X_1 \\ \vdots \\ X_k \end{pmatrix}^w$$

be the multivariate weighted version of $\boldsymbol{X}$, and let the weight function $w(\boldsymbol{x})$, where $w : \boldsymbol{X} \to A \subseteq R^+$, be non-negative with finite and nonzero expectation. Then, the multivariate weighted probability mass function corresponding to $P(\boldsymbol{x})$ is defined to be

$$Q(\boldsymbol{x}) = \frac{w(\boldsymbol{x})\, P(\boldsymbol{x})}{E[w(\boldsymbol{X})]} . \tag{43.25}$$

Mahfoud and Patil (1982), Patil, Rao, and Ratnaparkhi (1986), and Arnold and Nagaraja (1991) have all discussed the bivariate weighted distributions, while Jain and Nanda (1995) have recently studied multivariate weighted distributions in great length. The last authors have established several interesting properties of multivariate weighted distributions, some of which are presented below.

1. Let $\overline{F}_{\boldsymbol{X}}(\boldsymbol{x}) = \Pr[X_1 \geq x_1, \ldots, X_k \geq x_k]$ be the joint survival function of non-negative random vector $\boldsymbol{X}$, and let $r_{\boldsymbol{X}}(\boldsymbol{t}) = \nabla R_{\boldsymbol{X}}(\boldsymbol{t}) = (r_{1\boldsymbol{X}}(\boldsymbol{t}), \ldots, r_{k\boldsymbol{X}}(\boldsymbol{t}))$ be the corresponding multivariate hazard rate where $R_{\boldsymbol{X}}(\boldsymbol{t}) = -\ln \overline{F}_{\boldsymbol{X}}(\boldsymbol{t})$ is the joint cumulative hazard rate. Then, $\boldsymbol{X}$ is said to be *multivariate IFR* if $\frac{\overline{F}_{\boldsymbol{X}}(\delta\boldsymbol{1}+\boldsymbol{t})}{\overline{F}_{\boldsymbol{X}}(\boldsymbol{t})}$ is nonincreasing in $\boldsymbol{t}$ such that $\overline{F}_{\boldsymbol{X}}(\boldsymbol{t}) > 0$ for all $\delta \geq 0$ and $\boldsymbol{1} = (1, 1, \ldots, 1)'$, together with this condition on all the marginals [see Marshall (1975)].
   Let $A(\boldsymbol{x}) = E[w(\boldsymbol{X}) \mid X_1 > x_1, \ldots, X_k > x_k]$. Further, let $\frac{A(\delta\boldsymbol{1}+\boldsymbol{t})}{A(\boldsymbol{t})}$ be nonincreasing in $\boldsymbol{t}$. Then, $\boldsymbol{X}^w$ is multivariate IFR if $\boldsymbol{X}$ is multivariate IFR.
2. Let $\boldsymbol{Y}$ be another non-negative random vector with joint survival function $\overline{F}_{\boldsymbol{Y}}(\boldsymbol{x}) = \Pr[Y_1 \geq x_1, \ldots, Y_k \geq x_k]$ and with the corresponding multivariate hazard rate $r_{\boldsymbol{Y}}(\boldsymbol{t}) = \nabla R_{\boldsymbol{Y}}(\boldsymbol{t}) = (r_{1\boldsymbol{Y}}(\boldsymbol{t}), \ldots, r_{k\boldsymbol{Y}}(\boldsymbol{t}))$, where $R_{\boldsymbol{Y}}(t) = -\ln \overline{F}_{\boldsymbol{Y}}(\boldsymbol{t})$. Then, $\boldsymbol{X}$ is said to be larger than $\boldsymbol{Y}$ in *multivariate monotone likelihood ratio*

ordering ($\boldsymbol{X} \overset{MLR}{\geq} \boldsymbol{Y}$) if $P_{\boldsymbol{X}}(\boldsymbol{x})/P_{\boldsymbol{Y}}(\boldsymbol{x})$ is nondecreasing in $\boldsymbol{x} \geq \boldsymbol{0}$ [see Whitt (1982)]. $\boldsymbol{X}$ is said to be larger than $\boldsymbol{Y}$ in *multivariate failure rate ordering* ($\boldsymbol{X} \overset{MFR}{\geq} \boldsymbol{Y}$) if $r_{\boldsymbol{X}}(\boldsymbol{t}) \leq r_{\boldsymbol{Y}}(\boldsymbol{t})$ for all $\boldsymbol{t} \geq \boldsymbol{0}$ or, equivalently, if $\overline{F}_{\boldsymbol{X}}(\boldsymbol{x})/\overline{F}_{\boldsymbol{Y}}(\boldsymbol{x})$ is nondecreasing in $\boldsymbol{x} \geq \boldsymbol{0}$.

We then have:

(i) If $w(\boldsymbol{x})$ is nondecreasing in $\boldsymbol{x} \geq \boldsymbol{0}$, then $\boldsymbol{X}^w \overset{MLR}{\geq} \boldsymbol{X}$.

(ii) If $w(\boldsymbol{x})$ is nondecreasing in $\boldsymbol{x} \geq \boldsymbol{0}$ and $P_{\boldsymbol{X}}(\boldsymbol{x})$ is TP_2 (totally positive of order 2) in pairs, then $\boldsymbol{X}^w \overset{MFR}{\geq} \boldsymbol{X}$.

(iii) With $A(\boldsymbol{x})$ as defined in **1**, if $A(\boldsymbol{x})$ is nondecreasing in $\boldsymbol{x} \geq \boldsymbol{0}$, then $\boldsymbol{X}^w \overset{MFR}{\geq} \boldsymbol{X}$.

(iv) For any nonzero weight function $w(\boldsymbol{x})$, $\boldsymbol{X} \overset{MLR}{\geq} \boldsymbol{Y}$ if and only if $\boldsymbol{X}^w \overset{MLR}{\geq} \boldsymbol{Y}^w$.

3. A random variable $\boldsymbol{X}$ is said to be *multivariate totally positive of order 2* (MTP_2) if [Karlin and Rinott (1980a)]

$$P_{\boldsymbol{X}}(\boldsymbol{x} \bigvee \boldsymbol{y})\, P_{\boldsymbol{X}}(\boldsymbol{x} \bigwedge \boldsymbol{y}) \geq P_{\boldsymbol{X}}(\boldsymbol{x})\, P_{\boldsymbol{X}}(\boldsymbol{y})\ \forall\ \boldsymbol{x}, \boldsymbol{y}. \tag{43.26}$$

If the inequality is reversed in Eq. (43.26), then $\boldsymbol{X}$ is said to satisfy *multivariate reverse rule of order 2* (MRR_2); see Karlin and Rinott (1980b). $\boldsymbol{X}$ is said to be *associated* if $\text{cov}(h_1(\boldsymbol{X}), h_2(\boldsymbol{X})) \geq 0$ for any nondecreasing functions h_1 and h_2 for which $E[h_1(\boldsymbol{X})]$, $E[h_2(\boldsymbol{X})]$, and $E[h_1(\boldsymbol{X})h_2(\boldsymbol{X})]$ all exist.

Let the weight function $w(\boldsymbol{x})$ be MTP_2 [i.e., it satisfies the inequality in (43.26)]. Then $\boldsymbol{X}^w$ is MTP_2 if $\boldsymbol{X}$ is MTP_2. The same result holds for MRR_2. Furthermore, $\boldsymbol{X}^w$ is associated if $\boldsymbol{X}$ is MTP_2.

4. In the spirit of the concepts in **3**, Karlin and Rinott (1980a) have defined the ordering $P_{\boldsymbol{Y}} \overset{TP_2}{>} P_{\boldsymbol{X}}$ if

$$P_{\boldsymbol{Y}}(\boldsymbol{x} \bigvee \boldsymbol{y})\, P_{\boldsymbol{X}}(\boldsymbol{x} \bigwedge \boldsymbol{y}) \geq P_{\boldsymbol{X}}(\boldsymbol{x})\, P_{\boldsymbol{Y}}(\boldsymbol{y})\ \forall\ \boldsymbol{x}, \boldsymbol{y}. \tag{43.27}$$

If the inequality in (43.27) is reversed, we have the ordering $P_{\boldsymbol{Y}} \overset{RR_2}{>} P_{\boldsymbol{X}}$.

Now, let $P_{\boldsymbol{Y}} \overset{TP_2}{>} P_{\boldsymbol{X}}$ and $w_1(\boldsymbol{x})$ and $w_2(\boldsymbol{x})$ be two weight functions with $w_2 \overset{TP_2}{>} w_1$ [i.e., they satisfy the inequality in (43.27)]. Then $P_{\boldsymbol{Y}}^{w_2} \overset{TP_2}{>} P_{\boldsymbol{X}}^{w_1}$, where $P_{\boldsymbol{X}}^{w_1}$ is the probability mass function of the weighted distribution of $P_{\boldsymbol{X}}$ with weight function w_1, and $P_{\boldsymbol{Y}}^{w_2}$ is the probability mass function of the weighted distribution of $P_{\boldsymbol{Y}}$ with weight function w_2. The same result holds for RR_2.

Jain and Nanda (1995) have also presented some similar ordering results by considering the marginal distributions, regression functions and multiple correlation coefficients. Chapter 33 contains a discussion of orderings for continuous univariate distributions.

As an example of the multivariate weighted distributions, let us consider the case $\boldsymbol{X} = (X_1, \ldots, X_k)'$, where X_i's are independent Poisson random variables with X_i distributed as Poisson (θp_i) with $\theta > 0$, $0 < p_i < 1$, and $\sum_{i=1}^k p_i = 1$. Let us consider the weight function

$$w(x_1, \ldots, x_k) = \left\{ \prod_{i=1}^{k} \frac{\Gamma(x_i + \ell_i)}{\Gamma(\ell_i)} \right\} \left\{ \prod_{i=1}^{k} p_i^{x_i} \right\}^{-1} \times \frac{\Gamma(\ell_1 + \cdots + \ell_k)}{\Gamma(x_1 + \cdots + x_k + \ell_1 + \cdots + \ell_k)}, \quad \ell_1, \ldots, \ell_k > 0. \tag{43.28}$$

Then, the probability mass function of the weighted distribution is

$$P^w(\boldsymbol{x}) = \frac{w(\boldsymbol{x})\, P(\boldsymbol{x})}{E[w(\boldsymbol{X})]}, \tag{43.29}$$

where

$$P(\boldsymbol{x}) = \frac{1}{\prod_{i=1}^{k} x_i!}\, e^{-\theta}\, \theta^{x_1 + \cdots + x_k} \prod_{i=1}^{k} p_i^{x_i}. \tag{43.30}$$

Observing now that

$$\begin{aligned} E[w(\boldsymbol{X})] &= \sum_{x_1=0}^{\infty} \cdots \sum_{x_k=0}^{\infty} w(x_1, \ldots, x_k) P(x_1, \ldots, x_k) \\ &= \frac{\Gamma(\ell_1 + \cdots + \ell_k)}{\Gamma(\ell_1) \cdots \Gamma(\ell_k)}\, e^{-\theta} \sum_{x_1=0}^{\infty} \cdots \sum_{x_k=0}^{\infty} \left[\left\{ \prod_{i=1}^{k} \frac{\Gamma(x_i + \ell_i)}{x_i!} \right\} \times \frac{\theta^{x_1 + \cdots + x_k}}{\Gamma(x_1 + \cdots + x_k + \ell_1 + \cdots + \ell_k)} \right] \\ &= 1, \end{aligned}$$

we have from (43.29)

$$\begin{aligned} P^w(\boldsymbol{x}) &= \frac{\Gamma(\ell_1 + \cdots + \ell_k) \prod_{i=1}^{k} \Gamma(x_i + \ell_i)}{\Gamma(x_1 + \cdots + x_k + \ell_1 + \cdots + \ell_k) \prod_{i=1}^{k} \{\Gamma(\ell_i) x_i!\}}\, e^{-\theta}\, \theta^{\Sigma_{i=1}^{k} x_i} \\ &= \frac{\prod_{i=1}^{k} \binom{-\ell_i}{x_i}}{\binom{-\ell_1 - \cdots - \ell_k}{x_1 + \cdots + x_k}}\, \frac{e^{-\theta}\, \theta^{\Sigma_{i=1}^{k} x_i}}{(\Sigma_{i=1}^{k} x_i)!}. \end{aligned} \tag{43.31}$$

The distribution (43.31) is called the *multivariate Poisson-negative hypergeometric distribution*; for $k = 2$, it reduces to the bivariate Poisson-negative hypergeometric distribution discussed by Patil, Rao, and Ratnaparkhi (1986). Jain and Nanda (1995) have also shown that the distribution (43.31) can be obtained as a mixture of the joint distribution of independent random variables X_i $(i = 1, 2, \ldots, k)$ with X_i distributed as Poisson (θp_i), $p_i > 0$, $\sum_{i=1}^{k} p_i = 1$, and p_i's $(i = 1, 2, \ldots, k-1)$ distributed as multivariate beta with parameters $\ell_1, \ell_2, \ldots, \ell_k$.

Gupta and Tripathi (1996) have discussed *weighted bivariate logarithmic series distributions* with the weight function $w(x_1, x_2) = x_1^{(r)} x_2^{(s)}$. For the case $r = s = 1$, they have presented an explicit expression for the joint probability mass function and have also discussed some inferential issues.

6 DISCRETE DIRICHLET DISTRIBUTIONS

Rahman (1981) presented a discrete analogue of Appell's polynomials as

$$A_{m,n}(x, y; N) = \frac{1}{K_n P_N(x, y)} \Delta_x^m \Delta_y^n \Big[(x - m + 1)^{(\gamma + m - 1)} \times (y - n + 1)^{(\gamma' + n - 1)} (N - x - y + 1)^{(\alpha - \gamma - \gamma' + m + n)} \Big], \qquad (43.32)$$

where K_n is some constant, Δ is the forward difference operator given by $\Delta f(x) = f(x+1) - f(x)$ and $\Delta^n f(x) = \Delta[\Delta^{n-1} f(x)]$ for $n = 2, 3, \ldots,$ and

$$P_N(x, y) = \frac{1}{N+1} \frac{\Gamma(\alpha + 1)}{\Gamma(\gamma)\Gamma(\gamma')\Gamma(\alpha + 1 - \gamma - \gamma')} \times (x+1)^{(\gamma-1)} (y+1)^{(\gamma'-1)} (N - x - y + 1)^{(\alpha - \gamma - \gamma')},$$
$$x, y = 0, 1, 2, \ldots, N, \ x + y \le N. \qquad (43.33)$$

The restrictions on the parameters α, γ, γ' and the variables x, y have been described by Rahman (1981). The function $P_N(x, y)$ in (43.33) is the discrete analogue of the bivariate Dirichlet distribution [see Wilks (1962) and Johnson and Kotz (1972)]. Incidentally, Rahman (1981) has shown that the polynomials $A_{m,n}(x, y; N)$ in (43.32) are orthogonal.

By generalizing this result, Agrawal and Dave (1993) discussed a discrete analogue of generalized Appell's polynomial as

$$A_{m_1,\ldots,m_k}(x_1, \ldots, x_k; N) = \frac{1}{K_n\, P_N(x_1, \ldots, x_k)} \Delta_{x_1}^{m_1} \cdots \Delta_{x_k}^{m_k} \left[\left\{ \prod_{i=1}^{k} (x_i - m_i + 1)^{(\gamma_i + m_i - 1)} \right\} \times \left(N - \sum_{i=1}^{k} x_i + 1 \right)^{(\alpha - \sum_{i=1}^{k} \gamma_i + \sum_{i=1}^{k} m_i)} \right], \qquad (43.34)$$

where K_n is once again some constant and

$$P_N(x_1, \ldots, x_k)$$
$$= \frac{1}{N+1} \frac{\Gamma(\alpha+1)}{\Gamma(\gamma_1)\cdots\Gamma(\gamma_k)\Gamma(\alpha+1-\Sigma_{i=1}^k \gamma_i)} \left\{ \prod_{i=1}^k (x_i+1)^{(\gamma_i-1)} \right\}$$
$$\times \left(N - \sum_{i=1}^k x_i + 1 \right)^{(\alpha - \Sigma_{i=1}^k \gamma_i)}, \qquad (43.35)$$
$$x_i = 0, 1, \ldots, N, \ 0 \le \sum_{i=1}^k x_i \le N.$$

The function $P_N(x_1, \ldots, x_k)$ in (43.35) is the discrete analogue of the multivariate Dirichlet distribution [see Wilks (1962) and Johnson and Kotz (1972)].

7 DISCRETE NORMAL DISTRIBUTIONS

Dasgupta (1993) has discussed multivariate distributions with discrete normal distributions as marginals. A *discrete normal distribution*, with support $Z = \{0, \pm 1, \pm 2, \ldots\}$, has probability mass function

$$P(x) = c\, e^{-\beta x^2}, \qquad \beta > 0, \qquad (43.36)$$

where

$$c^{-1} = \sum_{x=-\infty}^{\infty} e^{-\beta x^2} < \infty. \qquad (43.37)$$

This distribution does not have the additive closure property unlike its continuous counterpart (see Chapter 13).

Dasgupta (1993) has proved that for a k-dimensional random variable $(X_1, X_2, \ldots, X_k)'$ with independent and integer-valued coordinates, if the joint distribution of $(X_1, X_2, \ldots, X_k)'$ depends only on $r_k^2 = X_1^2 + \cdots + X_k^2$ (radial symmetry) and if $\Pr[X_i = 0] > 0$, then for $k \ge 4$ the marginal distributions of X's are discrete normal as given in (43.36). This characterization does not, however, hold for $k = 2$ and 3.

The counterexample in the case of $k = 2$ is based on the fact that the equation

$$f(x_1^2 + x_2^2) = f(x_1^2) + f(x_2^2), \ x_1, x_2 = 0, 1, 2, \ldots \qquad (43.38)$$

has a solution $f_1(x^2) = -\beta\, x^2$ corresponding to the discrete normal probability mass function (43.36), but also

$$f_2(x) = \begin{cases} 0 & \text{if } x = 0 \pmod 4 \\ 1 & \text{if } x = 1 \pmod 4 \\ 2 & \text{if } x = 2 \pmod 4 \end{cases} \qquad (43.39)$$

satisfies (43.38). This will yield the marginal probability mass function

$$P(x) = d\, e^{-\beta x^2 + \gamma(x^2 \bmod 4)}, \ \beta > 0, x \in Z. \qquad (43.40)$$

Similarly, in the case of $k = 3$,

$$f_2(x) = \begin{cases} 0 & \text{if} \quad x = 0 \pmod 4 \\ 1 & \text{if} \quad x = 1 \pmod 4 \\ 2 & \text{if} \quad x = 2 \pmod 4 \\ 3 & \text{if} \quad x = 3 \pmod 4 \end{cases} \tag{43.41}$$

satisfies the extended form of (43.38).

However, for $k \geq 4$ the use of Lagrange theorem that every positive integer is the sum of four (or less) squares of integers [see, for example, Davis (1962)] assures the uniqueness of the discrete normal distribution as the solution. The result for higher dimensions follows since one may take the remaining coordinates to be zero while representing a positive integer by sum of more than four squares.

An analogous result for distances different than the Euclidean distance established by Dasgupta (1993) states that for a k-dimensional random variable $(X_1, \ldots, X_k)'$ with independent and integer-valued coordinates, if the joint distribution of $(X_1, \ldots, X_k)'$ depends only on $r_k^4 = X_1^4 + \cdots + X_k^4$ and $\Pr[X_i = 0] > 0$, then for $k \geq 36$ the marginal distributions of X's are

$$P(x) = c\, e^{-\beta^4}, \qquad \beta > 0,\ x \in Z\,, \tag{43.42}$$

where

$$c^{-1} = \sum_{x=-\infty}^{\infty} e^{-\beta x^4} < \infty\,. \tag{43.43}$$

This result is based on Dickson's theorem of the representation of an integer by sum of 35 biquadrates [see, for example, Davis (1962, p. 337)]. In two dimensions, however, uniqueness can be achieved by imposing additional restrictions; if the values of $r^2 = X_1^2 + X_2^2$ are equispaced, X_1 and X_2 have the same set of possible values A, and $\Pr[X_i = 0] > 0$, then under the assumption of independence of X_1 and X_2 along with the radial symmetry of the joint distribution of X_1 and X_2, the marginal distributions of X_1 and X_2 are both discrete normal on A.

Finally, we note that discrete normal distribution on Z is equivalent to a geometric distribution on an appropriate support. For example, if the joint probability mass function of X_1 and X_2 is of the form

$$P(x_1, x_2) = c\, e^{-\beta(x_1^2 + x_2^2)}, \qquad x_1, x_2 \in Z\,, \tag{43.44}$$

the random variable $r^2 = X_1^2 + X_2^2$ has a geometric distribution over the admissible values of $x_1^2 + x_2^2$ when $x_1, x_2 \in Z$. The quantity $r = \sqrt{X_1^2 + X_2^2}$ is called *eccentricity* and is utilized in quality control problems.

BIBLIOGRAPHY

Agrawal, B. D., and Dave, G. N. (1993). m-dimensional discrete orthogonal system corresponding to multivariate Dirichlet distribution, *Bulletin of the Calcutta Mathematical Society*, **85**, 573–582.

Arnold, B. C., and Nagaraja, H. N. (1991). On some properties of bivariate weighted distributions, *Communications in Statistics—Theory and Methods*, **20**, 1853–1860.

Burtkiewicz, J. (1971). Wykorzystanie wielomianowego rozkladu do k-wymiarowej alternatywnej klasyfikacji, *Archiv. Automat. Telemech.*, **16**, 301–309 (in Polish).

Charalambides, C. A., and Papageorgiou, H. (1981). Bivariate Poisson-binomial distributions, *Biometrical Journal*, **23**, 437–450.

Chen, W. C. (1979). On the infinite Pólya urn models, *Technical Report No. 160*, Carnegie–Mellon University, Pittsburgh, PA.

Dasgupta, R. (1993). Cauchy equation on discrete domain and some characterization theorems, *Theory of Probability and Its Applications*, **38**, 520–524.

Davis, H. T. (1962). *The Summation of Series*, Trinity University Press.

Dyczka, W. (1973). On the multidimensional Pólya distribution, *Annales Societatis Mathematicae Polonae, Seria I: Commentationes Mathematicae*, **17**, 43–53.

Fisher, R. A. (1934). The effect of methods of ascertainment upon the estimation of frequencies, *Annals of Eugenics*, **6**, 13–25.

Gupta, R. C., and Tripathi, R. C. (1996). Weighted bivariate logarithmic series distributions, *Communications in Statistics—Theory and Methods*, **25**, 1099–1117.

Jain, K., and Nanda, A. K. (1995). On multivariate weighted distributions, *Communications in Statistics—Theory and Methods*, **24**, 2517–2539.

Jeuland, A. P., Bass, F. M., and Wright, G. P. (1980). A multibrand stochastic model compounding heterogeneous Erlang timing and multinomial choice parameter, *Operations Research*, **28**, 255–277.

Johnson, N. L., and Kotz, S. (1972). *Continuous Multivariate Distributions*, New York: John Wiley & Sons.

Johnson, N. L., and Kotz, S. (1976). On a multivariate generalized occupancy model, *Journal of Applied Probability*, **13**, 392–399.

Johnson, N. L., and Kotz, S. (1977). *Urn Models and Their Applications*, New York: John Wiley & Sons.

Karlin, S., and Rinott, Y. (1980a). Classes of orderings of measures and related correlation inequalities. I. Multivariate totally positive distributions, *Journal of Multivariate Analysis*, **10**, 467–498.

Karlin, S., and Rinott, Y. (1980b). Classes of orderings of measures and related correlation inequalities. II. Multivariate reverse rule distributions, *Journal of Multivariate Analysis*, **10**, 499–516.

Kawamura, K. (1979). The structure of multivariate Poisson distribution, *Kodai Mathematical Journal*, **2**, 337–345.

Kemp, C. D., and Loukas, S. (1978). The computer generation of discrete random variables, *Journal of the Royal Statistical Society, Series A*, **141**, 513–519.

Kemp, C. D., and Papageorgiou, H. (1982). Bivariate Hermite distributions, *Sankhyā, Series A*, **44**, 269–280.

Kunte, S. (1977). The multinomial distribution, Dirichlet integrals and Bose-Einstein statistics, *Sankhyā, Series A*, **39**, 305–308.

Levin, B., and Reeds, J. (1975). Compound multinomial likelihood functions are unimodal, *Unpublished Report.*

Mahfoud, M., and Patil, G. P. (1982). On weighted distribution, in *Statistics and Probability: Essays in Honor of C. R. Rao* (Eds. G. Kallianpur, P. R. Krishnaiah, and J. K. Ghosh), pp. 479–492, Amsterdam: North-Holland.

Marshall, A. W. (1975). Multivariate distributions with monotone hazard rate, in *Reliability and Fault Tree Analysis* (Ed. R. E. Barlow), pp. 259–284, Philadelphia: Society for Industrial and Applied Mathematics.

Milne, R. K., and Westcott, M. (1993). Generalized multivariate Hermite distributions and related point processes, *Annals of the Institute of Statistical Mathematics*, **45**, 367–381.

Mosimann, J. E. (1962). On the compound multinomial distribution, the multivariate β-distribution and correlations among proportions, *Biometrika*, **49**, 65–82.

Mosimann, J. E. (1963). On the compound negative multinomial distribution and correlations among inversely sampled pollen counts, *Biometrika*, **50**, 47–54.

Patil, G. P., Rao, C. R., and Ratnaparkhi, M. V. (1986). On discrete weighted distributions and their use in model choice for observed data, *Communications in Statistics—Theory and Methods*, **15**, 907–918.

Rahman, M. (1981). Discrete orthogonal systems corresponding to Dirichlet distribution, *Utilitas Mathematica*, **20**, 261–272.

Rao, C. R. (1965). On discrete distributions arising out of methods of ascertainment, in *Classical and Contagious Distributions* (Ed. G. P. Patil), pp. 320–332, London: Pergamon Press; Calcutta: Statistical Publishing Society.

Steyn, H. S. (1976). On the multivariate Poisson normal distribution, *Journal of the American Statistical Association*, **71**, 233–236.

Taillie, C., Ord, J. K., Mosimann, J. E., and Patil, G. P. (1975). Chance mechanisms underlying multivariate distributions, in *Statistical Distributions in Ecological Work* (Eds. J. K. Ord, G. P. Patil, and E. C. Pielou), pp. 157–191, Dordrecht: Reidel.

Titterington, D. M., Smith, A. F. M., and Makov, U. E. (1985). *Statistical Analysis of Finite Mixture Distributions*, New York: John Wiley & Sons.

Whitt, W. (1982). Multivariate monotone likelihood ratio and uniform conditional stochastic order, *Journal of Applied Probability*, **19**, 695–701.

Wilks, S. S. (1962). *Mathematical Statistics*, New York: John Wiley & Sons.

Abbreviations

C	conforming
cgf	cumulant generating function
CEP	conditional even-points (estimation)
CMtn	compound multinomial
EP	even-points (estimation)
ESF	Ewens sampling formula
GEM	generalized Engen–McCloskey distribution
GMg	generalized hypergeometric distribution
IFR	increasing failure rate
MDEF	multivariate discrete exponential family
MED	multivariate Ewens distribution
MEG	multivariate extended geometric distribution
MEG_s	multivariate extended geometric distribution of order s
MENB	multivariate extended negative binomial distribution
MENB_s	multivariate extended negative binomial distribution of order s
MG	multivariate geometric distribution
MG_s	multivariate geometric distribution of order s
mgf	moment generating function
MLE	maximum likelihood estimator
MLS	multivariate logarithmic series distribution
MLS_s	multivariate logarithmic series distribution of order s
MoMIP	modified multivariate inverse Pólya distribution
$\text{MoMIP}_{s,I(II)}$	modified multivariate inverse Pólya distribution of order s [Type I(II)]
MoMLS	modified multivariate logarithmic series distribution
$\text{MoMP}_{s,I(II)}$	modified multivariate Pólya distribution of order s [Type I(II)]
MNB	multivariate negative binomial distribution

MNB_s	multivariate negative binomial distribution of order s
MP	multivariate Pólya distribution
MP_s	multivariate Pólya distribution of order s
MRR	multivariate reverse rule
MSSPSD	multivariate sum-symmetric power-series distribution
MTP_2	multivariate totally positive of order 2
MV(U)(E)	minimum variance (unbiased) (estimator)
NC	nonconforming
P	univariate Pólya distribution
pgf	probability generating function
PMP	polytomic multivariate Pólya
SMP	singular multivariate Pólya–Eggenberger distribution
SUMH	singular unified multivariate hypergeometric distribution
UMH	unified multivariate hypergeometric distribution
$UMH_{(1)}$	univariate unified hypergeometric distribution

Author Index

The purpose of this index is to provide readers with quick and easy access to the contributions (pertinent to this volume) of any individual author, and not to highlight any particular author's contribution.

Subject Index

For readers' convenience, lists of all discrete distributions that are referred to in this volume have also been included in this index.

WILEY SERIES IN PROBABILITY AND STATISTICS

ESTABLISHED BY WALTER A. SHEWHART AND SAMUEL S. WILKS

Editors

Vic Barnett, Ralph A. Bradley, Nicholas I. Fisher, J. Stuart Hunter, J. B. Kadane, David G. Kendall, David W. Scott, Adrian F. M. Smith, Jozef L. Teugels, Geoffrey S. Watson

Probability and Statistics

ANDERSON · An Introduction to Multivariate Statistical Analysis, *Second Edition*
*ANDERSON · The Statistical Analysis of Time Series
ARNOLD, BALAKRISHNAN, and NAGARAJA · A First Course in Order Statistics
BACCELLI, COHEN, OLSDER, and QUADRAT · Synchronization and Linearity: An Algebra for Discrete Event Systems
BARTOSZYNSKI and NIEWIADOMSKA-BUGAJ · Probability and Statistical Inference
BERNARDO and SMITH · Bayesian Statistical Concepts and Theory
BHATTACHARYYA and JOHNSON · Statistical Concepts and Methods
BILLINGSLEY · Convergence of Probability Measures
BILLINGSLEY · Probability and Measure, *Second Edition*
BOROVKOV · Asymptotic Methods in Queuing Theory
BRANDT, FRANKEN, and LISEK · Stationary Stochastic Models
CAINES · Linear Stochastic Systems
CAIROLI and DALANG · Sequential Stochastic Optimization
CHEN · Recursive Estimation and Control for Stochastic Systems
CONSTANTINE · Combinatorial Theory and Statistical Design
COOK and WEISBERG · An Introduction to Regression Graphics
COVER and THOMAS · Elements of Information Theory
CSÖRGŐ and HORVÁTH · Weighted Approximations in Probability Statistics
*DOOB · Stochastic Processes
DUDEWICZ and MISHRA · Modern Mathematical Statistics
DUPUIS · A Weak Convergence Approach to the Theory of Large Deviations
ETHIER and KURTZ · Markov Processes: Characterization and Convergence
FELLER · An Introduction to Probability Theory and Its Applications, Volume 1, *Third Edition,* Revised; Volume II, *Second Edition*
FREEMAN and SMITH · Aspects of Uncertainty: A Tribute to D. V. Lindley
FULLER · Introduction to Statistical Time Series, *Second Edition*
FULLER · Measurement Error Models
GHOSH · Sequential Estimation
GIFI · Nonlinear Multivariate Analysis
GUTTORP · Statistical Inference for Branching Processes
HALD · A History of Probability and Statistics and Their Applications before 1750
HALL · Introduction to the Theory of Coverage Processes
HANNAN and DEISTLER · The Statistical Theory of Linear Systems
HEDAYAT and SINHA · Design and Inference in Finite Population Sampling
HOEL · Introduction to Mathematical Statistics, *Fifth Edition*
HUBER · Robust Statistics
IMAN and CONOVER · A Modern Approach to Statistics
JUREK and MASON · Operator-Limit Distributions in Probability Theory
KASS and VOS · Geometrical Foundations of Asymptotic Inference: Curved Exponential Families and Beyond
KAUFMAN and ROUSSEEUW · Finding Groups in Data: An Introduction to Cluster Analysis

*Now available in a lower priced paperback edition in the Wiley Classics Library.

Probability and Statistics (Continued)

KOTZ · Leading Personalities in Statistical Sciences from the Seventeenth Century to the Present
LAMPERTI · Probability: A Survey of the Mathematical Theory, *Second Edition*
LARSON · Introduction to Probability Theory and Statistical Inference, *Third Edition*
LESSLER and KALSBEEK · Nonsampling Error in Surveys
LINDVALL · Lectures on the Coupling Method
MANTON, WOODBURY, and TOLLEY · Statistical Applications Using Fuzzy Sets
MARDIA · The Art of Statistical Science: A Tribute to G. S. Watson
MORGENTHALER and TUKEY · Configural Polysampling: A Route to Practical Robustness
MUIRHEAD · Aspects of Multivariate Statistical Theory
OLIVER and SMITH · Influence Diagrams, Belief Nets and Decision Analysis
*PARZEN · Modern Probability Theory and Its Applications
PRESS · Bayesian Statistics: Principles, Models, and Applications
PUKELSHEIM · Optimal Experimental Design
PURI and SEN · Nonparametric Methods in General Linear Models
PURI, VILAPLANA, and WERTZ · New Perspectives in Theoretical and Applied Statistics
RAO · Asymptotic Theory of Statistical Inference
RAO · Linear Statistical Inference and Its Applications, *Second Edition*
*RAO and SHANBHAG · Choquet-Deny Type Functional Equations with Applications to Stochastic Models
RENCHER · Methods of Multivariate Analysis
ROBERTSON, WRIGHT, and DYKSTRA · Order Restricted Statistical Inference
ROGERS and WILLIAMS · Diffusions, Markov Processes, and Martingales, Volume I: Foundations, *Second Edition;* Volume II: Îto Calculus
ROHATGI · An Introduction to Probability Theory and Mathematical Statistics
ROSS · Stochastic Processes
RUBINSTEIN · Simulation and the Monte Carlo Method
RUBINSTEIN and SHAPIRO · Discrete Event Systems: Sensitivity Analysis and Stochastic Optimization by the Score Function Method
RUZSA and SZEKELY · Algebraic Probability Theory
SCHEFFE · The Analysis of Variance
SEBER · Linear Regression Analysis
SEBER · Multivariate Observations
SEBER and WILD · Nonlinear Regression
SERFLING · Approximation Theorems of Mathematical Statistics
SHORACK and WELLNER · Empirical Processes with Applications to Statistics
SMALL and McLEISH · Hilbert Space Methods in Probability and Statistical Inference
STAPLETON · Linear Statistical Models
STAUDTE and SHEATHER · Robust Estimation and Testing
STOYANOV · Counterexamples in Probability
STYAN · The Collected Papers of T. W. Anderson: 1943–1985
TANAKA · Time Series Analysis: Nonstationary and Noninvertible Distribution Theory
THOMPSON and SEBER · Adaptive Sampling
WELSH · Aspects of Statistical Inference
WHITTAKER · Graphical Models in Applied Multivariate Statistics
YANG · The Construction Theory of Denumerable Markov Processes

Applied Probability and Statistics

ABRAHAM and LEDOLTER · Statistical Methods for Forecasting
AGRESTI · Analysis of Ordinal Categorical Data

*Now available in a lower priced paperback edition in the Wiley Classics Library.

Applied Probability and Statistics (Continued)

AGRESTI · Categorical Data Analysis
AGRESTI · An Introduction to Categorical Data Analysis
ANDERSON and LOYNES · The Teaching of Practical Statistics
ANDERSON, AUQUIER, HAUCK, OAKES, VANDAELE, and WEISBERG · Statistical Methods for Comparative Studies
ARMITAGE and DAVID (editors) · Advances in Biometry
*ARTHANARI and DODGE · Mathematical Programming in Statistics
ASMUSSEN · Applied Probability and Queues
*BAILEY · The Elements of Stochastic Processes with Applications to the Natural Sciences
BARNETT and LEWIS · Outliers in Statistical Data, *Second Edition*
BARTHOLOMEW, FORBES, and McLEAN · Statistical Techniques for Manpower Planning, *Second Edition*
BATES and WATTS · Nonlinear Regression Analysis and Its Applications
BECHHOFER, SANTNER, and GOLDSMAN · Design and Analysis of Experiments for Statistical Selection, Screening, and Multiple Comparisons
BELSLEY · Conditioning Diagnostics: Collinearity and Weak Data in Regression
BELSLEY, KUH, and WELSCH · Regression Diagnostics: Identifying Influential Data and Sources of Collinearity
BERRY · Bayesian Analysis in Statistics and Econometrics: Essays in Honor of Arnold Zellner
BERRY, CHALONER, and GEWEKE · Bayesian Analysis in Statistics and Econometrics: Essays in Honor of Arnold Zellner
BHAT · Elements of Applied Stochastic Processes, *Second Edition*
BHATTACHARYA and WAYMIRE · Stochastic Processes with Applications
BIEMER, GROVES, LYBERG, MATHIOWETZ, and SUDMAN · Measurement Errors in Surveys
BIRKES and DODGE · Alternative Methods of Regression
BLOOMFIELD · Fourier Analysis of Time Series: An Introduction
BOLLEN · Structural Equations with Latent Variables
BOULEAU · Numerical Methods for Stochastic Processes
BOX · R. A. Fisher, the Life of a Scientist
BOX and DRAPER · Empirical Model-Building and Response Surfaces
BOX and DRAPER · Evolutionary Operation: A Statistical Method for Process Improvement
BOX, HUNTER, and HUNTER · Statistics for Experimenters: An Introduction to Design, Data Analysis, and Model Building
BROWN and HOLLANDER · Statistics: A Biomedical Introduction
BUCKLEW · Large Deviation Techniques in Decision, Simulation, and Estimation
BUNKE and BUNKE · Nonlinear Regression, Functional Relations and Robust Methods: Statistical Methods of Model Building
CHATTERJEE and HADI · Sensitivity Analysis in Linear Regression
CHATTERJEE and PRICE · Regression Analysis by Example, *Second Edition*
CLARKE and DISNEY · Probability and Random Processes: A First Course with Applications, *Second Edition*
COCHRAN · Sampling Techniques, *Third Edition*
*COCHRAN and COX · Experimental Designs, *Second Edition*
CONOVER · Practical Nonparametric Statistics, *Second Edition*
CONOVER and IMAN · Introduction to Modern Business Statistics
CORNELL · Experiments with Mixtures, Designs, Models, and the Analysis of Mixture Data, *Second Edition*
COX · A Handbook of Introductory Statistical Methods
*COX · Planning of Experiments

*Now available in a lower priced paperback edition in the Wiley Classics Library.

Applied Probability and Statistics (Continued)

COX, BINDER, CHINNAPPA, CHRISTIANSON, COLLEDGE, and KOTT · Business Survey Methods
CRESSIE · Statistics for Spatial Data, *Revised Edition*
DANIEL · Applications of Statistics to Industrial Experimentation
DANIEL · Biostatistics: A Foundation for Analysis in the Health Sciences, *Sixth Edition*
DAVID · Order Statistics, *Second Edition*
*DEGROOT, FIENBERG, and KADANE · Statistics and the Law
*DEMING · Sample Design in Business Research
DILLON and GOLDSTEIN · Multivariate Analysis: Methods and Applications
DODGE and ROMIG · Sampling Inspection Tables, *Second Edition*
DOWDY and WEARDEN · Statistics for Research, *Second Edition*
DRAPER and SMITH · Applied Regression Analysis, *Second Edition*
DUNN · Basic Statistics: A Primer for the Biomedical Sciences, *Second Edition*
DUNN and CLARK · Applied Statistics: Analysis of Variance and Regression, *Second Edition*
ELANDT-JOHNSON and JOHNSON · Survival Models and Data Analysis
EVANS, PEACOCK, and HASTINGS · Statistical Distributions, *Second Edition*
FISHER and VAN BELLE · Biostatistics: A Methodology for the Health Sciences
FLEISS · The Design and Analysis of Clinical Experiments
FLEISS · Statistical Methods for Rates and Proportions, *Second Edition*
FLEMING and HARRINGTON · Counting Processes and Survival Analysis
FLURY · Common Principal Components and Related Multivariate Models
GALLANT · Nonlinear Statistical Models
GLASSERMAN and YAO · Monotone Structure in Discrete-Event Systems
GNANADESIKAN · Methods for Statistical Data Analysis of Multivariate Observations, *Second Edition*
GREENWOOD and NIKULIN · A Guide to Chi-Squared Testing
GROSS and HARRIS · Fundamentals of Queueing Theory, *Second Edition*
GROVES · Survey Errors and Survey Costs
GROVES, BIEMER, LYBERG, MASSEY, NICHOLLS, and WAKSBERG · Telephone Survey Methodology
HAHN and MEEKER · Statistical Intervals: A Guide for Practitioners
HAND · Discrimination and Classification
*HANSEN, HURWITZ, and MADOW · Sample Survey Methods and Theory, Volume 1: Methods and Applications
*HANSEN, HURWITZ, and MADOW · Sample Survey Methods and Theory, Volume II: Theory
HEIBERGER · Computation for the Analysis of Designed Experiments
HELLER · MACSYMA for Statisticians
HINKELMAN and KEMPTHORNE: · Design and Analysis of Experiments, Volume 1: Introduction to Experimental Design
HOAGLIN, MOSTELLER, and TUKEY · Exploratory Approach to Analysis of Variance
HOAGLIN, MOSTELLER, and TUKEY · Exploring Data Tables, Trends and Shapes
HOAGLIN, MOSTELLER, and TUKEY · Understanding Robust and Exploratory Data Analysis
HOCHBERG and TAMHANE · Multiple Comparison Procedures
HOCKING · Methods and Applications of Linear Models: Regression and the Analysis of Variables
HOEL · Elementary Statistics, *Fifth Edition*
HOGG and KLUGMAN · Loss Distributions
HOLLANDER and WOLFE · Nonparametric Statistical Methods
HOSMER and LEMESHOW · Applied Logistic Regression
HØYLAND and RAUSAND · System Reliability Theory: Models and Statistical Methods
HUBERTY · Applied Discriminant Analysis

*Now available in a lower priced paperback edition in the Wiley Classics Library.

Applied Probability and Statistics (Continued)

IMAN and CONOVER · Modern Business Statistics
JACKSON · A User's Guide to Principle Components
JOHN · Statistical Methods in Engineering and Quality Assurance
JOHNSON · Multivariate Statistical Simulation
JOHNSON and KOTZ · Distributions in Statistics
Continuous Univariate Distributions—2
Continuous Multivariate Distributions
JOHNSON, KOTZ, and BALAKRISHNAN · Continuous Univariate Distributions, Volume 1, *Second Edition*
JOHNSON, KOTZ, and BALAKRISHNAN · Discrete Multivariate Distributions
JOHNSON, KOTZ, and KEMP · Univariate Discrete Distributions, *Second Edition*
JUDGE, GRIFFITHS, HILL, LÜTKEPOHL, and LEE · The Theory and Practice of Econometrics, *Second Edition*
JUDGE, HILL, GRIFFITHS, LÜTKEPOHL, and LEE · Introduction to the Theory and Practice of Econometrics, *Second Edition*
JUREČKOVÁ and SEN · Robust Statistical Procedures: Aymptotics and Interrelations
KADANE · Bayesian Methods and Ethics in a Clinical Trial Design
KADANE AND SCHUM · A Probabilistic Analysis of the Sacco and Vanzetti Evidence
KALBFLEISCH and PRENTICE · The Statistical Analysis of Failure Time Data
KASPRZYK, DUNCAN, KALTON, and SINGH · Panel Surveys
KISH · Statistical Design for Research
*KISH · Survey Sampling
LAD · Operational Subjective Statistical Methods: A Mathematical, Philosophical, and Historical Introduction
LANGE, RYAN, BILLARD, BRILLINGER, CONQUEST, and GREENHOUSE · Case Studies in Biometry
LAWLESS · Statistical Models and Methods for Lifetime Data
LEBART, MORINEAU., and WARWICK · Multivariate Descriptive Statistical Analysis: Correspondence Analysis and Related Techniques for Large Matrices
LEE · Statistical Methods for Survival Data Analysis, *Second Edition*
LEPAGE and BILLARD · Exploring the Limits of Bootstrap
LEVY and LEMESHOW · Sampling of Populations: Methods and Applications
LINHART and ZUCCHINI · Model Selection
LITTLE and RUBIN · Statistical Analysis with Missing Data
LYBERG · Survey Measurement
MAGNUS and NEUDECKER · Matrix Differential Calculus with Applications in Statistics and Econometrics
MAINDONALD · Statistical Computation
MALLOWS · Design, Data, and Analysis by Some Friends of Cuthbert Daniel
MANN, SCHAFER, and SINGPURWALLA · Methods for Statistical Analysis of Reliability and Life Data
MASON, GUNST, and HESS · Statistical Design and Analysis of Experiments with Applications to Engineering and Science
McLACHLAN and KRISHNAN · The EM Algorithm and Extensions
McLACHLAN · Discriminant Analysis and Statistical Pattern Recognition
McNEIL · Epidemiological Research Methods
MILLER · Survival Analysis
MONTGOMERY and MYERS · Response Surface Methodology: Process and Product in Optimization Using Designed Experiments
MONTGOMERY and PECK · Introduction to Linear Regression Analysis, *Second Edition*
NELSON · Accelerated Testing, Statistical Models, Test Plans, and Data Analyses
NELSON · Applied Life Data Analysis
OCHI · Applied Probability and Stochastic Processes in Engineering and Physical Sciences

*Now available in a lower priced paperback edition in the Wiley Classics Library.

Applied Probability and Statistics (Continued)

OKABE, BOOTS, and SUGIHARA · Spatial Tesselations: Concepts and Applications of Voronoi Diagrams
OSBORNE · Finite Algorithms in Optimization and Data Analysis
PANKRATZ · Forecasting with Dynamic Regression Models
PANKRATZ · Forecasting with Univariate Box-Jenkins Models: Concepts and Cases
PORT · Theoretical Probability for Applications
PUTERMAN · Markov Decision Processes: Discrete Stochastic Dynamic Programming
RACHEV · Probability Metrics and the Stability of Stochastic Models
RÉNYI · A Diary on Information Theory
RIPLEY · Spatial Statistics
RIPLEY · Stochastic Simulation
ROSS · Introduction to Probability and Statistics for Engineers and Scientists
ROUSSEEUW and LEROY · Robust Regression and Outlier Detection
RUBIN · Multiple Imputation for Nonresponse in Surveys
RYAN · Modern Regression Methods
RYAN · Statistical Methods for Quality Improvement
SCHOTT · Matrix Analysis for Statistics
SCHUSS · Theory and Applications of Stochastic Differential Equations
SCOTT · Multivariate Density Estimation: Theory, Practice, and Visualization
SEARLE · Linear Models
SEARLE · Linear Models for Unbalanced Data
SEARLE · Matrix Algebra Useful for Statistics
SEARLE, CASELLA, and McCULLOCH · Variance Components
SKINNER, HOLT, and SMITH · Analysis of Complex Surveys
STOYAN, KENDALL, and MECKE · Stochastic Geometry and Its Applications, *Second Edition*
STOYAN and STOYAN · Fractals, Random Shapes and Point Fields: Methods of Geometrical Statistics
THOMPSON · Empirical Model Building
THOMPSON · Sampling
TIERNEY · LISP-STAT: An Object-Oriented Environment for Statistical Computing and Dynamic Graphics
TIJMS · Stochastic Modeling and Analysis: A Computational Approach
TITTERINGTON, SMITH, and MAKOV · Statistical Analysis of Finite Mixture Distributions
UPTON and FINGLETON · Spatial Data Analysis by Example, Volume 1: Point Pattern and Quantitative Data
UPTON and FINGLETON · Spatial Data Analysis by Example, Volume II: Categorical and Directional Data
VAN RIJCKEVORSEL and DE LEEUW · Component and Correspondence Analysis
WEISBERG · Applied Linear Regression, *Second Edition*
WESTFALL and YOUNG · Resampling-Based Multiple Testing: Examples and Methods for p-Value Adjustment
WHITTLE · Optimization Over Time: Dynamic Programming and Stochastic Control, Volume I and Volume II
WHITTLE · Systems in Stochastic Equilibrium
WONNACOTT and WONNACOTT · Econometrics, *Second Edition*
WONNACOTT and WONNACOTT · Introductory Statistics, *Fifth Edition*
WONNACOTT and WONNACOTT · Introductory Statistics for Business and Economics, *Fourth Edition*
WOODING · Planning Pharmaceutical Clinical Trials: Basic Statistical Principles
WOOLSON · Statistical Methods for the Analysis of Biomedical Data
*ZELLNER · An Introduction to Bayesian Inference in Econometrics

*Now available in a lower priced paperback edition in the Wiley Classics Library.

Tracts on Probability and Statistics

BILLINGSLEY · Convergence of Probability Measures

KELLY · Reversability and Stochastic Networks

TOUTENBURG · Prior Information in Linear Models